Advances in Underwater Technology, Ocean Science and Offshore Engineering

Volume 16

Oceanology '88

ADVANCES IN UNDERWATER TECHNOLOGY, OCEAN SCIENCE AND OFFSHORE ENGINEERING

CONFERENCE PLANNING COMMITTEE

D. Lennard, *D. E. Lennard & Associates Ltd*
S. Archer, *Miros Ltd*
M. Smith, *Ministry of Defence*
C. Green, *Shell Internationale Petroleum Mij BV*
A. Richards, *Consultant*
C. Johnston, *Institute of Offshore Engineering, Heriot-Watt University*
D. Wardle, *Society for Underwater Technology*
D. Ardus, *British Geological Survey*

Advances in Underwater Technology, Ocean Science and Offshore Engineering

Volume 16

Oceanology '88

Proceedings of an international conference (Oceanology International '88), organized by Spearhead Exhibitions Ltd, sponsored by the Society for Underwater Technology, and held in Brighton, UK, 8–11 March, 1988

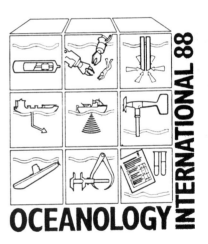

Published by

Graham & Trotman

A member of the Kluwer Academic Publishers Group
LONDON/DORDRECHT/BOSTON

First published in 1988 by

Graham & Trotman Limited   Graham & Trotman Inc.
Sterling House   101 Philip Drive
66 Wilton Road   Assinippi Park
London SW1V 1DE   Norwell, MA 02061
UK   USA

ISBN 0 86010 984 4 (Vol. 16)
ISBN 0 86010 922 4 (Series)

British Library Cataloguing in Publication Data

Oceanology International '88; *Conference*
  (Brighton, England)
  Oceanology '88
  1. Ocean engineering
  I. Title   II. Society for Underwater
  Technology
  620'.4162

Typeset, printed and bound in Great Britain
by Henry Ling Ltd, Dorchester

Contents

PART III
TECHNIQUES AND INSTRUMENTATION

Part I

Navigation and Hydrography

1

Starfix: Commercial Satellite Positioning

Lee Ott, Starfix Division of John E. Chance & Associates, Inc., 8200 Westglen, Suite A, Houston, Texas 77063, USA

Starfix is a privately owned and operated positioning system which became operational in early 1986 after three years of development effort. Starfix is a 24-hour-a-day positioning system encompassing the continental US and up to 500 miles offshore with a 5 metre 2 DRMS accuracy. At the present time Starfix has worked for every major oil company and seismic contractor, primarily in the Gulf of Mexico. This chapter will describe the operation of the Starfix system and compare Starfix with Syledis.

STARFIX SYSTEM DESCRIPTION

The major components of the Starfix system are a constellation of at least three satellites; uplink facilities for transmitting signals to each satellite; a nationwide tracking network for determining the position of each satellite; a central computing station known as the Master Site (Master Computing Facility); and passive user systems that provide the navigation function.

The Starfix positioning system makes use of geostationary communication satellites that are nearly motionless in relation to the earth 22 300 miles above the equator (Fig. 1).

Additional satellites may be used in the future to strengthen the geometry of the solution and provide redundancy. The purpose of the uplink facilities is to send data generated from the Master Site to the Starfix satellites for broadcast to the user. The data are being uplinked to the satellites at a frequency of 6 GHz. The Starfix tracking network is comprised of ten sites located throughout the United States. Each Remote Site (Tracking Station) has been precisely surveyed using state-of-the-art techniques to provide consistency of the network to the level of a few centimetres. The purpose of the network is to provide tracking data which allow the precise determination of the Starfix satellites in real time. The tracking stations also provide differential information that is necessary in the user position calculation.

The equipment at each tracking station consists of four Starfix receivers with antennas, a signal processing unit (SPU), a leased phone line modem to communicate with the Master Site, a back-up dial up modem to communicate with the Master Site, and a rubidium frequency standard.

Advances in Underwater Technology, Ocean Science and Offshore Engineering, Volume 16: Oceanology '88

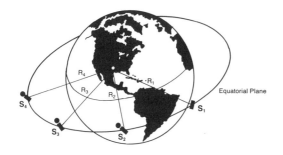

Fig. 1 Starfix geostationary satellite configuration.

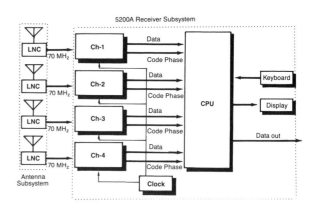

Fig. 2 Simplified block diagram of user receiving system.

The user equipment necessary in the Starfix positioning system consists of a 5200A receiver and antenna system (Fig. 2; Tables I and II).

TABLE I
Starfix receiver unit (model 5200A)

Weight:	100 lb
Size:	19″ rack mount—17.5″ high, 24″ deep 22″ rack depth
Power:	22 to 44 VDC 20 A
Operating temperature:	0 to +40°C
Storage temperature:	−20 to +75°C
Humidity:	Up to 95% non-condensing

TABLE II
Starfix antennae

Antenna

Weight:	200 lb
Size:	48″ diameter, 48″ high
Power:	Power supplied by azimuth controller
Operating temperature:	−10 to +50°C
Storage temperature:	−20 to +75°C

Azimuth Controller

Weight:	30 lb
Size:	19″ rack mount—5¼″ high, 22″ depth
Power:	115 VAC 50–60 Hz 4 A
Operating temperature:	0 to +40°C
Storage temperature:	−20 to +75°C
Humidity:	Up to 95% non-condensing

The purpose of the Master Site is to collect and format data for transmittal to the satellites, to calculate orbit coordinates of the satellites from all the tracking station data, and to provide a central facility for communication to the satellite uplink facilities. The Master Site is located in Austin, Texas and has leased phone lines to all the Remote Sites as well as the uplinks. A back-up Master Site is located in Houston, Texas with duplicate leased phone lines between all Remote Sites and the uplinks. The Master Site also monitors the Starfix system to identify problems, and to alert users if a potential problem exists by means of the Starfix message capability. A typical block diagram of a Master Site is shown in Figure 3.

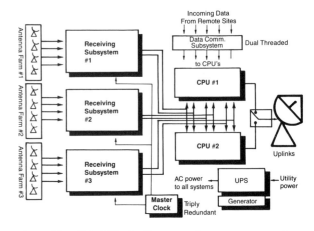

Fig. 3 Simplified block diagram of master computing facility.

USER POSITION DETERMINATION

In the Starfix system the satellites are used only as relays for information sent up by controllers on the ground. Unlike other navigational satellite systems, all timing is accomplished on the ground.

Starfix operates on the differential phase measurement (pseudo-ranging) principle. The Starfix satellites operate at that part of the radio frequency spectrum known as *C*-band (3.7–4.2 GHz).

The phase of the signal being sent from each satellite is measured by the user and at each differential site (Remote Site). All users and Remote Sites (which are essentially users at a known location) have clocks to maintain an accurate timebase. Since every time a user system is activated it will have a new timebase, the Starfix system is set up to solve for the space coordinates (X, Y, Z) of the user and the difference in the timebase of the user and each of the differential sites (Remote Sites). This last unknown is called the clock offset (δt). The basic equation to be solved is expressed as follows:

$$(0_u + \delta t) - 0_d = [(X_u - X_s)^2 + (Y_u - Y_s)^2 + (Z_u - Z_s)^2]^{1/2}$$
$$- [(X_d - X_s)^2 + (Y_d - Y_s)^2 + (Z_d - Z_s)^2]^{1/2}$$

Known variables

0_u Phase measurement at the user
0_d Phase measurement at the differential site
X_s, Y_s, Z_s Space coordinates of a satellite
X_d, Y_d, Z_d Space coordinates of the differential site

Unknown variables

δt Clock offset or δt (difference in time base of a differential site and the user)
X_u, Y_u, Z_u Space coordinates of user

As can be seen from the above equation, there are four unknowns to be solved $(X_u, Y_u, Z_u, \delta t)$. By observing three satellites, three independent equations can be written.

Owing to the geometry of the equatorial satellites (insufficient spatial volume), even if four or more satellites were observable, the solution would not be very precise. Thus, in order to position with Starfix, the height above the reference ellipsoid (geodetic height) of the user must be known. With this information a fourth independent equation can be written. This equation is the distance to the centre of the earth from the user's position and is expressed in the following form:

$$(X_u - 0)^2 + (Y_u - 0)^2 + (Z_u - 0)^2 =$$
$$(\text{distance to centre of earth})^2$$

With the four independent equations the four unknowns can be solved $(X_u, Y_u, Z_u, \delta t)$ and a user position determined.

SATELLITE ORBIT DETERMINATION

The calculation of each satellite's position (orbit determination) is made at the Master Site, *separate* from the user software.

Starfix is supported by a nationwide tracking network comprised of ten sites: a Master Site and nine Remote Sites located throughout the US to provide as long a baseline as possible. Each Remote Site is in effect a user at a precisely surveyed location. A set of phase measurements at each Remote Site is transmitted in real time back to the Master Site two times per second via leased telephone lines. The incoming phase measurements from the Remote Sites are differenced from the Master Site phase measurements to yield up to nine sets of phase differences that are used to determine the position of the satellites.

The basic equation to solve for the position of the satellite is:

$$(0_r + \delta t) - 0_m = [(X_r - X_s)^2 + (Y_r - Y_s)^2 + (Z_r - Z_s)^2]^{1/2}$$
$$- [(X_m - X_s)^2 + (Y_m - Y_s)^2 + (Z_m - Z_s)^2]^{1/2}$$

Known variables

0_r Phase measurement at Remote Site
0_m Phase measurement at Master Site
X_r, Y_r, Z_r Space coordinates of Remote Site
X_m, Y_m, Z_m Space coordinates of Master Site

Unknown variables

δt Clock offset or δt (difference in time base of Master Site and Remote Site)
X_s, Y_s, Z_s Space coordinates of the satellite

As can be seen by the above equation, there are four unknowns to be solved. In looking at the three satellite case, if one Remote Site is used, there will be three equations and ten unknowns $(X, Y, Z,$ of each satellite and $\delta t)$. Whenever a Remote Site is added to the solution, another unknown variable is also added (δt with respect to the new Remote Site). However, the number of equations available increases by three. Therefore, the minimum number of Remote Sites to determine the position of the three satellites is five (this also provides some redundancy). Any additional Remote Sites will increase the redundancy of the solution.

The Master Site contains three independent sets of receivers and antennas which generate three sets of phase measurements in order to achieve a majority vote on the phase measurement.

STARFIX SIGNAL FLOW

Starfix uses a form of transmission known as direct sequence spread spectrum modulation. A code known by the intended receiving station is modulated with the data to be transmitted. The receiving station tries to match its replica of the code with the signal being received. If the signal does not match, the code is shifted until a positive correlation is found. When the codes are matched, the receiver's code is phase-locked to the broadcast code (Costas Lock) and the data can be decoded. Once the receiver is locked, there is 41 dB of gain (process gain), effectively increasing the strength of the signal. By using the spread spectrum signal, the transmitter can be operated at low power and the antenna size can be small. However, the rate at which data can be sent is limited (150 Hz or 150 baud).

The signal is uplinked on a 6 GHz carrier wave with a chipping rate of 2.4576 MHz. The signal is frequency shifted in the satellite and retransmitted on a 4 GHz carrier wave.

Starfix uses a string of 16 384 0s and 1s known as the Pseudo Noise (PN) code. This PN code is modulated at a rate of 2.4576 MHz and the data are transmitted at a baud rate of 150 bits s^{-1}, or

$$(150 \text{ cycle s}^{-1}) \times (16\,384 \text{ chips cycle}^{-1}) = 2\,457\,600 \text{ chips s}^{-1}$$

Since the code repeats 150 times a second, the code has a wavelength equal to 1/150 the speed of light, or

$$(1/150) \times (299\,792\,485 \text{ m s}^{-1}) = 1\,998\,616.387 \text{ m}$$

The data signal being sent to the user from the satellites consists of the following: (1) time-tagged phase measurements from each tracking station (ranging information), (2) the health status of each tracking station, (3) time-tagged satellite positions (orbit information), (4) health status of each satellite, (5) base of time measurements (time transfer), and (6) user message information. The round trip for the signals to be generated from the Master Site, to the uplinks, to the satellites, back to the Master Site is approximately 1.8 seconds.

The phase measurements (ranging information) from each tracking station are sent every ½ second via leased phone lines, to the Master Site. The Master Site smooths the data through a Kalman filter and

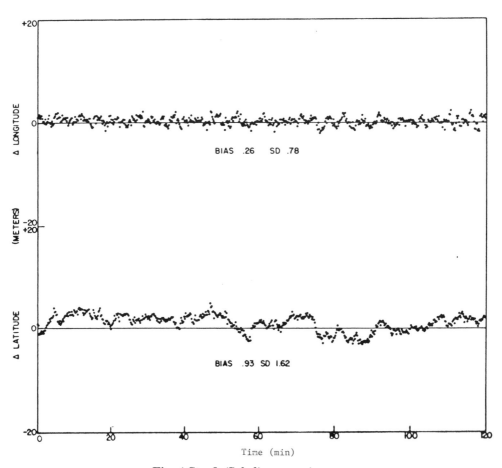

Fig. 4 Starfix/Syledis comparison.

best-fits the measurements to a fourth order polynomial. The polynomial consists of the position, velocity, acceleration and jerk of the phase measurements. The Master Site then sends the coefficients of the fourth order polynomial every 50 seconds to the users via the satellites.

The health status of each tracking station is determined at the Master Site and a flag is continuously sent to the user.

To determine the position of the satellites, the phase measurements from each tracking station must be acquired. These phase measurements are already being sent to the Master Site (see above ranging information). The Master Site calculates the position of the satellites via a 54 state Kalman filter. The Master Site then sends the coefficients of the second order polynomial every 180 seconds to the uplinks. The coefficients represent position and velocity of the satellite position in the X, Y, Z coordinate system.

Since the satellites do not change dramatically from an average position (nominal) in space, the data being sent are the difference between the current position and the nominal position. This cuts down on the amount of data being transmitted.

The timebase of all the measurements being taken (Starfix system time), is generated by the Master site. This timebase is sent to the uplinks every 11 seconds.

The message information service allows anyone to send short messages (up to 80 characters long) to specific receivers or groups of receivers. The alphanumeric message will appear on the display screen of the designated receiver(s). The Master Site generates the message to be sent.

OPERATIONAL RESULTS

Starfix has been tested and used at sea since the beginning of 1986. Most of the significant tests have been performed in the Gulf of Mexico, where extensive comparisons can be made against existing high-precision shore-based radionavigation systems used throughout the offshore oil industry.

After Starfix has been installed on the vessel at dockside, the antenna is precisely surveyed using standard techniques, and the surveyed coordinates are entered into the 5200A receiver to perform a calibration of the differential pseudo-ranges. Thus, the dockside point becomes the point of reference as described above, and generally it has been desirable to try and return to the same point after the sea trial to verify that the calibration remained consistent.

On a number of sea trials comparisons were made with state-of-the-art microwave radionavigation systems that exhibit high accuracy over short ranges (i.e. line-of-site to the transmitter). One such comparison made around 150 km offshore in the Gulf of Mexico is shown in Figure 4. The accuracy of the comparison system (Syledis) is generally accepted to be in the 5–7 m range at the transmitter distances employed about 25 km), so the results clearly indicate that Starfix accuracy was in that range also.

2

A Quality Control Index for Swathe Sounding Data

R. L. Cloet and C. R. Edwards, Bathymetrics Ltd

There are currently many aids to navigation which assist in making hazards visible and used sensibly they greatly enhance safety at sea. A conspicuous exception to this is the means to avoid the hazard of grounding. In this respect the navigator is still restricted by tunnel vision. Information about a water depth hazard only reaches him via his echosounder when the ship is already half way across it. Needless to say he should have consulted the chart and have avoided known hazards.

A chart contains a compilation of what is known about the seabed. It has never pretended to be a guarantee that no high spots are omitted or unrecorded. Indeed care is taken to spell this out by pointing out in the Sailing Directions that it is possible for a battleship to lie undetected between the sounding lines when even a large scale echo-sounding survey has been carried out.

The almost universal use of sidescan sonar as an adjunct search tool of the area between the lines while surveying has greatly reduced the possibility of large targets remaining undetected, but it has not yet permitted a statement on the likely margin of error of the depth measurements incorporated in surveys, and hence eventually in charts. This is because there is still a wide divergence between the depth sampling density along a track and across it. Typically this is every 0.5 m and possibly 125 m between lines—a ratio of 1:250.

Swathe sounding has made it possible to reduce the ratio to 1:1 and, if the tracks are run with a suitable overlap, to 2:1. It has therefore become possible to be as certain of discovering hazards across track as along it.

During the last two years, experience has been gained showing that high spots were found which had, or would have, gone undetected in an ordinary echo-sounding survey even had a sidescan sonar been used.

It is too early to say that no hazards will remain undetected by swathe sounding, but it is possible to be specific about their locations and possible dimensions on the basis of the distribution of coverage density achieved by a swathe sounding survey.

The scale of data gathering is now so large that its storage has to permit efficient, rapid access, in

Advances in Underwater Technology, Ocean Science and Offshore Engineering, Volume 16: Oceanology '88

order to process it. This accessibility can be made available on ships, so it is possible for the navigator to have more detailed information available for examination when entering formerly unfamiliar waters. This may lead to a fuller use of available water by reducing the sometimes very large under-keel tolerances now advocated in the light of current uncertainties, with the possible consequent reduction in congestion on busy routes, which is itself a potential navigation hazard.

CURRENTLY AVAILABLE SURVEY QUALITY INDICES

The chart is effectively a paper instrument abounding in safety features such as navigational aids information, only one of which is the selection of soundings shown. Particularly, though not only, in Port Approaches, care is taken by chart producers to monitor unstable seabed areas and incorporate new findings by means of small or large corrections, new editions or new charts. Annotations of such actions are provided as footnotes.

By means of an inset panel on the chart the user is also informed of the source, the ages, and scales of the survey data, originally used to compile it. Further notes indicate how the position determination has been harmonised and the basis used for correcting the survey soundings to tidal datum.

The variation in quality of the soundings is implied by the dates and scales of the source surveys, it being assumed the former is an indication of the degree of instrument sophistication which was available and refined over the years. The latter should represent the detail incorporated in the relevant survey. The basic principle which permits such an interpretation is straightforward. Sounding lines are normally plotted on the survey sheet 5 mm apart. Therefore if the survey scale is 1:10 000 the lines will have been 50 m apart and at 1:25 000, 125 m apart, and so on. These limitations are carefully pointed out in the supporting Admiralty texts, in which attention is drawn to the sizes of features which can remain undetected between sounding lines. It is salutary to have it stated that a battleship could lie undetected between sounding lines from a 1:12 500 scale survey if this rested on echo-soundings, because open sea surveys, even of areas as busy as Dover Strait, are mostly on scales of 1:25 000 or smaller.

Useful as survey dates and scales are, they do not adequately indicate the quality of the survey. The prudent surveyor, past or present, will have examined suspect areas between the lines, either on the basis of intelligence gathered over the years, his own experience, or even intuition. Nowadays the use of sidescan sonar can be a great help. It does, however, require interpretative skill and is better at identifying discrete objects on the seabed (provided there are not too many) than at finding the locations of mounds. Knowledge that an area has been examined by sidescan sonar does by no means guarantee the discovery of interline high spots. Hence an indication of sidescan sonar having been used is still a subjective quality evaluation.

The closeness of examination of an area depends on the surveyor's expectation of finding a possible danger, supplemented by any indications of changes of slopes seen from the depth profiles in the standard line pattern.

There is little agreement in fact about the optimum orientations for survey line patterns. If these are run in the direction of expected maximum slope, at least the major features should be charted, provided the slope direction has a reasonable measure of spatial persistence and the possible secondary features patterns are themselves not too large. Until recent years it was the orientation of the positioning system lines, rather than any characteristics of seabed features, which dictated the directions in which sounding lines were run. It is, and was, also a question of whether the survey launch or the ship was used in a particular area which, for navigational reasons, dictated the pattern orientation. Indeed so many are the variables that an indication of the name of the survey ship and of her captain often gives a better evaluation of the quality of a survey than anything else, but this kind of appreciation has no meaning for the user public.

The date of survey tends to be seen as meaning that the newer is also the better. A judgement of this kind would ignore changes in the conceived importance of surveying a particular sea area. A simple example of this is that before the Second World War the 3 to 6 fathom depth zone was invariably surveyed more intensively than greater or lesser depths because it was navigationally most critical. Since then the 6 to 11 fathom, or the 10 to 20 m, zone has received closest attention, and more recently the lower limit of the critical depth area has doubled. Not surprisingly it is found that the shallower, 3 to 6 fathom, area was often surveyed in greater detail a century ago than it is now. At times those surveys were also on a larger scale, than most recent surveys of the same area and may have been repeated more frequently.

We must not ignore the effects of budgetary constraints on the quality of surveys. Funds for hydrographic surveying have never been unduly generous, and this limitation has always been reflected in the chosen scale of the survey. That is in no way an indication of the degree of bed form complexity the surveyor anticipates exists in the area.

At one time, in the case of Port approach surveys, the scale of a survey appeared to be related to the distance from base of the area to be examined because, when day running, only a limited time was available each day for surveying in the area without the authorities incurring the expense of payment for overtime, or more, for the crew. Hence neither date nor scale of survey is an objective indicator of the accuracy or reliability of surveys or the charts based upon them. They are as likely to reflect a conceived valuation of the importance of an area rather than a reflection of the physical complexity of the seabed, and hence the degree of confidence which can be attached to them.

DESIRED QUALITY INDICES

Normally a set of measurements of any kind are qualified by a tolerance factor. It is reasonable to expect that when such a measurement is repeated its dimensions will not differ by more than the tolerance in the majority of cases.

The purpose of surveying is to make sufficient measurements of a surface such that these represent the entire surface adequately. Evidently there are practical limits to the amount of detail which can be, or finally need to be, represented on paper, but there is much concern that the shallowest depths should be shown.

On the whole natural features are repetitive or clustered, while man-made ones, at any rate at sea, more frequently tend to be singularities. Man-made features are generally also smaller than seabed features. There appears to be some evidence that natural agents frequently create linear forms and that these patterns are not randomly distributed.

The data produced by hydrographic surveys may need to be used for several purposes, ranging from the traditional navigational to seabed emplacement engineering, but there is a common requirement which is of interest to all users: the detection of obstacles.

Whereas it can be argued that natural seabed features frequently have linear characteristics and therefore that a linear search pattern, carefully orientated can take advantage of this fact, the same argument cannot apply to the shapes of obstacles. Therefore the quality index of the depth measurements in a survey is potentially very low indeed. The US Coast Pilot for an area on the Alaskan Coast

contains the following advice: 'As a measure of safety, it is considered advisable for vessels to avoid areas having depths no more than 30 feet greater than draft' (ref. 1). The area was surveyed by modern methods by a highly respected professional Government organization, but it clearly proved an impossible task. The caution is not intended as a safety margin which can be used with any assurance, more an indication of uncertainty. A recent accident in the area when a tanker was ripped open by a boulder indicates that the advice was either not heeded or was not sufficient. The environmental damage and that suffered by the polluting vessel are the cause of litigation which should put a financial valuation on the completeness of surveys.

The offending obstruction is said to have been one of many boulders known to occur in and near the channel. It is thought to have been about 10 m in diameter.

There are other cases of a similar character. Even a post dredge survey on a 25 m line spacing has been known in another part of the world to have missed the presence of a boulder left standing by the dredger, only to be found by a passing vessel.

It is evident that there is little merit in quoting a measurement tolerance for an echo-sounding depth as a meaningful indication of the worth of the survey to an eventual user. The position of any one sounding is uncertain, the tidal datum is a best guess at a distance offshore, and even the depth variability within the acoustic footprint of a single sounding is an unknown quantity, but above all the collection of soundings conveys next to nothing about the depths between one sounding and its neighbours.

Navigational safety can only be ensured by depicting all the highest spots which have been found wherever ships may operate. These must be in the correct positions and be shown as corrected depths, the values of which refer to a recoverable reference plane. To know where all the high spots are to an acceptable degree of certainty requires an assurance that all the remainder of any given area is deeper. No echo-sounding survey can or will ever make such a claim. Anything less than 'complete' coverage cannot provide such an assurance.

To measure the quality, and therefore the adequacy or the safety, of any hydrographic survey and the charts derived from it, a number of indices would be required, each based on the performance of several components.

POSITIONAL ACCURACY

Great strides have been made in the last few decades

in positioning a vessel at sea, and the accuracy claims which have been made are indeed impressive. It is a measurement which can be difficult to check unless reference can be made to a fixed target on the seabed.

Positioning data is processed to produce a set of coordinate values. The quality of the fix is expressed as an RMS value which should be reliable but need not be. Table I gives an example of 30 seconds of positioning data recorded at 1 second intervals. The increments have been added alongside the coordinates, while in the last column is the RMS value attributed to the latter as provided by the producers of the positioning processor. Because some of the

TABLE I
Interference in Syledis and the corresponding positioning error values

Time 1930 (h)	East 50 000 (m)	Difference	North 988 000 (m)	Difference	Error circle (m)
193010	350997.3		988978.4		1.9
11	94.5	−2.8	81.7	3.3	1.9
12	94.4	−0.1	84.6	2.9	2.0
13	93.6	−0.8	87.3	2.7	2.0
14	94.1	0.5	89.6	2.3	1.8
15	93.2	−0.9	92.3	2.7	1.8
16	91.7	−1.5	96.1	2.8	2.0
17	90.8	−0.9	98.8	2.7	2.0
18	89.3	−1.5	9001.6	2.8	1.9
19	88.4	−0.9	04.3	2.7	1.9
20	87.6	−0.8	06.9	2.7	1.8
21	59.6	*−28.0*	01.8	*−5.1*	*1.8*
22	58.8	−0.8	04.4	2.6	2.0
23	56.6	*−2.2*	12.6	*−8.2*	*2.0*
24	55.8	−0.8	15.3	2.7	2.0
25	54.9	−0.9	17.9	−2.6	1.6
26	56.5	1.6	16.8	−1.1	1.6
27	1146.7	*190.2*	23.6	*6.8*	*1.6*
28	45.8	−0.9	26.3	2.8	1.7
29	45.8	0	28.6	2.7	2.5
30	44.9	−0.9	31.3	2.7	2.5
31	43.4	−1.5	35.6	4.3	1.6
32	42.5	−0.9	38.3	2.8	1.6
33	0978.2	−164.3*	34.1	*−4.2*	*2.1*
34	77.3	−0.9	36.8	2.7	2.1
35	76.4	−0.9	39.4	2.6	3.3
36	76.1	−0.3	41.3	2.9	3.3
37	75.3	−0.8	43.9	2.6	2.9
38	74.4	−0.9	46.6	2.7	2.9
39	73.5	−0.9	49.3	2.7	2.1
40	71.3	*2.2*	59.6	*10.3*	*2.1*

Area: Cleeton "A"
Line: 3004

errors are large, they are easily spotted in the list of coordinates, but some of the starred values would have been accepted were it not that the high interrogation rate showed them to be erroneous.

There is clearly something suspect about the algorithm which produced the optimistic RMS values. The error was not identified in the echosounder survey which was made in parallel with the swathe sounder survey, and this must at least in part be due to its less frequent sampling rate. It became evident while surveying from the footprint plot of the Bathyscan survey that position errors were occurring because of the distinctive pattern of apparent data loss. Track increments are continually monitored and replaced by DR values for as long as 5 seconds before the system complains that no positioning is available. Corrections can be made during post processing.

It is evidently important to obtain a reliable measure of the quality of the positioning data, but it is also immensely reassuring to be able to monitor its reliability. An attractive feature of Bathyscan swathe sounding is that it can do so. Even well proven position processing systems cannot guarantee against human error. In a recent case the abeam offset of the swathe sounding tow fish was inadvertently inserted with the wrong sign and made to appear on the port instead of the starboard side. This only came to light when large depth misclosure errors were found during the post processing stage. On gridding the overlapping swathes individually it was seen that entire features had been displaced and from this the size of the erroneous offset was correctly estimated.

When swathe surveying it is the practice to overlap with the previous swathe because it ensures as far as is possible that shadow areas on one run will be filled by the next one. For this to work good relative positioning is needed, otherwise the fill in appears in the wrong place. In the process of merging adjacent swathes a great deal of common ground is sounded twice. Because the swathes are wide it is easy to recognize common features and from this comes an independent test of relative positioning. A wreck sitting in a scour hole provides such a feature. It is by means of swathe comparisons that tow fish abeam offsets and laybacks can be monitored and, if need be, corrected.

Swathe comparison therefore enhances the resolution of the positioning system to a fraction of a grid bin. Since any survey is likely to contain several swathes, this can be checked many times if necessary. Whereas kinks in tracks may cause suspicion when echo-sounding, swathe sounding produces an independent cross check (ref. 2).

TIDAL DATUM

Surveying is carried out throughout several tidal cycles in the same area. It is universal survey practice to subtract the height of the tide from the recorded depths to convert them into soundings which represent the minimum depth likely to be encountered, provided the seabed remains stable.

The virtue of using a datum such as LAT or MLLWS is that the surveyor or end user need only apply positive corrections for the time of local passage to the soundings to obtain the measure of available water, it being an established fact that if both positive and negative corrections were needed according to the state of tide someone would be bound to use the wrong sign, with possibly fatal consequences.

Both surveys and charts are treated in this way, and once a survey has been so corrected, the fair sheet would need to be recompiled to alter any of it.

Even in our busy domestic waters there exist insufficient offshore tidal measurements to determine the tidal phase and amplitude. Their values are extrapolated by means of proportionality factors from a Standard Port obtained from a co-tidal chart. In an area like the S. North Sea the correction factors largely ignore the effects of the East Anglian sandbanks on the tidal phase and amplitude, although it is generally acknowledged that both can be significantly different even on either side of one sand bank. This was well illustrated by an extensive investigation of the tides in the Outer Thames Estuary, carried out by the Ministry of Defence Hydrographic Department about twenty years ago, which resulted in detailed reduction zones for various tidal conditions.

Recently when processing a swathe sounding survey and using the overlap technique to merge adjacent swathes, it became evident that the misclosures in the overlap areas were time dependent rather than due to positioning errors and that in order to obtain a better fit, the amplitude and the phase of the tide had to be altered by a significant amount. On further investigation it was found that this conclusion accorded with unpublished data collected at nearby platforms.

Figure 1 shows the sites of the Standard Port of Immingham and the temporary recording stations labelled T1 and T2, as well as the area of the survey. Figure 2 shows the predicted spring curves at the three sites, the curve predicted for the survey area by the co-tidal chart, and the interpolated curve for the survey area, based on stations T1 and T2 and corroborated by swathe misclosure data.

It is characteristic of the method of determining the level of LAT offshore that the latter is assumed to lie below low water by the same proportion as the ratio applied to the tidal amplitude. Therefore errors in the estimate of the amplitude of the tide have a greater effect on depths recorded around high tide than near low tide. In the present case the high tide corrections were in excess of 1 m whereas the low tide reading differed by barely 0.2 m.

Since there is a sensible tendency to prefer sounding in shoal water, near high rather than low water, it follows that the largest errors in tidal reduction, due to the presence of shoal areas, are likely to have occurred in some potentially critical areas.

An echo-sounding survey carried out independently from Bathymetrics and simultaneously with the swathe sounding survey did not identify this time related misclosure.

Figure 3 is a composite three-dimensional diagram which shows on the right a series of swathes merged before correction of the tidal datum error, and on the left the same area after correction. A series of sandwaves have been numbered to show how the shapes of features can become obscured by datum errors.

It would be foolish to recommend ignoring coastal tidal observations when surveying some distance offshore, even though it is possible to reconstitute segments of the tidal curve from the misclosures between the overlapping swathes. Nevertheless it cannot but be reassuring that the swathe survey procedure provides an independent check on a variable which affects the accuracy of a survey and which has hitherto been unverifiable.

MEASURING THE DEPTH OF THE WATER

Obviously, a ship at sea is an unstable platform from which to make any kind of measurement. Therefore determining the depth at a given point, if it is to have any meaning, is bound to be a composite of several measurements.

When swathe sounding with a Bathyscan, the system is mounted in a tow fish, and depth profiles are measured abeam in terms of angle and range. The accuracy of the roll angle is therefore very important, particularly because its effect increases with range. It is an error source which is detectable to very fine angles, subject to positional accuracies, by using a relatively wide overlap, in the misclosure measurements between swathes. Such individual swathe, roll related, misclosures could be ±0.2 m at ranges of 100 m along well run lines, but this could increase to ±0.9 m if sudden course corrections had to be made. The latter could last for a few seconds and would

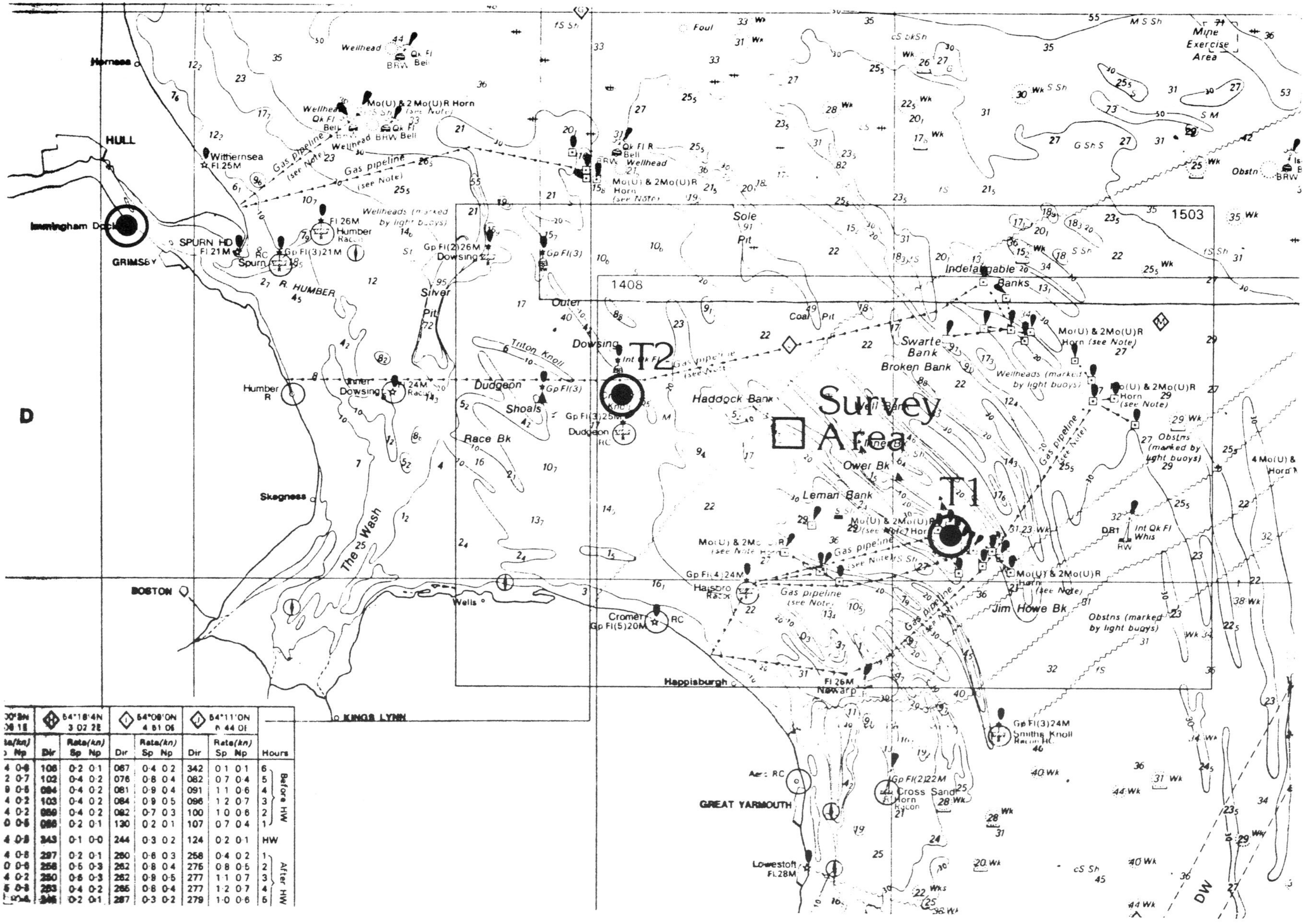

Fig. 1 Location of additional tidal data and survey area.

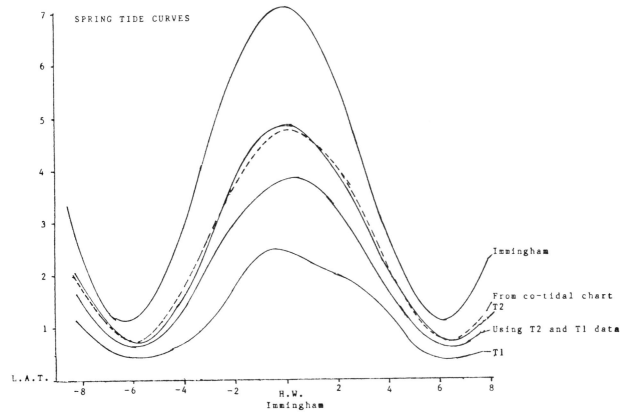

Fig. 2 Comparative spring and neap tides for Immingham, stations T2 and T1, and the survey site.

thus produce a small 'feature' which would misclose on the adjacent swathe.

A depth error will also result if the heave meter or the pressure depth sensor do not read correctly. To apply the heave correction future information is needed. The phase retardation which has to be applied is a function of the amplitude of the heave. A mean time correction can be applied on line, but the true heave value is obtained only during the post processing stage.

The pressure sensor is used as an indicator of the flight level of the tow fish. When deployed only a few metres below water level, about 1.5 min to 2 min are needed to eliminate the effect of surface waves. There is thus a response period between 20 seconds sampled by the vertical accelerometer, and 1.5 min, which cannot be observed directly by attitude sensors. This can be the duration of a course correction while surveying. Experience has shown that the amplitude of such a slow heave can be about 0.5 m.

The mean flight level of the tow fish is affected by the motion of the tow point on the ship and by the speed through the water which can vary in the course of the day, but also when moving from deep to shallow water.

The accumulation of these errors, although individually small, is not acceptable. To improve upon them instrumentally would be extremely costly. This is another reason why the practice of swathe overlapping has been adopted. It is assumed that it is most unlikely that an equal dimensioned but opposite error will be found on the adjacent run whose swathe overlaps it. Therefore, on the principle that each position has a unique depth value, the overlap area can be used to reduce the effect of the cumulated errors. The limitation of this postulate is that there is also an inevitable uncertainty attached to each position determination, so that it is necessary to think in terms of position areas rather than positions. From this it follows that the principle should be restated

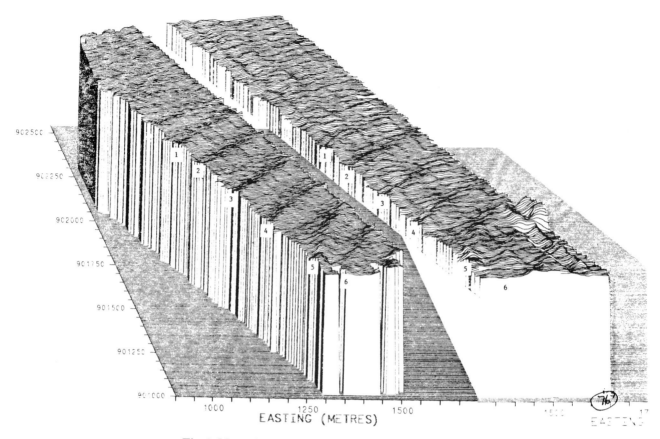

Fig. 3 Merged swathes before and after datum correction.

as being that the mean depth in each position area approximates to a single value and that deviations from it are observational or instrumental.

THE CALCULATION OF THE DEPTH MATRIX

The density of the depth matrix should ideally be such that the features occurring on the seabed are fully represented without aliasing. However, reality dictates that within a single swathe the maximum density cannot be more than the ground spacing between profiles. Swathes, when merged, cannot retain detail of features which have lengths smaller than the positional accuracy, because of the probability that the relevant measurements from adjacent swathes are posted into different grid bins. In fact an excessively small grid mesh could degrade the bedforms more than the possible smoothing effect of larger mesh sizes, where many small features occur. Fortunately there often appears to be a systematic distribution of the positional error dimensions so that the errors in relative positions in adjacent swathes are mostly reasonably small. It is possible to survey some considerable distance offshore and effectively use a 5 m matrix.

When operating at 5 knots an average bin population of about 25 depth values should be attained in one swathe. There can be large variations on that, and it is the purpose of the footprint plot to show, online, the adequacy of the coverage density which has been achieved. Low counts are caused by shadow areas, simple loss of signal, or rejection of data when post processing. The deviation from the bin mean depth value is affected by the population density. It can amount to 0.3 m with a count of as little as 5 and drops to 0.1 m when the count exceeds 10. If larger variations are met, an instrumental malfunction can normally be suspected. There are exceptions to this generalization, the principal being rocky or stony seabed or wreckage. Such targets tend to declare their presence also by being associated with shadow areas and can result in causing overlapping swathes to be unmergeable because a 5 m grid bin is then too coarse, while a smaller mesh size exceeds the position resolution. However, the pre-merging individual swathe grids can be compared and this provides a good indication of the local positioning error. This analytical approach, using local fitting of bed forms, has on several occasions been used to identify and correct unsuspected errors in the operation of positioning systems. It can also add considerably to the working-up time. Regrettable as that is, it must be recognized that these errors are generally overlooked in echo-sounding surveys and were pursued principally to establish the reliability of the swathe sounder.

Confirmation of the depths derived from the overlapping swathe increases the confidence in those measurements. This confidence must, however, be moderated by taking account of the population densities. Not only is a depth derived from a low count potentially less accurate than the others, but there

TABLE II

```
DYNAMIC SYNOPSIS:
*****************************************************************************************************
* MEAN  * AVERAGE * AVERAGE * AVERAGE    * NO. OF    * MISCLOSURE    * ERROR ABOUT *
* PING  * FLYING  * HEAVE   * SWATHE     * MISCLOSURE * CORRECTED FOR * AV. CORRECTED *
* NO.   * DEPTH   *         * MISCLOSURE * SAMPLES   * FISH DEPTH    * MISCLOSURE  *
*****************************************************************************************************
*  152 *  4.0  *  0.00  *   0.06    *   421   *   -0.04   *    0.05
*  452 *  4.0  *  0.00  *   0.05    *   465   *   -0.05   *    0.04
*  752 *  4.0  *  0.00  *   0.30    *   444   *    0.20   *    0.29
* 1052 *  4.1  *  0.00  *  -0.13    *   425   *   -0.13   *   -0.04
* 1352 *  4.1  *  0.00  *  -0.07    *   416   *   -0.07   *    0.02
* 1652 *  4.1  *  0.00  *  -0.04    *   392   *   -0.04   *    0.04
* 1952 *  4.1  *  0.00  *  -0.09    *   409   *   -0.09   *   -0.00
* 2252 *  4.1  *  0.00  *  -0.06    *   459   *   -0.06   *    0.03
* 2552 *  4.1  *  0.00  *  -0.06    *   506   *   -0.06   *    0.03
* 2852 *  4.1  *  0.00  *  -0.03    *   445   *   -0.03   *    0.06
* 3152 *  4.1  *  0.00  *  -0.51    *   509   *   -0.51   *   -0.42
* 3452 *  4.1  *  0.00  *  -0.02    *   520   *   -0.02   *    0.07
* 3752 *  4.1  *  0.00  *  -0.08    *   462   *   -0.08   *    0.01
* 4052 *  4.1  *  0.00  *  -0.23    *   670   *   -0.23   *   -0.14
* 4352 *  4.1  *  0.00  *  -0.45    *   717   *   -0.45   *   -0.36
* 4652 *  4.1  *  0.00  *  -0.24    *   665   *   -0.24   *   -0.15
* 4952 *  4.1  *  0.00  *  -0.06    *   706   *   -0.06   *    0.03
* 5252 *  4.1  *  0.00  *  -0.09    *   726   *   -0.09   *    0.00
* 5552 *  4.1  *  0.00  *   0.01    *   651   *    0.01   *    0.10
* 5852 *  4.1  *  0.00  *   0.06    *   611   *    0.06   *    0.15
* 6152 *  4.1  *  0.00  *   0.17    *   598   *    0.17   *    0.26
* 6452 *  4.1  *  0.00  *   0.07    *   328   *    0.07   *    0.16
* 6669 *  4.1  *  0.00  *   0.00 !! *     0   *    0.00   *   ****
*****************************************-------------************------------*******************
          AVERAGES:                   -0.08          ***         -0.09
*****************************************-------------************------------*******************

!! CORRECT FOR FLYING FISH DEPTH ONLY
   (not used in average calculation)

AVERAGE FISH TOW DEPTH=    4.1




TRIAL LINE:    58
  SWATHE DIFFERENCE FOR (   57B -      58)
        SWATHE    58 IS SHALLOWER
MERGE THIS LINE?
Y
                    INPUT FROM TEMP GRIDFILE: SYS:58    .TGR
  DO YOU WANT DYNAMIC CORRECTION?
Y
```

is then also a greater possibility that a small discrete object, such as a plate on edge, or a moderately small boulder, has escaped discovery. The responsible surveyor will have done his best to ensure high counts where depths are critical or where other indications lead to the suspicion that small discrete targets occur. It is fortunately easy to identify low count bins because that information is retained with the grid. The practice is to plot either the population density as a quality control guide, or to eliminate soundings, with cut-offs at different minimum density counts. This results in plots with data holes where low counts, or no counts, occur.

GRID MERGING

Making due allowance for shadow areas and possibly randomly erroneous 'depths' which may need closer scrutiny, the same surface surveyed from different aspects, in the swathe overlaps, should be the same. Individual swathe profiles cannot be matched directly to each other because the positioning data are inadequate for that degree of detail and the profiles almost certainly do not cover the identical portion of seabed. Moreover it is not possible, when towing shallow, to obtain an instantaneous measure of the tow fish flight level. The practice has therefore developed at the post processing stage, before merging adjacent overlapping swathes, of examining the misclosures between pairs of corresponding platelets each consisting of 300 profiles, the equivalent of 1 to 1.5 min. Table II shows the mean ping number, the platelet flying height, and a check whether there is drift on the heave meter. It then calculates the misclosures before the flying height difference of the offered-up swathe is applied. The number of samples in each platelet of the latter is used as weighting of the

correction. The platelet with the highest count, out of each pair contributes the largest proportion to the adopted merged depth.

When the average of all misclosures is small and not systematic, the correction is applied, leaving a residual error, shown in the last column. In this way unacceptably large residuals are highlighted and examined further if need be.

It is conceivable that this interactive post processing is unattractive to processing speed merchants. The surveyor still has to apply his judgement and skill, but the workload has been reduced to manageable proportions despite the vast quantity of data involved. As to the quality of the product of swathe sounding it will have become clear that the residual errors shown in Table II provide a workable index of its worth and could be converted into a global statistic for an entire survey. It represents an amalgam of all the contributory error sources including a measure of the persistence of the surveyor in tracking down the causes of unacceptable misclosure errors.

CONCLUSION

Swathe sounding has moved hydrographic surveying into the realm of quantifiable mensuration. It requires massive intervention from the computer and the use of rapid access storage devices. It has not become a substitute for the skilled surveyor although it does allow him to shed much drudgery.

The advent of a system which produces so much, more accurate, data is already leading to demands for an even better ability to visualize the seabed which after all still compares poorly with overland vision. To achieve this it is probable that the navigator's chart will have to become a very different product from today's essential paper instrument.

REFERENCES

1. United States Coast Pilot Vol 9. (1987). p. 99. Pacific and Arctic Coast Alaska Cape Spencer to Beaufort Sea.
2. Cloet, R. L. and Read, R. V. B. (1981). Cleaning the data of a Start Bay sandwaves survey. *Hydr. J.* No. 22, pp. 23–28.

3

The Challenge of Digital Data in Hydrography

Patrick Hally, Canadian Hydrographic Service, Québec Region, Institut Maurice-Lamontagne, 850, Route de la Mer, C.P. 1000, Mont-Joli (QC) G5H 3Z4, Canada

The recent advent of digital technology has had major impacts on acquisition, processing, storage and utilization of hydrographic data. These impacts are described and placed in context to show the explosion of the possibilities opened up by this technology. An overview of digital hydrography with the links to related techniques and sciences is made in order to demonstrate the future challenges facing us. New users and new ways to see and feel the bottom are only examples of what is now a reality with the digital data.

CHARACTERISTICS OF HYDROGRAPHIC DATA

We can describe a typical hydrographic operation as being the flow illustrated in Figure 1. The first step is the data acquisition stage, where the data is collected using a broad variety of sensors. The collected data then pass on to the data processing stage where it is cleaned, concatenated, computed, adjusted, etc. Once the data have reached a certain quality level, they are kept and managed at the

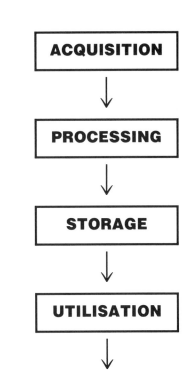

Fig. 1

Advances in Underwater Technology, Ocean Science and Offshore Engineering, Volume 16: Oceanology '88

storage stage. Finally comes the application (utilization) stage where the information is built with the data.

The major element of a hydrographic data set is bathymetry. Composed of positions associated with related depths, it is characterized as being point data, with no relation to the representation scale in itself. It is the information that requires the major portion of the efforts of a survey. The unknown aspect of the data collection, the physical considerations such as the sensors that have to be used to measure the depths (acoustics), and the practical platforms necessary to pursue surveys render this task very difficult compared to land data gathering.

The shoreline and the line data from the topography are scale related information forming the second significant data group of hydrographic data. This information is much more easily gathered using well proven land survey techniques, and its volume in hydrography is relatively minor compared to the previous group.

Other types of data, such as navigation aids, limits, bottom characteristics, form together the rest of the set in a minor proportion. Their collection requires a minimum of efforts compared again to the bathymetry.

THE PAST

What did we do 15 years ago in hydrography? We were very limited in our exploration of the subwater environment. We could only guess at what it would look like. Our tools based on mechanical, basic electronic or electrical 'technologies' could give us some spot information but we had to use our hands and our skills to compute, draw and analyse the results. The results seemed impressive at the time.

Data acquisition was, to say the least, tedious. Geodetic computations—for the localization of the ground-based stations—were done by hand, log tables or mechanical calculator. Every single piece of information on the position—even from an electronic positioning system—had to be recorded by hand on a note pad. The soundings were recorded on an echogram in analog form by the echosounder, with hand-written annotations. We had to record the time from our watches for the tide values, and the patterns that our launches or ships were following were the best sampling we could get of a designated area.

Data processing was also tedious manual interpretation of the echograms, manual plotting of position and soundings on a manually calculated projection. If there was an error, there was no alternative but to start over again. Data interpretation was based on the graphical method. Trying to see more—other scales, other geodetic projections, other data densities—was a question of redoing the processing steps—computations, drafting.

The storage of the data and the supporting documents was a question of archiving. Trying to keep the precious medium away from all the physical disturbances that could have ruined its accuracy, reliability and its own existence was a major preoccupation.

The utilization of such data was also restricted by its form. A graphic plot could be reproduced, cut, enlarged or reduced a little. They were all mechanical and manual processes with many limitations.

However, a revolution has arrived and is gaining momentum.

THE REVOLUTION

This revolution is the advent of digital technology now rendered possible with the introduction of computers with ever increasing capabilities. As in virtually every area of the modern society, the computer has widely spread in the hydrographic community in the last ten years. The impact on all facets of hydrography is not about to fade; instead it is bringing us an innovating array of tools to better discover, understand and utilize the underwater world.

ACQUISITION IN THE DIGITAL ENVIRONMENT

At the acquisition stage, the impact is major. Digital information is gathered in a huge quantity with a quality never before achieved.

Positioning is done with such tools as electronic distance measurement (EDM) systems that include built in computer facilities, enabling digital data to be outputed. Complex computations are performed inside these machines for multi-station interrogation, filtering, averaging, smoothing and data formating before the data is transferred to the outside of the instrument via standard interfaces.

New positioning systems such as the global positioning system (GPS) can achieve amazing results in dynamic mode feasible only because of the presence of digital technology. Computations such as online differential corrections, phase comparison techniques, and data analysis are performed in order to reach these results.

The navigation techniques used in hydrographic data collection to stay on line during survey operations are not the same anymore. The primary functions of the past have been replaced by efficient ways of operating; ranging from efficient straight line patterns with automatic piloting capabilities to complex customized functions like shoal searching with memory. Historical track information and smart online indicators are other applications. Radar information can now be digitized and processed for filtering and enhancement. A vast variety of sensors such as ship parameters providing data on digital form can now be interfaced to give an array of informations to the navigation during survey operations.

The depth acquisition has also been revolutionized by the digital data. The echo from the echosounder can now be digitized and various possibilities are now offered. The data can be compensated for heave, roll and pitch, enabling sounding operations in much more rough seas with a better quality. Multiple sounding sensors such as sweep systems are now able to handle up to several dozens of simultaneous echos, providing full bottom coverage. Similarly, swath systems with a complex array of transducers can achieve approximately the same results in deeper waters.

Accoustic imaging of much better quality with quantitative information is now possible with the digital data. Side scan sonars have been improved to provide data in digital form for later processing.

Remote sensing techniques have also taken their share in the depth data acquisition in hydrography. Analytical photogrammetry enables photobathymetry to be performed on incomplete models—something impossible to realize with analog stereo plotters. Multispectral imaging assimilated to fully digital photography is another example. Laser bathymetry flown from aircraft is acquiring data at a rate of 20 Hz, recording the whole signal history in order to later derive the depth information. Electromagnetic profilers flown from helicopters are gathering reflections of emitted electromagnetic signals produced at tremendous rates.

The most important aspect of all these applications is the standardization of the digital output and the interfacing to a central computer. Almost any sensor in digital form can be interfaced for data logging. The mass volume of data can now be entirely recorded during the costly time in the field and used integrally later on.

Other hydrographic parameters have also benefited from the advent of digital data. Tide measurements can now be sampled, recorded and broadcasted as well during survey operations.

Current meters can now be moored for long periods and record the data *in situ*.

One important aspect of the data acquisition in digital form is the conversion of old usable graphical data. Digitizers of all kinds, ranging from simple digitizing tablets to sophisticated scanners enable data to be rendered computer readable. Character recognition techniques are coming along, reading faster and improving the quality of the transformation work.

We can see the impacts of the digital age on the acquisition stage. Vast volumes of data of an incredible quality help us to take a better picture of the sub-bottom. A variety of attributes can be collected at this stage that will give the data several new dimensions to interact and build upon. These new dimensions in the data allow us to go beyond its primary functions.

PROCESSING IN THE DIGITAL ENVIRONMENT

Once the data has been collected, a certain number of functions have to be performed. Even with all its qualities, the data has to be cleaned. This task is easily done in digital form by means of filtering algorithms or interactive editing.

Complex precision analysis can be performed in order to assess data quality and validity. Survey coverage is computed; corrections for deformation are made; merging of complementary data is accomplished. These are easy tasks for a computer for data in digital form. Data reduction routines are performed in order to get more manageable data sets. Selection routines are performed, to choose the best representation of the data in order to evaluate the primary results. Automatic contouring is a first class tool for hydrographic work. This first product allows us to 'visualize' the bottom while still on site.

One of the main evolutions can be seen at this stage. The work done in the field, aided by the digital revolution, enables the hydrographic data collector to leave the area after having 'cleared', so to speak, the bottom. With all these tools, he is no longer the 'bottom sampler' who would have to guess what the bottom would look like. Almost in an exploratory fashion, he would have confirmed it only a long time after he was back in the office.

STORAGE IN THE DIGITAL ENVIRONMENT

The amount of data collected, further multiplied by the various steps involved at the processing stage, must be orderly and structured in order to extract the best of it. The filing concept of the past graphical world no longer applies because of the new nature of the data. It has acquired more than one dimension that can be efficiently accessed only through a digital data base management system (DBMS). Complex relations can be built in the data storage structure for fast and easy access with ever-demanding flexibility to adapt to new kinds of data and formats at the input and at the output.

The key to the hydrographic data in a DBMS environment naturally becomes naturally its earth position. To render it more universal, the geographical coordinates are taken as the working base to build on. The geographical information system (GIS) is a system that can perform the particular DBMS functions required by this kind of data.

All the attributes that were previously lost or accessed only with difficulty in the graphical form are now available digitally and linked to the key, the geographic coordinates (spatial position). At the same time a new dimension can now be added to the data in the form of temporal information for bathymetric data. The historical aspect can be quickly accessed bringing in a total new vision for sedimentation studies for example.

Data exchange can take a fully new approach since it can readily take the form of any of the incomer to the DBMS. Data exchange formats can be developed to further standardize these exchanges between DBMS of various agencies and systems. Satellite communications with their speed and globality can let us foresee the global links of a general information centre for hydrographic data, furthermore linked to other related physical parameters of above and under water.

UTILIZATION IN THE DIGITAL ENVIRONMENT

The data can now be utilized in a various array of applications. The first obvious one is the computer assisted cartography. All the information being in digital form, sophisticated hardware and software can help the operator (the former cartographer) to perform interactive data selection. Screen drawing with all the elements (attributes, references) and functionalities (colours, zooming, panning and powerful commands) are available to make digitally graphical products. Performance and quality are greatly improved compared with traditional manual methods. Hypotheses and trials can be performed in data interpretation of information,

and later dropped in favour of another test almost at the fingertips.

Furthermore, cartography can now be even more automated with the help of expert systems (ES) and artificial intelligence (AI). The basic rules, exceptions and standards can be taught to the computer and most of the cartographic operations can then be performed automatically.

However, trade-offs and compromises have to be made compared with traditional cartography. All the elements of a graphical cartographic product can be reproduced perfectly at a very high cost at times and it might not be useful to do so. New ways to render the bottom are accessible and we have to take advantage of them to reconsider the old products that were solely based on the old graphical and manual world considerations.

Electronic charting is a prime example. The paper chart with its limitations can now be replaced by an integration of several technologies. Radar display, positioning and digital hydrographic data can be combined to give a much more powerful tool to the end user. Integration of the tide and current information can lead to data representation where only the danger (red) and no danger (green) zones are represented on the screen as a map for a given ship. The ship itself is properly and dynamically positioned on top, complemented with the digital radar echo showing the instantaneous dangers such as other ships.

A whole new world of applications no longer linked to the traditional products can now be derived. One can think of digital terrain modelling (DTM) capabilities that can help other sciences or techniques (oceanography, dredging, coastal engineering) to model the data to best suit their requirements. The data density enable us to use techniques similar to 'bottom photogrammetry and remote sensing' to see and feel the bottom.

CONCLUSION

This brief description of the impact of digital data on hydrography demonstrates its enormous implications and its potential.

The challenges we face are:
- We must maintain the quality and integrity of the data in an ever more complex and less controllable system.
- We must cease to think and act 'graphically' in a digital environment. We have to stop the reproduction of our former graphical products (little digits on a sheet for example).

- We must start to think 'digitally', by ensuring a clear separation between storage (more global) and application (more specific) of the data. The graphic medium is not the information anymore but just a graphical representation of it.
- We must push the limits of the applications even further.

REFERENCES

1. Burrows, K. G. (1987). Information Management in the Modern Hydrographic Office: a Challenge for the 21st Century. Hydrographic Symp. XIIIth Int. Hydrographic Conf., Monaco, May 12, 1987.
2. DMR and Associates Ltd. (1986). *Development of a CHS Data Base Management System,* Department of Fisheries and Oceans Internal Contract Report, Ottawa.
3. Evangelatos, V. Timothy (1985). Evolving Communications Standards in the Charting and Mapping World: a Report and a Proposal. Department of Fisheries and Oceans Internal Contract Report, Ottawa.
4. Department of Fisheries and Oceans (1986). DFO Energy R&D Bibliography. Department of Fisheries and Oceans internal report, Ottawa.
5. Kerr, A. J. and Verma, H. P. (1987). Hydrography in the Digital Era. Hydrographic Symp., XIIIth Int. Hydrographic Conf., Monaco, May, 12, 1987.
6. Kleins, John. (1986). Proposed Strategy for Large Scale Scientific Information Processing at Fisheries and Oceans Canada. Department of Fisheries and Oceans Internal Contract Report, Ottawa.
7. Van Opstal, L. H. *et al.* (1987). Draft Specifications for Electronic Chart Display and Information Systems (ECDIS). Report by an ad-hoc Group of Experts of the IHO committee on ECDIS, The Hague.
8. Verma, H.P. (1984). An Interactive Graphic Editor for Hydrography. *Lighthouse, Journal of the Canadian Hydrographic Association,* Number 30.
9. Eaton, R. M. *et al.* (1987). Progress with an Electronic Chart Testbed. CHS Conference, Burlington (Ont.).

4

New Echosounding Methods for Shallow Water and Deep Sea Surveying

H. F. Wentzell and Rolf Ziese, Krupp Atlas Elektronik, Federal Republic of Germany

Results from two advanced complementary research echosounding systems are examined and discussed.

The more complex system is a Hydrographic Wide Swath Sweeping Survey Echosounder for shallow and deep water applications (Hydrosweep) for accurate bathymetric surveys and terrain-following navigation. The second system (Parasound) is a hull-mounted Dual Channel Parametric Narrow Beam Deep Sea Survey and Sub Bottom Profiling Echosounder with the functions of a wide beam (20° at 18 kHz, 12° at 33 kHz), narrow beam (2.5° at 33 kHz, 2° and 4° at 18 kHz), and sub-bottom profiler (4° at 2.5–5.5 kHz).

Both systems have been developed under sponsorship of the Ministry of Research and Technology of the Federal Republic of Germany for the new research vessel Meteor.

OVERALL CONCEPT OF THE HYDROACOUSTIC RESEARCH SYSTEM ON BOARD RV METEOR (see Fig. 1)

Both systems described are part of an overall hydroacoustic research system consisting of: hydrographic sweeping survey system (Hydrosweep), parametric survey echosounder (Parasound), standard survey echosounder (Deso), pinger signal and slave survey recorder (Deso), multi-beam fishery sonar, fishery net-sounder, navigation echosounder (461), sonar doppler log (Dolog), sound velocity sounder (C-Sonde), various slave recorders (Deso), various slave colour scopes (320), various digital depth indicators (Filia), gyro-stabilized heave/roll/pitch sensor (Heco).

All sub-systems of the hydroacoustic research system are connected to a central echosounding control system (Echocontrol) which cross-connects trigger and pre-trigger signals; calculates and distributes blanking delay signals; converts all echosounding signals to a standard carrier frequency for distribution to slave recorders and colour scopes; records or replays a signal to/from a tape deck; distributes common draught, sound velocity, heave (roll/pitch), paper speed, time, and annotation text data to all survey echosounders, slave recorders and slave colour scopes; distributes all digital depth values to remove indicators; provides an instant monitor display of all echosounder status data and of used sounding signals in the various laboratories (Fig. 2).

Advances in Underwater Technology, Ocean Science and Offshore Engineering, Volume 16: Oceanology '88

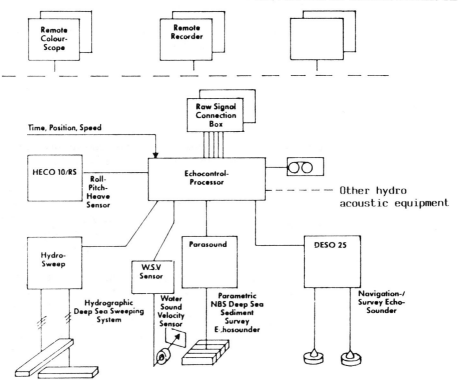

Fig. 1 Hydroacoustic research system on board RV *Meteor*.

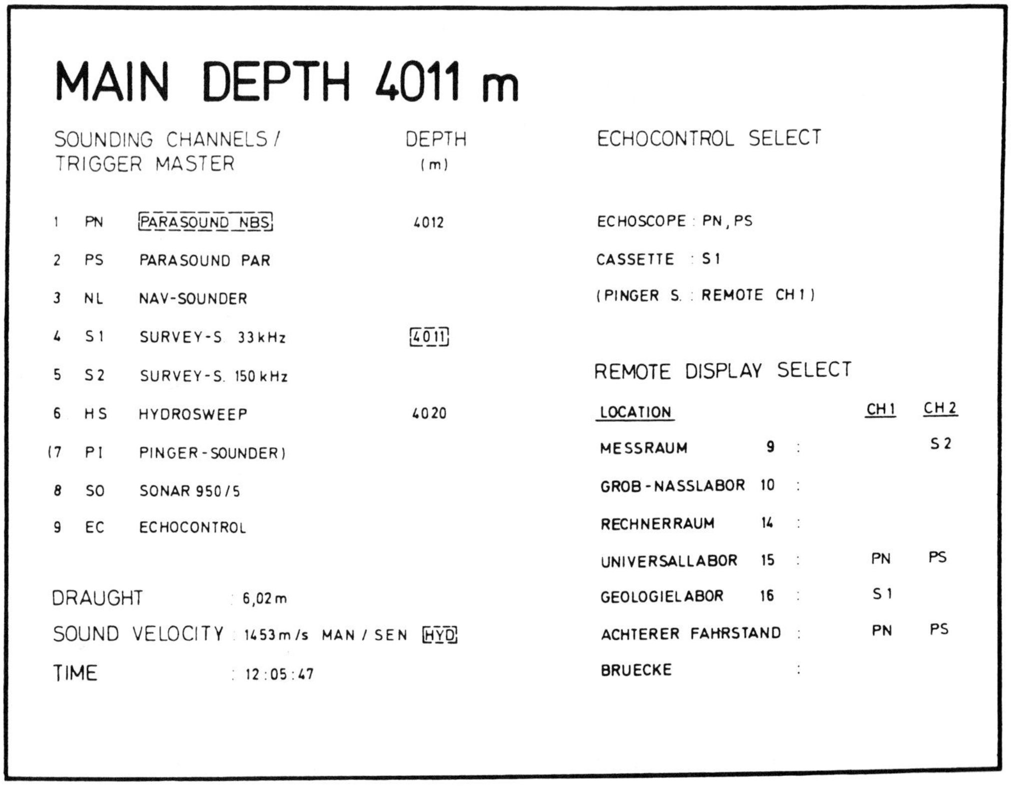

MAIN DEPTH 4011 m

SOUNDING CHANNELS / TRIGGER MASTER		DEPTH (m)
1	PN	PARASOUND NBS	4012
2	PS	PARASOUND PAR	
3	NL	NAV-SOUNDER	
4	S 1	SURVEY-S 33 kHz	4011
5	S 2	SURVEY-S 150 kHz	
6	HS	HYDROSWEEP	4020
(7	P I	PINGER-SOUNDER)	
8	SO	SONAR 950/5	
9	EC	ECHOCONTROL	

ECHOCONTROL SELECT

ECHOSCOPE : PN , PS

CASSETTE : S 1

(PINGER S. : REMOTE CH 1)

REMOTE DISPLAY SELECT

LOCATION		CH 1	CH 2
MESSRAUM	9 :		S 2
GROB-NASSLABOR	10 :		
RECHNERRAUM	14 :		
UNIVERSALLABOR	15 :	PN	PS
GEOLOGIELABOR	16 :	S 1	
ACHTERER FAHRSTAND	:	PN	PS
BRUECKE	:		

DRAUGHT   : 6,02 m

SOUND VELOCITY 1453 m/s MAN / SEN HYD

TIME   : 12 : 05 : 47

Fig. 2 Echosounder control system—monitor.

It was for us as a manufacturer an exceptional situation: we could supply all hydroacoustic sub-systems, partly as existing standard products, but the survey echosounders all had to be developed and produced anew in parallel (and in time) within a con-tract period of only 18 months. Technically, this gave us the chance to introduce a very high amount of commonality throughout all products by hardware components such as silent and dust-free thermo-recorders, high resolution colour displays, and—most importantly—a very high commonality of spare parts.

THE HYDROGRAPHIC SWEEPING SURVEY SYSTEM (Fig. 3)

The main characteristic of the Hydrosweep equip-ment is the large sector of 90° covered by a fan of 59 pre-formed beams. This is made possible by a patented real time calibration method without which refraction effects would lead to large measurement errors in the case of the outer sound beams. With this method the fan of beams from the echosounder is transmitted in the ship's longitudinal direction at regular intervals. The 'oblique profile' measured in this way is stored. During the distance that has already been sailed, a 'vertical profile' was measured with the centre beam.

The two profiles are compared. In an iterative pro-cess, the value of the mean sound velocity in the water is changed repeatedly in both formulas until maximum similarity is obtained between the two profiles, i.e. until the sum of the error squares is a minimum. The mean sound velocity that is necessary for this is output as a 'quasi-measured value'. The Hydrosweep produces on a colour graphical display a real time bottom chart (Fig. 4) or a cross-sectional view (Fig. 5) and an isoline strip chart via an ink-jet printer (Figs 6, 7).

The system will be operated from the operating console. All inputs will be done by menu technique via a function keyboard and controlled from an alphanumerical display.

We have implemented an almost fully digitized sig-

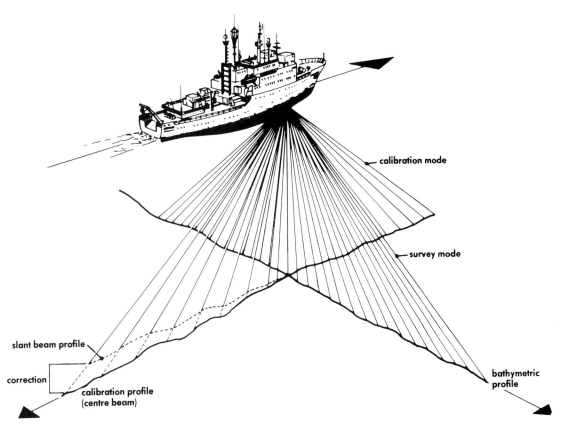

Fig. 3 Hydrosweep: crossed multi-beam configuration for self-calibration beam deflections.

nal processing technique, which allows a high flexibility in transmit/receive schemes. The consequences of this technology are multiple:

- Identical flat and solid state (ice resistant) transducers for transmission and reception;
- 90° rotation of beam pattern to perform calibration sounding;
- Omni-directional transmission-mode for shallow depth, 10–100 m, allowing the operation even in extremely shallow water without changing frequency or transmitter/receiver configurations;
- Sectoral directional transmission-mode for larger depth, 100–10 000 m, allowing the use of a higher transmit frequency than usual, resulting in a higher resolution and accuracy, and smaller transducer dimensions;
- 'Intelligent' (processor controlled) adaptive amplification to improve signal to noise ratio and to minimize side lobe echo interference;
- Adaptive swath coverage from 90° to 45° at depths from 5000 to 10 000 m.

SUMMARY OF OTHER PERFORMANCE CHARACTERISTICS

(1) Real time colour coded bottom map display, true to vessel speed for terrain following navigation (Fig. 4).
(2) Real time cross-sectional depth and signal strength display for quick-look performance analysis (Fig. 5).
(3) Quiet, true to scale, fully annotated isoline strip chart record (Figs 6 and 7).
(4) Quick-Tab menu dialogue via separate VDU.
(5) Complete data logging functions to accept/merge position data with depths etc., magnetic tape storage, ~100 h capacity.
(6) Automatic self calibration of slant beam deflections (Fig. 3).
(7) Automatic determination and output of the 'mean water sound velocity' over the whole water column, in time/distance intervals corresponding to approximately twice the water depth.

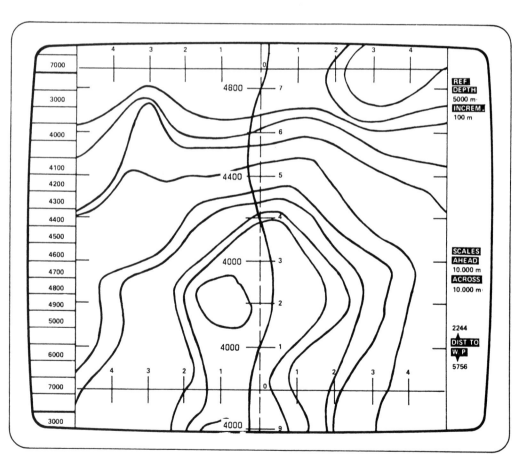

Fig. 4 Hydrosweep: bottom map display.

8) Roll-, pitch-, and heave compensation. (Heave compensation is very important for high resolution bathymetry analysis in shallow water)

(9) Built-in Operator Training Program.

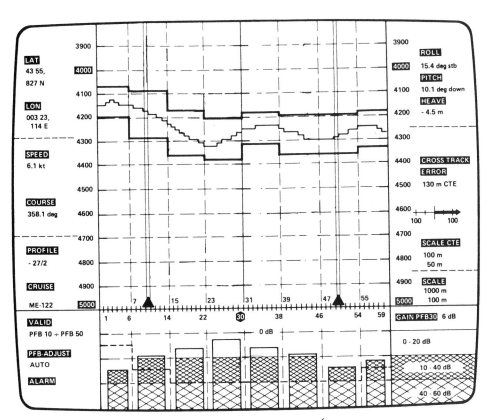

Fig. 5 Atlas Hydrosweep: profile and performance monitoring display. The display is 'steady'; its contents are overwritten anew with each receiving cycle.

Left Annotation of the sailing parameters

Middle

Upper part Cross profile display (during calibration: longitudinal profile
> (1) For each PFB, the measured depth of water is displayed with a vertical linear scale.
> (2) The 'search window' for the depth digitizer is repositioned automatically.
> (3) The lateral boundary of the beams of which the data are to be recorded can be selected manually.
> (4) The numbers of the beams are marked on this line.

Lower part For each of eight amplifier channels of the side beams, a bar graph is shown for the 'attenuation', i.e. the relative reflectivity of the bottom.

Right

Upper part Annotation of dynamic parameters of the ship at the instant of transmission.

PRESENTATION AND DISCUSSION OF RESULTS

The worst problem was that *Meteor* caused severe problems to all hydroacoustic sensors by generating an unbelievable amount of aerated water under the hull, approximately one-fifth of the length behind the bulbous bow.

The disturbances were so severe that it was decided to cut the bulb and to replace it with a new, almost conventional, bow. After this conversion was finished in October 1986, extensive sea trials of the Hydrosweep System took place in the Bay of Biscay, off the coast of Brest.

Although much more recent data have been obtained with the system, the data from these trials will be used as they have been analysed most thoroughly by at least three independent organizations: the company's own bathymetric post processing system (a shipborne version of it is also installed on board *Meteor*); the Alfred Wegener Institute for Polar and Marine Research, FRG (Dr Schenke); the NECOR Sea Beam Development Center, the University of Rhode Island USA (Dr Robert Tyce). (Both intend to publish results of their work separately, but references were not available at the deadline for the manuscript of this paper.)

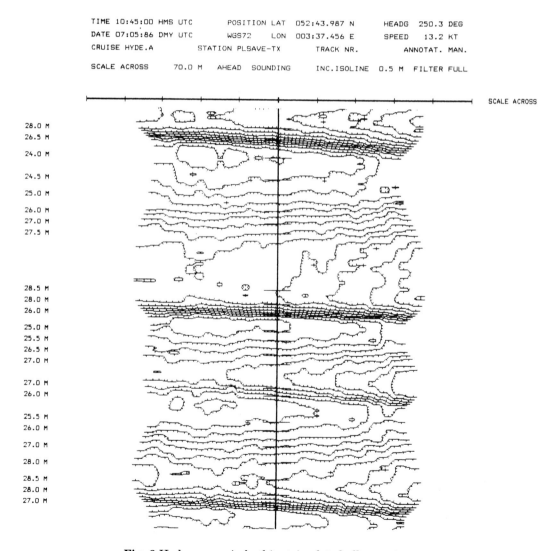

```
TIME  10:45:00 HMS UTC        POSITION LAT  052:43.987 N      HEADG   250.3 DEG
DATE  07:05:86 DMY UTC            WGS72  LON  003:37.456 E      SPEED    13.2 KT
CRUISE HYDE.A            STATION PLSAVE-TX       TRACK NR.          ANNOTAT. MAN.

SCALE ACROSS      70.0 M    AHEAD   SOUNDING    INC.ISOLINE  0.5 M  FILTER FULL
```

Fig. 6 Hydrosweep: isobathic strip plot shallow water example, giant ripples in the English Channel.

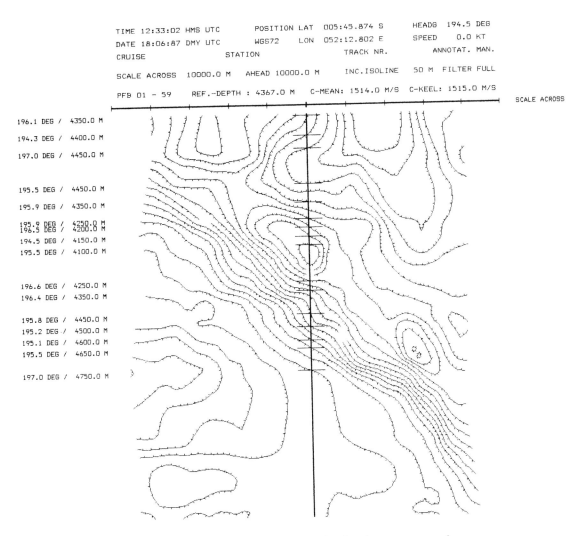

Fig. 7 Hydrosweep: isobathic strip plot deep sea example
off Madagascar.

Figure 8 shows a 3-dimensional view of the test area; Figure 9 shows a plan view with the test pattern; Figure 10 shows an isobathic strip-plot from Hydrosweep; Figure 11 shows the corresponding isobathic plot after post processing by the University of Rhode Island; Figure 12 shows a tabular summary of quantitative test results.

The data sets provided to the NECOR centre originally showed unusually bad data quality and repeatability. It was found, however, that the raw data storage philosophy of the Hydrosweep deviated considerably from the method used in the Seabeam system, which stores filtered data, and to which most bathymetric postprocessing systems are adapted.

As the raw analogue echosounding signals of the test area were still available from analogue video tapes, we could reprocess the raw data in the Hydrosweep sensor system and apply cross-track filtering before storing the digital depth patterns on to magnetic tape, this resulted in considerable improvement of data quality and repeatability.

In the NECOR post processing system, the data were then averaged along the track to produce comparable averaging parameters between the data points. Since averaging filters depth noise in the data, the cross-track averaging applied to the Hydrosweep data, together with the along-track averaging, gives it depth filtering comparable to that (according to Dr Tyce) generally applied to Seabeam Data.

The new Hydrosweep data still showed some rela-

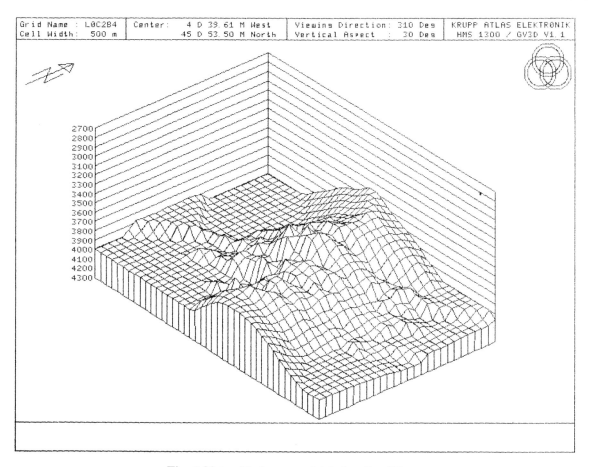

| Grid Name : LOC2B4 | Center: 4 D 39.61 M West | Viewing Direction: 310 Deg | KRUPP ATLAS ELEKTRONIK |
| Cell Width: 500 m | 45 D 53.50 M North | Vertical Aspect   : 30 Deg | HMS 1300 / GV3D V1.1 |

Fig. 8 *Meteor*/Hydrosweep trials location 2B.

tively modest side lobe interference effects; such errors do occur also with the previous multi-beam system when starting down a steep slope and the tracking gate fails to follow the actual bottom, and one such manifestation is referred to as the 'omega effect' in the literature.

This effect results from side lobe energy reflected back upslope with intensity comparable to that returning from the reduced backscattered return beneath the ship. In the meantime the transmission pattern of Hydrosweep has been modified to include downward as well as port and starboard transmissions so as to provide increased signal beneath the ship. This method has proved to be efficient, and the 'omega effect' is no longer a problem.

Before proceeding with gridding of data for comparison, navigation of the various tracks was adjusted to provide the most consistent bathymetry. Lines with GPS navigation generally do not need much

adjustment, since our experience is that this navigation is accurate, provided changes in the constellation utilized do not occur. Tracks with Transit navigation and dead reckoning were adjusted at the beginning and end of track only, giving the best fit with minimal adjustment. Tacks were numbered first from north to south on east–west tracks, then west to east on north–south lines from the original data. Line 6 from the original data was not included, but replaced with a new westernmost line labelled 9. This was rather arbitrary, since the time sequence actually follows line 9, 7, 8, 1, 2, 3, 4, 5. There is no line 6 in the new data. Owing to some obvious navigation errors in line 2, lines 1 and 2 were not used in the data comparison. Line 4 was chosen as the reference line and all other lines adjusted to match line 4. For the remaining lines the navigation adjustments were less than $\frac{1}{4}$ nautical mile.

All lines were plotted in original form. The data

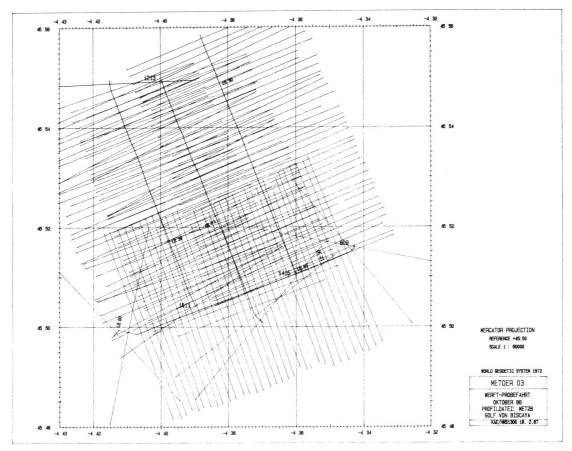

Fig. 9 *Meteor*/Hydrosweep trials location 2B: tracks with coverage 'spokes'.

were then gridded in a latitude/longitude grid with spacing of approximately 150 m, provided grid cell elements totalling 13 464 for the survey area. Grid cells were then examined to determine when more than one line had data in a cell for comparison. Pairs of lines, parallel lines, and all lines together were compared for depth agreement, with standard errors.

A small Seabeam survey of comparable size and depth, though with reduced overlap, was examined in the same way as the Hydrosweep data for comparison, the results are included in Figure 12.

This analysis shows that the quality of Hydrosweep data is at least comparable to previous systems. This is considered an excellent result since the Hydrosweep swath is more than twice as wide, and consequently the outer beams are much more susceptible to deflections. Clearly, the self-calibration method described above is very efficient in compensating those effects.

THE PARAMETRIC SURVEY ECHOSOUNDER

When the tender specification for the research vessel *Meteor* was issued in 1984, two separate echosounders were specified; a hull-mounted 3.5 kHz subbottom profiler (SBP) and a mechanically roll/pitch stabilized narrow beam 12/24 kHz deep sea echosounder (NBS). The budget was such that the decision to combine both pieces of equipment in one was a logical consequence, because at that time (for other purposes) we were experimenting with the 'parametric' or 'non-linear' phenomena of hydroacoustic systems.

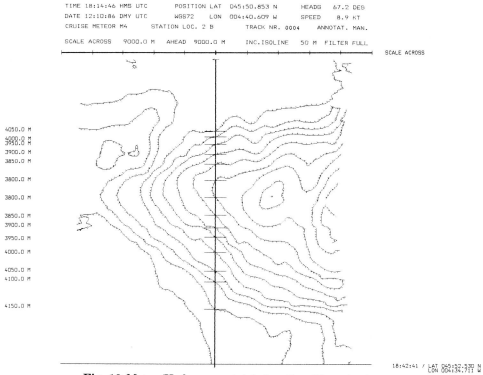

```
TIME 18:14:46 HMS UTC        POSITION LAT  045:50.853 N        HEADG    67.2 DEG
DATE 12:10:86 DMY UTC           WGS72   LON  004:40.609 W       SPEED    8.9 KT
CRUISE METEOR M4       STATION LOC. 2 B       TRACK NR. 0004    ANNOTAT. MAN.

SCALE ACROSS   9000.0 M    AHEAD   9000.0 M     INC.ISOLINE   50 M  FILTER FULL
                                                                    SCALE ACROSS
```

Fig. 10 *Meteor*/Hydrosweep trials location 2B: strip chart
plot.

	Elements used		Standard deviation (m)			
Lines	Only 1	More than 1	Minimum	Maximum	Average	Deviation
Hydrosweep data 136×99 grid (13 464 total elements)						
3,4	3031	1856	0	59	9	8
3,5	4255	948	0	65	14	11
3,7	2580	1896	0	79	11	9
3,8	2724	1545	0	55	12	9
3,9	2837	1551	0	84	12	11
4,5	3096	1620	0	52	13	11
4,7	3379	1589	0	81	12	9
4,8	3799	1100	0	56	11	9
4,9	3600	1262	0	47	10	8
5,8	4385	511	0	45	10	9
5,9	3962	785	0	63	13	10
7,8	2556	1536	0	72	14	10
7,9	2539	1607	0	76	14	12
8,9	3501	847	0	62	11	9
3,4,5	2987	2944	0	65	12	10
5,6,7	3927	1019	0	52	13	10
7,8,9	2525	2512	0	72	14	10
3,4,5, 7,8,9	2203	5004	0	81	13	9
Seabeam data 124×124 grid (15 376 total elements)						
All lines		1024	0	100	15	14

Fig. 11 *Meteor*/Hydrosweep trials location 2B:
post-processed isobathic plot.

Fig. 12 *Meteor*/Hydrosweep trials location 2B: quantitative
analysis.

Describing the phenomena in detail goes beyond the purpose of this paper, some information must be summarized to explain the performance characteristics of such an echosounder:

Wide beam SBP

A typical hull-mounted 3.5 kHz transducer for sub-bottom profiling has a surface of 0.5 to 1 m² and generates a beam with an opening angle of ~ 25–40°. Such a wide beam width 'illuminates' a large area on the sea bottom, and all reflections from there are interfering with each other in the receiver, causing a barely legible echogram, especially when the morphology is hilly (Fig. 13(a)). The problem is also shown in Figure 14.

Because of the indicated shortcomings, there is a trend to use fish-mounted SBPs, but this method is economically inefficient as it slows down the vessel speed to about 2–3 knots. It is therefore not practical for cruise operations.

Narrow beam SBP

If two signals with adjacent frequencies are radiated into the water at *very high energy*, then—and only

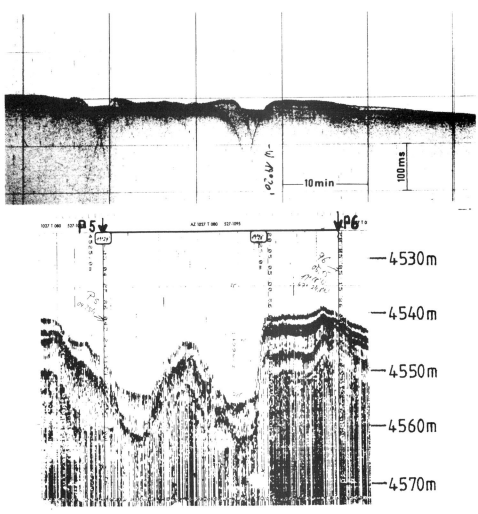

Fig. 13 (*a,b*) Comparison of two recordings made by a conventional 3.5 kHz sediment echosounder installed in the ship and the Atlas Parasound system, showing an example of sediment waves in water of depth 4550 m approximately. The recordings have a parallel displacement of about 1 nm relative to one another. Recorded on the Iberian deep-sea plane (research vessel *Meteor*).

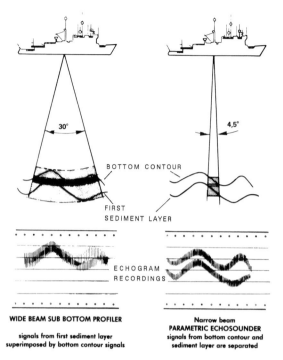

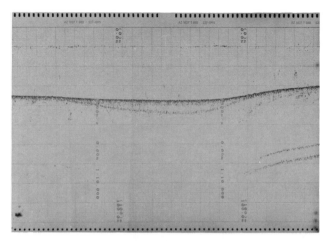

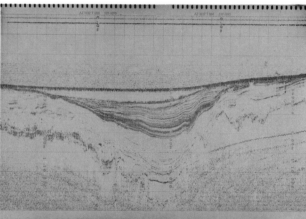

Fig. 14 Influence of beam angle on vertical resolution.

Fig. 15 (*a,b*) The acoustic parametric effect. Comparison of the 18 kHz recording of the sea bed (*a*) with the 3.5 kHz recording (*b*) made by mixing the frequencies of 18 and 21.5 kHz. Recorded in the Kattegat (research vessel *Meteor*, Parasound).

then—the difference between both 'primary' frequencies is generated in the first few metres of the water column below the transducer—the 'secondary' frequency. The secondary frequency is only generated within the main beam of the primary frequencies (only there is enough energy), resulting in a beam practically free from any side lobes and as narrow as the narrowest beam of the primary frequencies. Such a narrow beam 'illuminates' only a small area on the bottom, and the echogram obtained shows much more detail of both the morphology and the sediment structure (Fig. 13(*b*)). (The disturbances in this echogram are caused by aerated water under the transducer due to heavy seas.)

Figures 15(*a*) and (*b*) illustrate the acoustic parametric effect by a recording made in the Kattegat. The recordings show that the parametric frequency of 3.5 kHz selected here penetrates the sediment for a distance of up to 60 m, while the 18 kHz frequency penetrates only through the first 7 m of fairly liquid sediments. In addition to the large depth of penetration, the fine horizontal and vertical resolution of the sediment structure which can be achieved by the Atlas Parasound is also visible.

Additional NBS-function

Because the beam widths of the primary frequencies must be narrow in any case in order to generate

enough energy in the water column and because of the energy itself, the narrow beam deep sea function (NBS) is a consequential effect from a parametric echosounder, if the primary frequencies are chosen low enough. In our case a first primary frequency of 18 kHz and a second primary which is variable between 20.5 and 23.5 kHz (resulting in SBP-frequencies between 2.5 and 5.5 kHz) were chosen.

UNIVERSAL RESEARCH/SURVEY ECHOSOUNDER

To cover as wide a range of applications as possible, a number of features has been designed into the system which makes it a universal research/survey echosounder:

- Narrow beam, deep sea operation (NBS):
  18 kHz, 4° beam width, 10 000 m
- Extreme narrow beam width:
  18 kHz, 2° beam width (reduced side lobes) for extremely steep terrains
- Narrow beams, survey operation:
  33 kHz, 2.5°
- Wide beam, fishing operation and for shallow water:
  18 kHz, 20° or 33 kHz, 12°
- Narrow beam, SBP operation:
  frequency selectable in steps of 0.5 kHz from 2.5 to 5.5 kHz, 4°
- Multi Pulse Operation, synchronization and rate automatically controlled by a 'pilot—NBS signal'
- Dual channel operation and parallel recording of NBS and SBP signals, with automatic phase control for 'watch-free' overnight operations.

Automatic start control of bottom TVC (TVG)

The travel time of sound in water is different from that in sediments. To achieve optimum penetration and sediment layer resolution, the 'bottom TVC' for the SBP-signal can be adjusted separately. Although this is also possible in other SBP-sounders, it was always difficult and tedious to set and maintain the start time for the bottom TVC with changing water

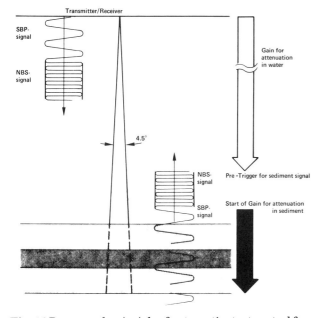

Fig. 16 Parasound: principle of automatic start control for the bottom TVG.

depth. Because Parasound uses a relatively high primary frequency with a precise depth digitizer in parallel, the start depth for the bottom TVC is always automatically and precisely determined and maintained (Fig. 16).

Horizontal resolution

For a good horizontal resolution at greater depth, it is not sufficient to send out the next sounding pulse only after the previous one has arrived. Below 2000 m, Parasound can be set to an automatic multi-pulse operation (sometimes also called pulse–train operation), i.e. more than one sound pulse is travelling in the water column at any moment.

Obviously, the number of pulses must be carefully controlled, as the pulse length and time distance between them are predetermined. The possible number of pulses is therefore a function of the depth. Again the same SBP systems provide a multi-pulse function, but only with manual control. Parasound provides this function automatically.

First a pilot pulse is sent out and the bottom depth is determined. The processor now automatically decides how many pulses can fit into the travel time and sends out a burst of these pulses. Next, the depth is determined again, and so on.

Vertical resolution

Compared with conventional SBPs, the parametric principle not only provides a sharper beam, but also results in a better vertical resolution. This is a consequence from the 6-times-higher primary frequency, which allows an approximately 6-times-shorter pulse length to be sent out. Furthermore the technology of Parasound allows the transmission of primary pulse length, which result in the generation of SBP pulses with precise wave periods. The operator can select to transmit SBP pulses of 1 through 8 wave periods.

The above measures result in a considerable improvement of vertical resolution of up to 10 or 30 cm, even in deep sea sediments.

Heave/roll/pitch compensation

Because of the high vertical resolution, the heaving of the research vessel would compromise display/recording of the sediment structures at high resolution phase settings, therefore a precision real-time heave-compensator Heco 10 is connected, which also contains Roll/Pitch pick-ups for vertical beam control.

Resolution of the sediment structure (examples of uses)

The fact that sediment layer resolution is possible up

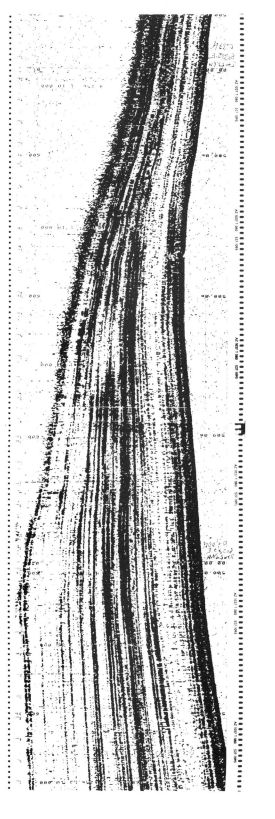

Fig. 17 Layer conditions along a north–south section with penetration depths of up to 100 m in water of mean depths 520 m. The sound velocity setting was constant: $c = 1469\,\mathrm{m\,s^{-1}}$. (Recorded in the Skagerrak on German research vessel *Meteor*, Parasound.)

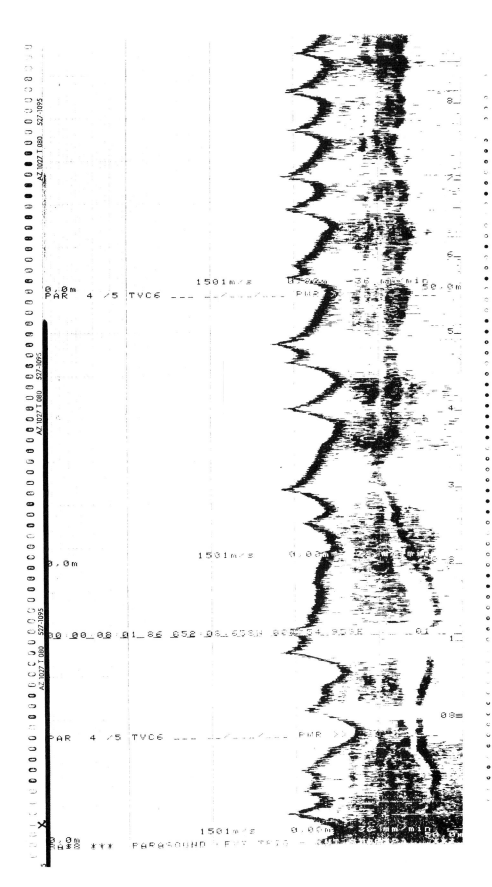

Fig. 18 Giant active sand waves up to 10 m deep in water of depth 30 m approximately. The static reflector can still be recognized clearly in this depth of sandy sediment. (Recorded in the English Channel on German research vessel *Meteor*, Parasound.)

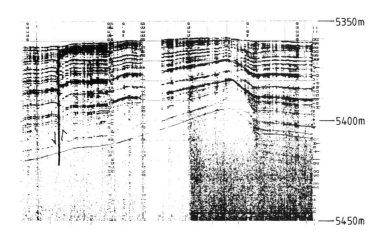

Fig. 19 Resolution of the sediment structure with a depth of 50 m in water of depth 5360 m. The fault in the layer sequence can clearly be seen in the left half of the recording. (Recorded in the Iberian deep-sea plane on research vessel *Meteor*.)

to 100 m is illustrated by the recording in Figure 17 which shows a north–south sectional view from the Skagerrak about 10 kM in length. The average depth of the water is about 520 m. In fact the depth of penetration in this example is even greater, because the analogue recording is scaled with the set sound velocity of $c = 1469\,\mathrm{m\,s^{-1}}$. In general, a sound velocity of about $1800\,\mathrm{m\,s^{-1}}$ can be assumed in 'sandy' sediments.

The high horizontal resolution of the morphology on the sea bed which can be achieved with the Parasound system is illustrated by an example from the English Channel (Fig. 18), showing an impressive view of giant active sand waves with a depth of up to 10 m. Here, a parametric frequency of 4 kHz was used.

An example of a deep-sea application is shown in Figure 19. The recording was made with the Parasound system on board the research vessel *Meteor* over the Iberian deep-sea plane. At a mean water depth of 5630 m, penetration depths of up to 50 m in the sediment were achieved. The vertical resolution is in the metre range. The fault in the layer sequence can clearly be seen in the left-hand half of the picture, and the associated 'arching' of the sediment layers in the right-hand half.

The relevance of high resolution sediment analysis to the task of determining the distribution of physical soil characteristics is clear from Figure 20. This figure shows the results of clear strength measurements of a core sample in comparison with a sediment echogram obtained with Parasound from the same location.

CONCLUSION

From the qualitative and quantitative analysis we can state:

(1) Sweeping system

The new hydrographic sweeping system has demonstrated a substantial improvement in the technology and performance of multi-beam swath survey systems:

- Transducers are smaller and less susceptible to mechanical damages.
- The swath coverage has been more than doubled.
- The accuracy has been slightly improved despite of the wider coverage due to the self calibration function.
- The real-time 'bottom map display' enhances the terrain following navigation capability of a research vessel.

(2) Parametric echosounder

The parametric narrow beam sub-bottom profiler/survey sounder has demonstrated the capability of high resolution sediment surveys with hull mounted equipment, with comparable quality to towed SBP-systems for the horizontal resolution, and with superior vertical resolution.

The additional NBS functions make the sounder usable also for general precision bathymetric surveys in deep and shallow water.

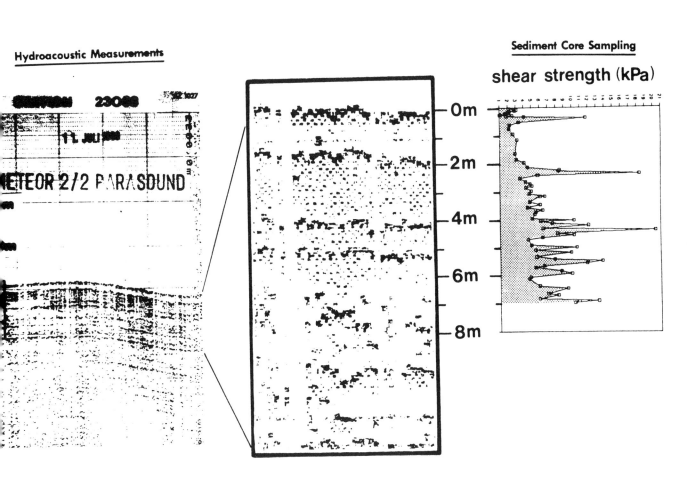

Fig. 20 Comparison between Atlas Parasound registration in the Norwegian Sea (RV *Meteor*) and shear strength measurements at a sediment core, sampled in the same region (Kassens, Geol. Palaeontol Inst., University of Kiel/Fed. Rep. of Germany).

Part II

Oceanography

5

The Canadian Meteorological Buoy Network

Andrew L. Wood, Seakem Oceanography Ltd, Sidney, British Columbia, Canada
and *Gary E. Wells,* Pacific Weather Centre, Atmospheric Environment Service, Environment Canada

From 1952 to 1981 the Canadian government operated two ships off the West Coast, primarily for meteorological observations but also for collecting a wide range of oceanographic data. These vessels sailed from Victoria to Ocean Station 'P' (PAPA) at 50°N 145°W (Fig. 1). In June 1981 they were decommissioned and sold because of budgetary constraints. For several years the only direct source of marine meteorological data was from commercial vessels, from drifting buoys, and from US buoys moored in the Gulf of Alaska. Fishermen constantly lobbied for reinstatement of the weatherships and for other programs to improve the marine forecasts.

During summer a high pressure system persists off the British Columbia coast (Fig. 1). The Hawaii High, as it is called, results in generally stable and calm conditions. In 1987 for example, Victoria, at the southern tip of Vancouver Island, had a record 153 days without significant rainfall from May to October. When the Hawaii High moves south in October, low pressure systems flow in off the north Pacific bringing frequent storms to the entire coastline. Figure 2 shows the difference in seasonal rainfall and the tracks of winter storms. In addition, British Columbia's rugged coastline is indented by

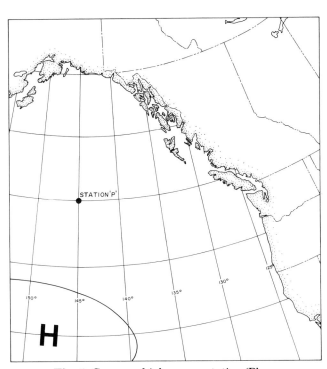

Fig. 1 Summer high, ocean station 'P'.

Advances in Underwater Technology, Ocean Science and Offshore Engineering, Volume 16: Oceanology '88

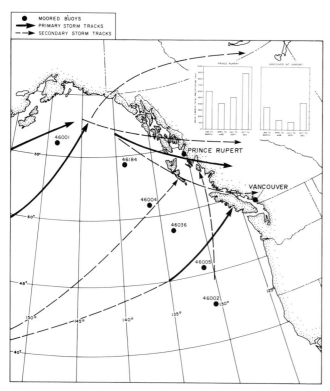

Fig. 2 Winter storms and seasonal rainfall on the British Columbia coast (September–March).

long narrow fjords, resulting in very localized weather conditions, especially during the winter when cold air flows down the valleys from the interior.

Major systems can be tracked with little difficulty by the Atmospheric Environment Service. On occasion, however, rapidly deepening low pressure systems move in off the Pacific. These storms, often referred to as 'bombs' can be highly destructive if adequate advance warning is not provided. On the night of October 11, 1984, a bomb struck the west coast of Vancouver Island sinking seven vessels and drowning five seamen. An inquiry was immediately launched to investigate the adequacy of marine weather forecasting on the West Coast. The subsequent report by Dr Paul LeBlond of the University of British Columbia made several key recommendations:

- The general level of weather services provided to mariners should be enhanced towards that available to aviators.
- Specialist marine forecasting positions should be created in coastal forecast offices.
- Sea state forecasting be included in the services provided to mariners.

- Development of a number of marine meteorological data acquisition systems should be pursued (e.g. anemometers for drifting buoys).
- Research on the physics of the rapid deepening process of storms should be carried out.

In a cooperative effort between the Atmospheric Environment Service, the Canadian Coast Guard, and Fisheries and Oceans Canada, steps are being taken on several fronts to improve marine sea state and weather forecasts and search and rescue (SAR) response capabilities. One of these is the establishment of a network of permanently moored buoys in the coastal zone and in the deep offshore.

THE OFFSHORE BUOY PROGRAM

The United States maintains, and is constantly upgrading and expanding, the world's largest network of weather buoys (Fig. 3). Its inventory includes 10 and 12 metre discus buoys, 6 metre Navy oceanographic meteorological automated devices (NOMADs), and more recently 3 metre discus buoys. The Canadian government turned to the US National Data Buoy Centre (in Bay St Louis, Mississippi) for advice and in 1986, awarded a contract to Seakem Oceanography Ltd (of Victoria, British Columbia) to build five NOMAD buoys. The aluminium hulls are the same as those used by the NDBC but the payloads are based on a ZENO designed by Coastal Climate Company (of Seattle, Washington) and built under licence by Seakem in Canada (Figs 4 and 5); NDBC buoys use a Magnavox DACT electronics package. The NOMADs, which

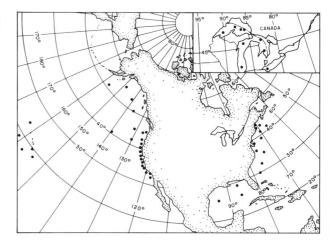

Fig. 3 NDBC buoy location map.

Fig. 4 NOMAD buoy.

Fig. 5 NOMAD buoy

weigh 10 tons fully ballasted, are equipped with two anemometers, two barometers, a water temperature sensor, an air temperature sensor, a heave sensor, and a 5 mile visibility navigation light. The sampling period is one hour, with processed data transmitted every hour through the geostationary operational environment satellite (GOES) to a downlink at AES's Pacific Weather Centre in Vancouver and the Global Telecommunications System through NOAA NESDIS. The data are also relayed by a backup ARGOS transmitter to AES's Edmonton, Alberta,

office. Power is supplied by air depolarized primary cells and solar panels.

The mooring design is a reverse catenary (Fig. 6) using a combination of $1\frac{1}{2}$ inch 12-strand braided polypropylene in the lower section and negatively buoyant $1\frac{1}{4}$ inch 12-strand polyester/olefin in the upper section to prevent fouling with the chain immediately under the buoy. A 12 ton anchor and 360 feet of $1\frac{1}{2}$ inch chain are used on the bottom.

In September, 1987, two of these Canadian buoys were deployed in about 3400 metres of water from a Canadian Coast Guard vessel at the locations shown in Figure 7. Coverage is also provided by a US NOMAD at the site west of Queen Charlotte Sound; this will be replaced by another Canadian Buoy in 1988. The 4th and 5th NOMADS will be used as spares during routine servicings every 2 years while the moorings will be replaced about every 6–10 years. On October 4, two weeks after their deployment, a bomb passed near the northern station. The plots in Figure 8 show the rapid and dramatic change at the site.

THE INSHORE PROGRAM

The Canadian databuoy program actually started in May 1986 with the deployment of a 3 metre discus (3D) buoy in Douglas Channel north of Prince Rupert (Fig. 9). In June 1987 another was moored in Howe Sound near Vancouver. These smaller buoys have the same payload as the NOMADs (but with only one pressure sensor and no ARGOS transmitter) and also report through the GOES satellite. The aluminium hull was originally designed at Woods Hole Oceanographic Institution and later modified by the NDBC. A recently completed one year test demonstrated that it can survive the severest storms in the north Pacific but for the near future it will only be used nearshore. AES has ordered two more 3D buoys for the west coast in 1988 and will add others in subsequent years.

THE GREAT LAKES PROGRAM

At present there is one Canadian 3D buoy moored in Lake Erie and three more will be deployed in the Great Lakes in 1988. Water levels in the Lakes were exceptionally high in 1986 and 1987, resulting in considerable property damage and coastal erosion. During the ice-free summer and fall, these buoys not only provide routine data for weather forecasts but also enable the AES to provide warnings of the high waves that cause the shoreline damage.

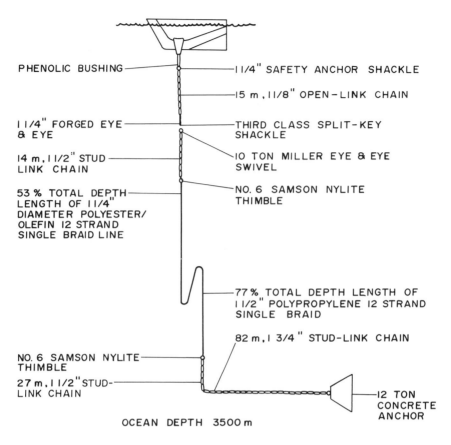

PHENOLIC BUSHING

1 1/4" SAFETY ANCHOR SHACKLE

15 m , 1 1/8" OPEN – LINK CHAIN

1 1/4" FORGED EYE & EYE

THIRD CLASS SPLIT-KEY SHACKLE

14 m , 1 1/2" STUD LINK CHAIN

10 TON MILLER EYE & EYE SWIVEL

53 % TOTAL DEPTH LENGTH OF 1 1/4" DIAMETER POLYESTER/ OLEFIN 12 STRAND SINGLE BRAID LINE

NO. 6 SAMSON NYLITE THIMBLE

77 % TOTAL DEPTH LENGTH OF 1 1/2" POLYPROPYLENE 12 STRAND SINGLE BRAID

82 m , 1 3/4" STUD-LINK CHAIN

NO. 6 SAMSON NYLITE THIMBLE

27 m , 1 1/2" STUD-LINK CHAIN

12 TON CONCRETE ANCHOR

OCEAN DEPTH 3500 m

Fig. 6 NOMAD buoy: inverse catenary mooring.

EAST COAST

Although the north Atlantic is renowned for its severe storms (as Britain and France know from their experience in October, 1987), the AES had not failed to predict the advent of a major system for many years. Much of the success can be attributed to the relatively dense terrestrial data network upstream from the weather centre in Halifax, Nova Scotia. Consequently, initial financial resources available for the Canadian buoy network were allocated to the West Coast.

Nevertheless, resources permitting, an East Coast buoy network will be established to assist the fishing and other marine industries which generally must work much farther offshore than the West Coast. One 2.5 metre hexoid buoy is being instrumented for deployment in 1988 and the AES is conducting a test in which an automatic ARGOS-reporting weather station has been mounted on a fisheries patrol vessel that operates at the edge of the continental shelf off

Nova Scotia. To date, this system (designed by Coastal Climate Company) has operated successfully.

OTHER MARINE DATA COLLECTION SYSTEMS

Automated shipboard aerological program (ASAP)

ASAP was designed to provide accurate surface and upper atmospheric meteorological data from commercial ships-of-opportunity operating between Canada's West Coast and the Orient. Conceived by AES, there are now four ships carrying ASAP trailers (Fig. 10): three off the West Coast and one (on loan to the British government) operating between Canada and the United States. Every 12 hours a sonde is released from the trailer through a specially designed tube. Data are relayed by radio down to the ship, automatically processed on board, and transmitted through

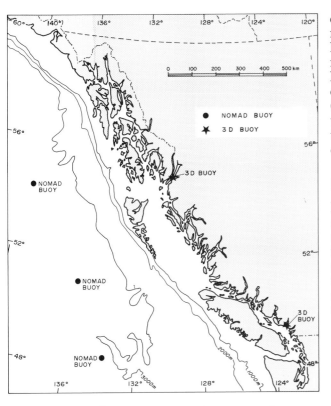

Fig. 7 Canadian West Coast buoy network.

GOES. The only human interaction is in preparing the balloon for launching and monitoring the data processing and transmission.

Shore Stations, Lighthouses, and MAREP

AES maintains a number of automated meteorological shore stations along the east and west coasts that report through GOES. In addition, the lighthouses operated by the Canadian Coast Guard provide routine meteorological forecasts.

Mariner Reporting Program (MAREP) is a volunteer program intended to supplement routine meteorological data collection, particularly in coastal waters. Within given areas mariners can report to designated receiver stations using radiotelephone, and the data are then relayed to the Pacific Weather Centre in Vancouver for use in upgrading the next forecast or for issuing a marine weather warning. In return the marine forecaster provides updated information to the mariner who supplied the report.

Supplementary programs

Fisheries and Oceans Canada is assisting with marine safety by research into wave patterns on both coasts and the Great Lakes. The Marine Environmental Data Service (MEDS) operates several wave heave and wave directional buoys reporting automatically via satellite to lighthouses or through radio link into the telephone system. Using the recently developed Canadian Search and Rescue Program (CANSARP) which models the drift of floating objects under given atmospheric and oceanographic conditions, the Coast Guard is upgrading its SAR capabilities during emergencies.

SUMMARY

In the last three years, Canada has implemented several important programs to improve its marine sea state and weather forecasting service and to assist in search and rescue operations. All of these programs are being constantly expanded and upgraded using the latest available technologies. With the longest coastline in the world, but with very low population density and limited resources, Canada has been able

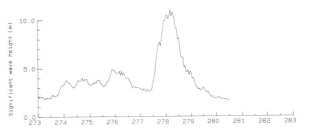

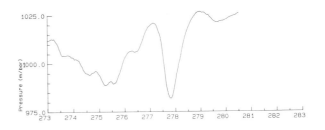

Fig. 8 Data from Canadian NOMAD at 53°55′N, 138°52′W.

to achieve a great deal in a very short time. In cooperation with the United States, Canada is rapidly achieving a world class reputation in this field to the benefit of all North Americans.

Fig. 9 3 metre discus buoy.

Fig. 10 Automated shipboard aerological program (ASAP) releasing a sonde.

REFERENCES

1. LeBlond, Paul H. (1984). Final Report of the Investigation on the Storm of October 11–12, 1984 on the West Coast of Vancouver Island. Report to the Minister of Environment Canada.
2. National Data Buoy Centre. (1987). Canadian 3-Metre Buoy Data Evaluation. US Department of Commerce. National Oceanic and Atmospheric Administration. No. F-842-4.

6

Airborne Surveys of Ocean Current and Temperature Perturbations Induced by Hurricanes

Peter G. Black, National Oceanic and Atmospheric Administration, Hurricane Research Division, Miami, Florida 33149, USA and *Russell L. Elsberry* and *Lynn K. Shay,* Department of Meteorology, Naval Postgraduate School, Monterey, California 93943, USA

As part of a Hurricane Planetary Boundary Layer Experiment, Airborne Expendable Current Profilers (AXCPs) were deployed during the passage of East Pacific Hurricane Norbert and Atlantic Hurricanes Josephine and Gloria on September 23, 1984, October 11, 1984 and September 26, 1985, respectively. The experiment was a joint effort between federal government and private industry which resulted in the first detailed current profile measurements below a hurricane. A total of 92 AXCPs were deployed in these storms, of which 45 transmitted useable data to the NOAA WP-3D aircraft. Most of the failures occurred in the highest wind and wave regions according to Sanford *et al.* (ref. 1). A total of seven satellite-tracked drifting buoys were also deployed in Hurricanes Josephine and Gloria. The Josephine buoys (3) were the Polar Research Lab (PRL) type and were equipped with an anemometer, pressure sensor, and a thermistor chain with sensors at the 40-, 60-, 100- and 200-metre levels. Initial results from analysis of this data set are provided in Black *et al.* (ref. 2). The Gloria buoys (4) were the Horizon Marine type and were instrumented only for position and sea surface temperature (SST).

The objective of the experiment was to measure the SST decrease induced by these hurricanes together with the storm-generated surface and subsurface current field in order to specify the dynamical processes which are involved. It is hoped that improved parameterizations of the SST decreases and storm-induced currents can be derived from this data set. It has been shown (refs 3 and 4) that the asymmetric SST distributions induced by hurricanes may contribute to their asymmetric structure and reduce air–sea sensible and latent heat fluxes.

The Norbert observations revealed a divergent cyclonic circulation within the mixed layer similar to that predicted by recent numerical model simulations. Maximum mean mixed layer currents of $1.2 \, \mathrm{m \, s^{-1}}$ were observed. Below the thermocline, a

Advances in Underwater Technology, Ocean Science and Offshore Engineering, Volume 16: Oceanology '88

weaker anticyclonic circulation was observed, which resulted in 180° phase shifts in current vectors across the thermocline in most quadrants of the hurricane. Local Richardson numbers were less than 0.2 in the right-rear quadrant of Norbert, where sea surface temperature decreases of 2°C were observed in a crescent-shaped pattern centered at a radius 1.5 times the radius of maximum winds. It is uncertain whether these strong currents and large shears are a result of enhanced surface stresses due to fetch-limited seas, as recently proposed by several authors, or to a resonant interaction of the hurricane wind field with inertially rotating currents.

The Josephine and Gloria observations revealed a more complex eddy pattern induced by the interaction of the storm with the ocean Subtropical Front. Maximum inertial current amplitudes of $0.6 \, \text{m s}^{-1}$ were observed, superimposed on the eddy currents. The Gloria buoy observations were not capable of resolving inertial motions, but did reveal strong, divergent surface currents in the right-rear quadrant, in agreement with the Norbert observations.

Large surface wave-induced currents were observed in the right quadrants of these storms. The largest values of nearly $2 \, \text{m s}^{-1}$ were observed in Gloria with a period of 12 s. When added to the mean currents of $1 \, \text{m s}^{-1}$, this resulted in a total peak current of $3 \, \text{m s}^{-1}$ every 12 s.

RELEVANT AIR–SEA PARAMETERS AND SCALES

The initial upper ocean response, or spin-up, to hurricane forcing is governed by the parameters of the storm. Linear theory of Geisler (ref. 5) has shown that the important parameters for estimating the type of response to be expected are the ratio of the storm translation speed to the first mode internal wave speed (U/C_1) (internal Froude number) and the ratio of the storm forcing scale (scale of positive wind stress curl region) to the baroclinic deformation radius ($2R_\text{m}/L_\text{b}$, where R_m is the maximum wind radius). The baroclinic radius, L_b, is simply the product of the first mode internal wave speed, C_1, and the inertial period, IP where $IP = 2\pi/24f$, and f is the Coriolis parameter. If the non-dimensional forcing scale is less than one, the response is primarily barotopic. If it is much larger than one, the response is largely baroclinic. If the Froude number is less than $2^{0.5}$, large mean currents are generated and upwelling is the dominant response. If the Froude number is larger than about 5, internal, inertial waves are the dominant type of response. The wavelength of the internal wave response, L_I, is pro-

portional to the product of the baroclinic radius and the Froude number, while the elevation of the resulting baroclinic ridge, ΔH, is proportional to the rate of vorticity input to the ocean, i.e. the wind stress curl. The amplitude of the internal wave response will be small if the non-dimensional time scale, $(2R_\text{m}/L_\text{b})/(U/C_1)$, is small compared to $\pi = 3.1416$ according to Veronis (ref. 6) and Geisler and Dickinson (ref. 7).

The observed ocean response parameters for Hurricanes Norbert, Josephine and Gloria are intercompared in Table I. Norbert was a small, intense storm, while Josephine and Gloria were large storms with broad, flat wind profiles. Curiously, all three storms exhibited secondary wind maxima inside of the primary wind maxima. These were remnants of an earlier eyewall, which was contracting and dissipating with time. The evolution of inward propagating convective rings, and their associated wind maxima, is discussed by Willoughby *et al.* (ref. 8). These double wind maxima give rise to considerable difficulty in interpreting the response of a hurricane as a simple Rankin vortex. Does this structure excite a non-linear response or higher harmonics?

One also sees from Table I that Norbert and Josephine were moving slowly with respect to C_1 while Gloria was moving faster. This, together with the fact that the forcing scale ($2R_\text{m}$) is of the same order as the baroclinic radius, makes it questionable whether or not linear theory would apply to these

TABLE I
Comparison of ocean response parameters (see text for definitions)

Parameter	Norbert	Josephine	Gloria
R_m, primary (km)	34	52	46
R_ms, secondary (km)	17	28	25
τ_m, primary (N m^{-2})	5.7	3.4	3.4
τ_ms, secondary (N m^{-2})	4.4	1.4	2.4
curl τ_m, primary ($\times 10^{-3}$)	0.40	0.14	0.02
curl τ_ms, secondary ($\times 10^{-3}$)	0.50	0.06	0.10
U (m s^{-1})	3.7	4.2	6.8
C_1 (m s^{-1})	2.8	3.0	3.0
U/C_1	1.4	1.4	2.3
IP (d)	1.6	1.0	1.0
f (s$^{-1} \times 10^4$)	0.50	0.73	0.73
L_b (km)	56	41	41
$2R_\text{m}/L_\text{b}$	1.1	2.5	2.2
$(2R_\text{m}/L_\text{b})/(U/C_1)$	0.8	1.8	1.0
L_I (km)	270	250	515
H (m)	40	55	45
ΔH (m)	12	12	60

cases. A mixed barotropic and baroclinic response is likely. The barotropic response cannot be addressed with AXCP data since these probes measure only relative currents. The combined barotropic and baroclinic response has been recently analysed for another similar case using current meter data by Shay and Elsberry (ref. 9). The baroclinic response for Norbert and Josephine has recently been analysed by Shay *et al.* (ref. 10), who found that the first four baroclinic modes explained 80 and 60%, respectively, of the current variability. We can also anticipate that the first upwelling peak, displaced one-quarter inertial wavelength ($\frac{1}{4} \times L_I$) behind the storm, would lay just beyond R_m for Norbert and Josephine where wind mixing is anticipated to be a maximum. Thus an interaction between mixing and upwelling would be anticipated for these storms, further complicating the ocean response processes.

MIXED LAYER RESPONSE

A typical SST response pattern with respect to a moving hurricane, shown in Figure 1 as derived by Black (ref. 11), illustrates the crescent-shaped pattern of SST decreases centred at about $2R_m$ to the right-rear of the storm. The SST patterns for Norbert and Josephine are shown in Figures 2 and 3. These fit the pattern of Figure 1 reasonably well with the largest SST decreases being 2.6 and 1.9°C, respectively.

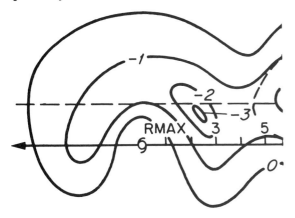

Fig. 1 Schematic SST change in Celsius (°C) induced by a hurricane. The distance scale is indicated in multiples of R_m = RMAX. Storm motion is to the left. Horizontal dashed line is at $1.5R_m$.

The shallowest mixed layer observed in response to Norbert (Fig. 4) occurred about 60 km to the rear of the centre ($\frac{1}{4} \times L_I$), where the mixed layer depth

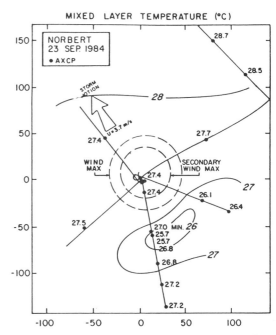

Fig. 2 SST patterns in °C within Hurricane Norbert. Maximum wind radii (primary and secondary) are indicated by quasi-circular dashed lines. Storm motion is indicated by the fat arrow. The SST is shown at AXCP locations (dots) along the aircraft flight path (line).

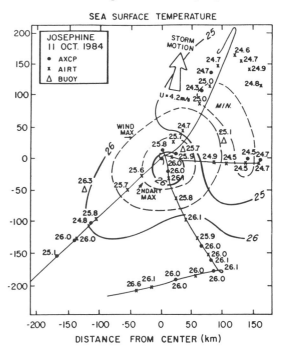

Fig. 3 Same as Fig. 2, except for Hurricane Josephine. Airborne infrared radiation thermometer (AIRT) values of SST are indicated by **x**.

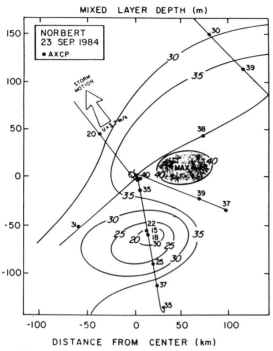

Fig. 4 Mixed layer depth (m) within Hurricane Norbert. Maximum depth of greater than 40 m is shaded.

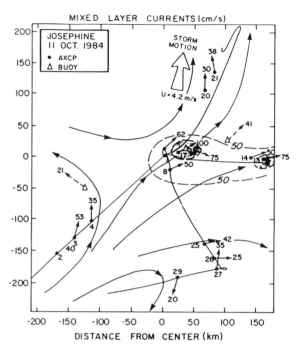

Fig. 6 Same as Fig. 5, except for Hurricane Josephine. Shading indicates currents greater than 75 cm s^{-1}. Open triangles with dashed vectors are buoy motions.

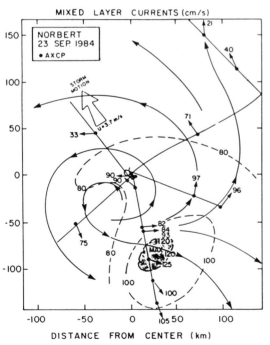

Fig. 5 Streamlines (direction toward) and isotachs of mean mixed layer currents (cm s^{-1}) for Hurricane Norbert. Shading indicates maximum currents of greater than 120 cm s^{-1}.

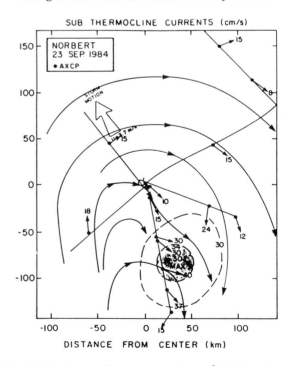

Fig. 7 Sub-thermocline currents (cm s^{-1}) in Hurricane Norbert. Shading indicates currents greater than 40 cm s^{-1}.

(MLD) was only 15 m. The SST minimum in this region appears to be a result of efficiently entrained cooler water from below the shallow mixed layer. Away from this region of strong vertical advection, the mixed layer is deeper (notice the region of greater than 40 m MLD to the right of the storm). The effect of the mixed layer shallowing due to the upwelling is also evident in the mean mixed layer currents (Fig. 5). The Stokes drift associated with large waves (swell) was removed from the AXCP data using the model discussed in Sanford *et al.* (ref. 1). Maximum mean currents exceed $1 \, \mathrm{m \, s^{-1}}$ near $2R_\mathrm{m}$ to the rear of the centre and are directed away from the storm track, which is consistent with the divergent currents associated with inertial pumping. This maximum is superimposed on a general divergent, cyclonic mixed layer current pattern with magnitudes on the order of $80 \, \mathrm{cm \, s^{-1}}$, as expected from various numerical models such as in Price (ref. 12). Undoubtedly, larger currents exist at the surface, but could not be measured by AXCPs.

In Hurricane Josephine, the pattern of mixed layer currents is not as well defined due to the presence of the oceanic Subtropical Front (Fig. 6). The maximum currents of $1 \, \mathrm{m \, s^{-1}}$ were located just to the right of the storm at about R_m, somewhat closer to the centre than in the case of Norbert. The displacements of the satellite-tracked drifting buoys in Josephine yield a maximum buoy mean speed of $0.5 \, \mathrm{m \, s^{-1}}$ according to Black *et al.* (refs 2 and 13). A recent study by Large *et al.* (ref. 14) suggests that for the Josephine hull type, the mean mixed layer currents have a weighting factor of 0.5 to 0.6 on the speed of the buoy, giving a mean mixed layer current estimate of $1 \, \mathrm{m \, s^{-1}}$, in agreement with that from the AXCPs. The buoys exhibited widely different trajectories, which is consistent with the highly variable pre-storm currents suggested by Figure 6.

THERMOCLINE RESPONSE

The maximum currents just below the seasonal thermocline (60–160 m) in the rear of Hurricane Norbert were somewhat in excess of $0.4 \, \mathrm{m \, s^{-1}}$ (Fig. 7). In this region of strong upwelling, the difference between the sub-thermocline and mixed layer current directions were 40–60°C. However, the sub-thermocline currents in the remainder of the domain were nearly opposite in direction to those in the mixed layer, giving rise to large vertical shears in the thermocline. These shears are shown in Figure 8 together with the bulk and local Richardson numbers (R_i). A substantial area of near critical R_is exist to the right of the storm near the region of the largest SST de-

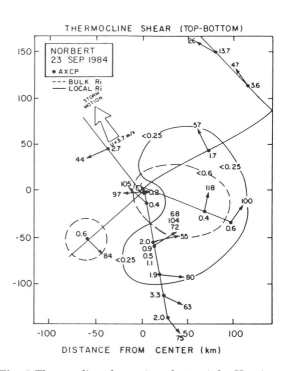

Fig. 8 Thermocline shears (top–bottom) for Hurricane Norbert are indicated near the arrow heads showing the vector shear. Areas exceeding critical bulk and local Richardson numbers (R_i) are outlined in dashed and solid lines, respectively. The local R_i are labelled near the dots.

creases. These direct calculations of R_i are the first confirmation of this physical process that has generally been assumed in numerical models of the oceanic response to hurricanes.

SURFACE WAVE RESPONSE

It is of some interest to examine the wave-induced motions that were subtracted from the AXCP profiles. These wave motions had a period of about 10 s for Norbert and, according to a side-looking airborne radar (SLAR), had a wavelength of 200 m. The wave-induced current is maximum at the surface and decreases exponentially with depth according to e^{-kz}, where $k = 0.0083$ and z is the depth in metres. The maximum surface currents induced by the swell (Fig. 9) were $80 \, \mathrm{cm \, s^{-1}}$ and located in the right quadrant of the storm, which is consistent with previous observations showing that highest waves and longest fetches are in the right quadrant. Figure 9 also shows that the pattern of swell propagation is nearly identical to the mean current flow pattern shown in Figure 5. This means that the wave-induced currents are nearly parallel to the mean currents. In the right

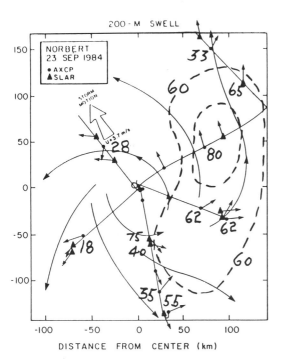

Fig. 9 Streamlines indicating swell propagation direction near Hurricane Norbert. Surface, wave-induced Stokes drift velocity (cm s^{-1}) is contoured with a dashed line. Triangles indicate swell directions obtained from side-looking airborne radar (SLAR) images.

quadrant, the mean current is modulated by 110% at the surface with a 10 s period. The total current thus varies from zero to 1.5 m s^{-1}. In the rear quadrant, there is only a 25% modulation, resulting in a maximum total current of about 1.7 m s^{-1}. It is conceivable that this wave-induced motion may also modulate the shear at the base of the mixed layer, producing intermittent mixing with a 10 s period, and playing an important role in the mixing processes.

HURRICANE GLORIA RESULTS

During the AXCP deployment in Hurricane Gloria, four Horizon Marine mini-drifting buoys were deployed in the storm along a line extending from the radius of maximum wind to a point 190 km northeast of the centre. The buoys contained the Service ARGOS transmitters for communicating the buoy's position and sea surface temperature data via NOAA polar orbiting satellites. The buoys were launched from the NOAA WP-3D research aircraft via the internal launch chute, which was the first such deployment of its kind. The buoys were deployed from an

altitude of 3 km and fell freely to the surface, where a drogue was deployed from each, descending to a depth of 25 m. Fortuitously, a fifth drifting buoy deployed 9 months earlier by oceanographers from NOAA was also in the area traversed by Gloria.

The primary purpose for deploying the buoys in Gloria was to map the background circulation features present in the region in order to place the AXCP measurements in proper perspective with respect to the prevailing currents. The second objective in deploying the buoys was to provide independent confirmation of surface mean currents in the vicinity of the hurricane for comparison with AXCP derived mean surface currents.

The buoy trajectories for the first 12 days after hurricane passage (Fig. 10) reveal the location of anticyclonic and cyclonic eddy centres indicated by an 'A' and 'C', respectively, and hence substantiate the existence of fairly energetic eddy currents in the semi-permanent Subtropical Front first described by Voorhis (ref. 15). The anticyclonic centre northeast of the storm track most likely existed prior to the passage of the storm. However, it is unclear whether the cyclonic centre along the storm track was induced by the storm or whether it existed prior to storm passage.

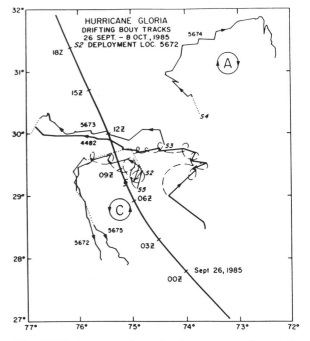

Fig. 10 Mini-buoy trajectories for a two-week period following the passage of Hurricane Gloria. Anticyclonic (A) and cyclonic (C) eddy centres are indicated, as are storm 3-hourly positions and track.

The AOML buoy was undrogued and hence exhibited the largest deflection to the right of the track by the storm-induced currents. An average buoy velocity over the 12 hours between fixes of 1.4 m s^{-1} was calculated during the time of the rightward deflection. This is probably representative of true mixed layer currents since drag on the mini-buoy was much smaller than drag on the larger PRL drifting buoys used in Josephine. However, instantaneous velocities, associated with hypothesized inertial period currents (dashed lines), were probably somewhat higher.

Figure 11 illustrates the average mixed layer current field derived from the AXCP data. Also indicated on Figure 11 by triangles are the current vectors derived from the drifting buoys which agree, to within 10 cm s^{-1}, with the AXCP currents. Note that the location of the cyclonic centre of circulation (strongly divergent and indicative of upwelling), corresponds to the cyclonic eddy centre inferred from the buoys. Both the buoys and the AXCPs indicate that the strongest currents are located in the right rear quadrant of the storm, in agreement with the findings from Norbert. In the Gloria case, the mean current

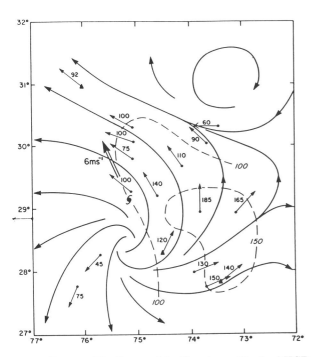

Fig. 11 Same as Fig. 5, except for Hurricane Gloria. AXCP positions are indicated by dots with current vector and magnitude shown. Triangles indicate mini-buoy locations and the arrow is their motion vector.

speeds are somewhat larger than Norbert, peaking at 1.85 m s^{-1}.

The wave-induced motions at the surface were also larger in Gloria than in Norbert, with amplitudes of 1.5 to 2.0 m s^{-1} and periods of 12 s in the right-front quadrant. This modulated the mean currents in this region by 200%. Maximum total currents in this region thus approached 3 m s^{-1}.

In conclusion, it can be said that the concurrent use of drifting buoys to provide a time history and AXCPs to provide a spacial snapshot greatly enhances the data base for a hurricane–ocean response experiment. Buoys with surface meteorological sensors, although more expensive, also aid in defining the surface wind field which forces the ocean response. It is highly recommended that the two tools—AXCPs and mini-drifting buoys—be used together in future experiments.

CONCLUSIONS

As shown above, the ocean response to a hurricane is primarily governed by the parameters of the forcing (Table I). The most crucial parameters seem to be the storm's translational motion and its maximum wind radius. The ocean dynamics can also be quite complicated, especially in areas of pre-existing ocean eddies, such as the Subtropical Front. Future ocean response experiments should be conducted away from these features to simplify the interpretation of the data.

Such features as secondary wind maxima (associated with 'double eye' features on radar displays of precipitation), exhibited by all three storms in this study, have not been treated in ocean response models. Such secondary maxima may introduce other scales into the problem than predicted by linear theory. Future experiments should accurately measure these mesoscale features so that their impact on the ocean currents and associated SST changes can be assessed and models to predict them correctly formulated.

It appears that storms with the right combination of storm speed, size and/or secondary wind maxima which generate upwelling just beyond the radius of maximum wind create conditions favourable for rapid and large decrease in SST. With this condition shown to be a maximum in the right-rear quadrant, where inflowing air from outside the storm must pass, considerable modification of inflowing air to the storm can take place as well as modification of vertical air–sea fluxes, both of which can lead to a modification of the storm intensification rate.

ACKNOWLEDGEMENTS

This project was made possible by the collaboration of the National Oceanic and Atmospheric Administration (NOAA), Hurricane Research Division and a consortium of oil companies. The work of the consortium steering committee co-chaired by Dr James Haustein, Mobil Research and Development Corporation, and Dr George Forristall, Shell Development Corporation, is especially appreciated. Air-launch capability for the AXCPs, mini-drifting buoys, logistics and AXCP data recording and processing were provided by James Feeney, Horizon Marine. The AXCP probe design, much of the processing software, and the surface wave elimination model were developed by Tom Sanford, University of Washington. Manufacture of the AXCPs was provided by Sippican Corporation. Assistance in buoy data processing was provided by Ron Kozak, Ray Partridge, and Glenn Hamilton of National Data Buoy Center. The AXCP and mini-buoy deployment were conducted by the Hurricane Research Division of NOAA using the NOAA-Office of Aircraft Operation (OAO) WP-3D aircraft. The assistance of OAO flight crews and engineers is greatly appreciated.

REFERENCES

1. Sanford, T. B., Black, P. G., Haustein, J., Feeney, J. W., Forristall, G. Z. and Price, J. F. (in press). Ocean response to hurricanes, Part I: Observations. *J. Geophys. Res.* Gill Memorial Issue.
2. Black, P. G., Elsberry, R. L., Shay, L. K., Partridge, R. and Hawkins, J. (in press). Hurricane Josephine surface winds and ocean response determined from air-deployed drifting buoys and concurrent research aircraft data. *J. Oceanic Atm. Tech.*
3. Holland, G. J. and Black, P. G. (in press). The boundary layer of Hurricane Kerry (1979), I: Mesoscale Structure. Submitted to *J. Atm. Sci.*
4. Black, P. G., Holland, G. J. and Powell, M. D. (in press). The boundary layer of Hurricane Kerry (1979), II: heat and moisture budgets. Submitted to *J. Atm. Sci.*
5. Geisler, J. E. (1970). Linear theory on the response of a two layer ocean to a moving hurricane. *Geophys. Fluid Dyn.* **1**: 249–272.
6. Veronis, G. (1956). Partition of energy between geostrophic and nongeostrophic oceanic motions. *Deep Sea Res.* **3**: 157–177.
7. Geisler, J. E. and Dickinson, R. E. (1972). The role of variable Coriolis parameter in the propagation of inertia-gravity waves during the process of geostrophic adjustment. *J. Phys. Oceanogr.* **2**: 263–272.
8. Willoughby, H. E., Marks, F. D. Jr. and Feinberg, R. W. (1984). Stationary and moving convective bands in hurricanes. *J. Atm. Sci.* **41**: 3189–3211.
9. Shay, L. K. and Elsberry, R. L. (1987). Near-inertial ocean current response to Hurricane Frederic. *J. Phys. Oceanogr.* **17**: 1249–1269.
10. Shay, L. K., Elsberry, R. L. and Black, P. G. (in preparation). Near-inertial ocean current response to Hurricanes Norbert and Josephine.
11. Black, P. G. (1983). Ocean Temperature Changes Induced by Tropical Cyclones. PhD dissertation, the Pennsylvania State University, University Park, PA, 278 pp.
12. Price, J. F. (1981). Upper ocean response to a hurricane. *J. Phys. Oceanogr.* **11**: 153–175.
13. Black, P. G., Elsberry, R. L., Shay, L. K. and Partridge, R. (1985). Hurricane Josephine surface winds and ocean response determined from air-deployed drifting buoys and concurrent research aircraft data. Extended Abstracts, 16th Conf. Hurr. and Trop. Met., Amer. Met. Soc., Boston, MA, pp. 22–24.
14. Large, W. G., McWilliams, J. C. and Niiler, P. P. (1986). Upper ocean thermal response to strong autumnal forcing of the Northeast Pacific. *J. Phys. Oceanogr.* **16**: 1524–1550.
15. Voorhis, A. D. (1969). The horizontal structure of thermal fronts in the Sargasso Sea. *Deep Sea Res.* **16** (Suppl), 331–337.

7

Thermal Structure and Remote Sensing Measurements in a Major Frontal Region

John C. Scott, Norman R. Geddes, Nichola M. Lane and
Anne L. McDowall, Ocean Science Division, Admiralty Research
Establishment, Portland, Dorset, UK

Remote sensing images of the ocean surface have a great potential for allowing monitoring of ocean dynamics. This chapter reports and compares in-water thermal structure data from three instruments—a long thermistor chain, a short thermistor spar, and an acoustic doppler current profiler—with data from a near-coincident infrared (AVHRR) image. The comparison underlines our need for a detailed knowledge of deep structures and of their interaction with the surface before remote sensing images can be used reliably to enhance what is a generally inadequate provision of in-water data.

THE IMPACT OF MODERN OBSERVING TECHNIQUES ON OCEANOGRAPHY

The major impact of technology on oceanography over the past twenty years has mainly been through advances in electronics, particularly in the electronics of physical measurement and data storage. Although this trend will no doubt continue, it is apparent that space technology is already having an increasing impact through its growing ability to provide detailed synoptic measurements from large areas of ocean surface. The combination of these two trends will allow major advances in our understanding of the oceans, when the space measurements are adequately supported by detailed in-water measurements.

Although this possibility has been recognized for some time now, examples of direct comparison of co-incident data, provided at comparable sampling scales by in-water and remote sensing techniques, have been relatively few. This chapter describes some recent results in which this comparison has been made.

Although satellites, and other orbiting space plat-forms such as space stations, can also measure sea surface elevation and surface waves along relatively narrow tracks, this chapter rather considers the possibilities offered by space-borne sensors which give large-area images. Passive infrared systems are

Advances in Underwater Technology, Ocean Science and Offshore Engineering, Volume 16: Oceanology '88

Fig. 1 An infrared AVHRR image of the Iceland–Faeroes
region obtained on June 21, 1986, at 1334Z. The warmer
water appears darker in this image. The contrast has been
stretched to give a good indication of the horizontal
variability in the region.

considered here, but the conclusions drawn can be expected to apply more generally, across the whole range of future remote sensing possibilities.

This chapter also draws on data obtained with two in-water techniques that have been developed quite recently: thermistor arrays and the acoustic doppler current profiler. These techniques exemplify the impact of modern electronic technology on the subject.

SATELLITE INFRARED IMAGES

High-quality images of the earth are provided routinely by the NOAA-TIROS series of satellites, carrying the advanced very high resolution radiometer (AVHRR) sensor at an altitude of 840 km. Detecting radiation in the visible and near infrared, this sensor package can give 1 km resolution images of ocean surface thermal radiation. Such images are invaluable sources of information about the spatial position and extent of ocean features that involve water masses of different temperature, ocean fronts being a good example.

Ideally, such images should allow us to make detailed synoptic extrapolations of necessarily sparse and spatially limited in-water data, with the aim of completely specifying the water mass structure of very large regions. Figure 1 shows an infrared image of the Iceland–Faeroes region, where the cold waters of the Norwegian Sea meet the warmer, generally northwards-flowing waters of the Atlantic Ocean. Regions such as this are vitally important oceanographically, but they are continually moving and are impossible to sample adequately with in-water measurements. Space-derived synoptic images offer a potentially invaluable means of extending these data.

There are three main areas of difficulty in using AVHRR images in this way: (1) the algorithm which is used to infer true surface temperatures from the four infrared and visible channels of the sensor is not completely verified; (2) the surface temperatures, as measured, may not reliably represent the temperatures of the underlying deep water masses; and (3) sea surface images are simply not obtainable when the region is covered by cloud.

Of these problem areas (1) attracts much attention and may eventually be removed by diligent comparative measurement under a range of different atmospheric conditions. Problems (2) and (3), however, are much more serious as obstacles to the routine oceanographic use of infrared images.

The seriousness of the cloud cover problem depends heavily on the particular region being investigated,

and we are powerless to affect it in any region. It is much more serious in the area shown in Figure 1 than, for example, in the Mediterranean and in the vicinity of the Gulf Stream. Figure 2 gives an approximate indication of the probability of obtaining cloud-free AVHRR images in the Iceland–Faeroes region. The only practical way of avoiding this problem is to make the radiometry measurements in a part of the spectrum, such as the microwave region, less affected by cloud (Shutko, 1985).

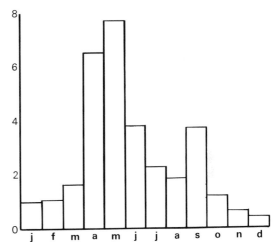

Fig. 2 A subjective impression of the probability of obtaining cloud-free images in the North Atlantic region. The data here are semi-quantitative, in that they incorporate statistical information over a long period of satellite imaging, but use an arbitrary subjective criterion of quality. (N. Ward, Southampton University, unpublished work, 1985.)

This paper principally addresses problem (2)—the reliability of the measured surface temperature distribution as an indicator of the spatial structure of the underlying water masses. This problem is particularly well illustrated by the fact that an AVHRR image of the sea surface can be obtained only under clear skies, and it is precisely under these conditions that the surface temperature will be changing most rapidly.

Only a thin surface skin of the ocean contributes to its infrared radiation, and in conditions of strong sunshine and low winds this skin can be rapidly warmed by several degrees, relative to the temperature even a few metres below.

Although crucial for problem (2), this effect need not be too significant for our present purpose of synoptic extension of limited in-water information. If we ignore the absolute values of radiometric temperature and concentrate on what the image tells us of the spatial variation of temperature across a region, then perhaps we can still derive valuable information on the horizontal disposition of frontal features.

We must nevertheless be careful, since it may not be reasonable to assume that the whole region has been exposed historically to the same pattern of heating (or cooling) conditions for the same length of time. Particularly in frontal regions, where the distribution of air temperature and cloud cover may be significantly influenced by sea surface temperature, such an interpretation may not be valid.

The problem extends beyond considerations of thin diurnally heated layers on single, isolated cloud-free days. Longer-term cumulative effects, in which the whole depth of the upper ocean boundary layer is affected by sub-seasonal time-scale biases in the regional distribution of solar radiation, may also be important.

Such factors relate only to the atmospheric and solar influences on the dynamics of the upper layer. In a frontal region such as that shown in Figure 1 we also require a knowledge of the deeper ocean structures. It is with the hope of deducing the position and dynamics of these structures that we examine the surface images. Ideally, we seek a relationship— even a qualitative one—between the surface layer temperature distribution and the underlying water masses.

Considered in this light, the problem becomes that of how much the above-water influences discussed above disturb the natural influence of the lower layers on the upper. To assess this it is obvious that we need to measure the water column to a sufficient depth, with greater detail in the upper layer.

To some extent, the problem is less pressing if we intend to use image data simply to extend existing in-water data. We may still reasonably use an image to give horizontal extensions of whatever interpretations we can base on point and line section thermal data. In this case, problem (2) reduces to that of how far away from a position or track of in-water data can we extend the data, both spatial and temporal extensions being relevant.

All the questions identified so far can only be addressed by experimental investigation, involving the direct correlation of images with detailed simultaneous in-water data.

IN-WATER DATA: THERMISTOR ARRAYS AND ACOUSTIC DOPPLER CURRENT PROFILER

Two techniques have been developed, in recent years, for the direct in-water measurement of water properties.

Towed undulating CTDs, such as Batfish and Sea-Soar, cover long vertical sections through the ocean, taking continuous temperature and conductivity data as they 'fly' up and down through the ocean behind the towing platform. The horizontal sampling rate is of the order of the depth of the flight, which can be down to 300–400 m.

Thermistor chains, on the other hand, carry more or less vertical arrays of sensors, sampling regularly as they are towed steadily through the ocean. As the name implies, the predominant sensors carried are temperature-measuring thermistors, although conductivity sensors may also be carried. These, however, are more prone to errors from fouling, and the reliable data obtained by thermistor-only chains make them an acceptable working compromise.

Although undulating CTDs have the advantages of simultaneous conductivity (and hence salinity) measurement, higher tow speed (4–5 m s^{-1} compared with 2–3 m s^{-1}), and also denser sampling in the vertical compared with the fixed sensor spacing of a thermistor chain, chains have the advantage of much denser sampling in the horizontal. In frontal regions, where horizontal coherence of water mass structures is little known, the higher sampling rate of thermistor chains can be an important, if not crucial, advantage (Fedorov, 1986).

IF '86: A Survey of the Iceland–Faeroes Frontal Region

MV *Sea Searcher* was making detailed thermal measurements in the Iceland–Faeroes region (the IF '86 survey) at about the same time as the image in Figure 1 was obtained. Two thermistor arrays were deployed on long tracks through the region: the ARE digital thermistor chain (Lane and Scott, 1986) and the IOS thermistor spar (Thorpe and Hall, 1980).

The ARE chain has 100 temperature sensors spaced over a 400 m length. It overcomes the problems associated with long individual analogue signal wires between the in-water sensors and the above-water recording equipment by making the measurements close to the sensors and passing the data to the surface in a robust digital form. The system can thus be modular and of adaptable length and variable sensor spacing. In the IF '86 configuration, a proportion of the sensor 'pods' also contained pressure sensors which indicate the depth and shape of the towed chain, and some conductivity sensors.

The chain uses a six-wire electrical cable (two power, two address, and two signal lines), and the pods are polled sequentially, at about 1 Hz, giving 2 m horizontal sampling at its preferred 2 m s^{-1} tow speed. This is found adequate for ocean structural studies.

The modular design of the chain allows up to 100 pods to be distributed at a mixture of 2, 4 or 8 m spacings for chain lengths between 100 and 400 m, the maximum length of chain so far towed. A 400 m chain gives a sampled depth around 300 m at 4 knots. It would be possible to make the chain longer, without difficulty, but 400 m gives an excellent practical compromise between pod separation, sampled depth, and tow speed. More pods can be carried, and more could be sampled, with suitable software modification.

The IOS Thermistor Spar is fundamentally similar in principle, except that it is designed to sample only the upper 10 m of the ocean. It is an analogue system, and its 11 thermistors are rigidly attached to a global-mounted neutrally buoyant spar, which hangs freely from a stable towed catamaran platform. In this design the catamaran, towed ahead of the ship wake, rides the surface waves, the spar remaining at about the same depth relative to the free surface. The spar thermistors were generally spaced 0.5 m apart, and their temperatures were sampled every 0.25 s.

The spar was developed specifically to investigate the detailed thermal development of the upper ocean boundary layer under different solar radiation and atmospheric conditions. Thorpe and Hall (1987) report related data from the survey reported here. It is an ideal system for remote sensing verification work in that it also allows us to see in detail the way that deeper thermal structures appear at the surface.

The third instrument whose data are drawn on here is the acoustic doppler current profiler (ADCP). This is essentially a sonar with four downward-pointing, mutually inclined beams. The backscatter from each depth range bin is assessed for relative doppler shifts which indicate movement of the water at that depth relative to the ship. Current profiles are calculated from the averaged ensembles of signals backscattered from a series of range bins down to about 500 m. The RDI instrument used in 1986 operated at 150 kHz, and 100 s averages were recorded.

FRONTAL STRUCTURE DATA

Three features of the Figure 1 image will be discussed here, to show how the in-water data, interpreted together with the image, can indicate the behaviour of the water in the frontal region.

To reduce the information contained in Figure 1 into a manageable form, the image has been processed to give contours of equal sea surface temperature, and the result is shown in Figure 3. The isotherm contours, separated by 0.5°C intervals, have been chosen here to delineate the features under consideration. All the tracks of the IF '86 survey are shown here; those for which data are presented are shown as bold dashed lines.

A cold eddy

The position of the ship at the time of the image is marked on Figure 3 with an arrow. It was heading South at that time, having passed through the cold surface feature marked 'A' between 0400Z and 0800Z, sufficiently close to the satellite image time for the correspondence to be good. Meteorological data recorded on board at the time indicates a strong wind, 20–25 knot, coming from 220°.

The appropriate sections of the coincident thermistor chain and thermistor spar records are shown in Figure 4. The considerable horizontal compression of these data from their 2 and 0.5 m sampling rates should be noted. Although in the chain data this leads to a confused representation, this form of summary highlights the gross features, and the over-sampling enhances confidence in the validity of the data. Even with the strongest horizontal gradients encountered in regions such as this, successive profiles show a high degree of correlation. The same cannot be said of CTD station, undulating CTD, or XBT data.

The chain data (Fig. 4(a)) are given here as 0.2°C isotherms down to 200–250 m depth, and they clearly show the strongly layered, asymmetrical, and apparently turbulent structure of the feature, which is reasonably interpreted as a cold core eddy. Figures 1 and 3 would indicate that this cold water is completely detached from the parent cold water mass situated to the north of the main frontal boundary.

Although it cannot be deduced from the figure, the data show that 5.0°C water, which is observed just outside the eddy at a depth around 350 m, has risen to about 50 m within it. The temperature at a constant 200 m depth decreases within the feature from 7.6 to around 3.0°C.

The spar data (Fig. 4(b)) are given here as the temperature outputs of the upper four thermistors, covering the thoroughly wind-mixed top 2.5 m of the water column..

The eddy feature is clearly asymmetrical, with the horizontal temperature gradients very much more abrupt in the Northern boundary than they are in the Southern boundary. The Northwards displacement of the upper layer 'signature' of the eddy is also clear, particularly in the spar record. The displacement of the southern boundary is considerably greater than that of the northern boundary, which has lead to the signature being considerably smaller than the eddy itself.

This 'overwashing' of the cold eddy by the warmer water situated upwind is a highly significant phenomenon from the point of view of image interpretation, and possibly also concerning the decay of the eddy and the mixing of its water.

The great detail visible in Figure 4(a) is one of the most important attributes of thermistor chain data. It is apparent here that a wide range of horizontal coherence scales is present. This proves to be the case in many important oceanographic situations. Even if these horizontal scales are not important in themselves for a given interpretation of the data, this density of detail is still useful for defining the sampling rates necessary to avoid the misinterpretation of less well sampled data.

The doppler current profiler data taken during the eddy transect show a more or less symmetrical variation of the profiles, consistent with interpretation as an eddy. Shear currents of the order of $0.5\,\mathrm{m\,s^{-1}}$ were found near the eddy boundaries.

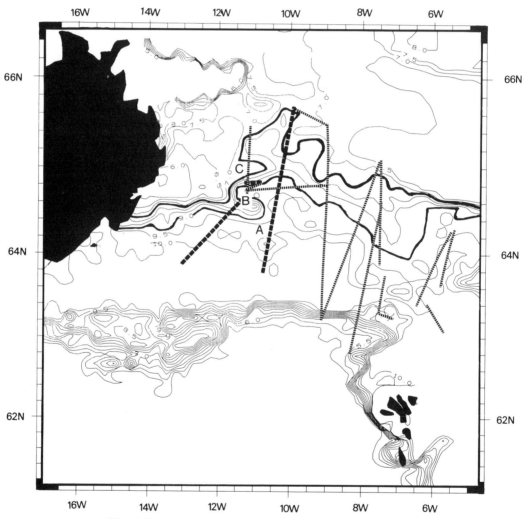

Fig. 3 Surface temperature contours derived from the Figure 1 image using standard algorithms with a nominal atmospheric correction. Contours are shown at 0.5°C intervals, and the 7.5 and 9.0°C contours, drawn thicker, outline the principal features examined in this paper. The tracks from which in-water data are presented here are shown as dashed lined. The arrow points to the ship location at the time the image was made.

A 'Hook' in the Frontal Boundary

The feature marked as B in Figure 3 is commonly called a 'hook' in the informal taxonomy of frontal features, and such features are thought to be extreme examples of the meandering instability of the major frontal boundary of this region. An indication of the time development of this particular feature is obtained from this image and two others, one obtained the previous day and the other obtained about two hours later. Two examples of less extreme meanders—also traceable in the other images—can be seen to the West of feature B in Figure 3.

The track shown in Figure 3, although made about two days after the Figure 1 image, did cut across a cold water mass in approximately the expected position, between 0800 and 1300 on 23 June 1986. The identification of the in-water structure with the

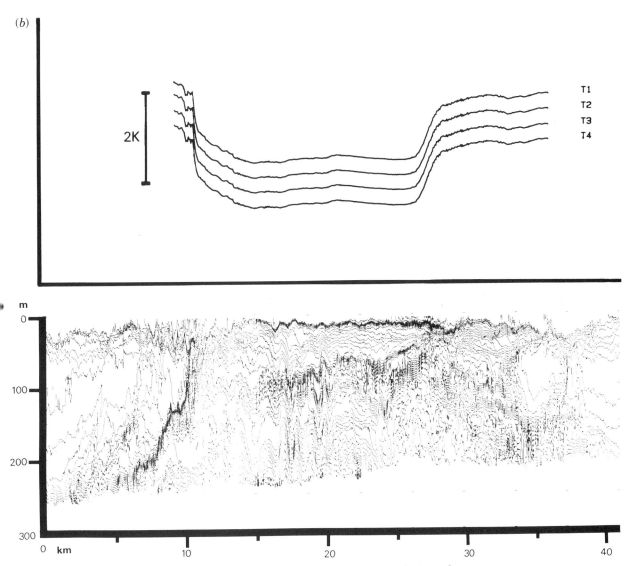

Fig. 4 A section of coincident thermistor chain and thermistor spar data for the passage through the feature marked 'A' in Figure 3, to the same horizontal scales: (*bottom*) the thermistor chain data, with 0.2°C isotherms; (*top*) the outputs of the top four thermistors of the spar, spanning the top 2.5 m of the water.

image may therefore be reasonable, although the width of the cold water region crossed was about 20 km, rather larger than in the image.

Figure 5 shows a small part of the thermistor spar record for the Northern boundary of the cold water in the hook, as the spar passed back into warm water. Here, the outputs of all 11 of the spar thermistors are shown. Note that the horizontal scale of this section is greatly expanded compared with Figure 4(b).

The horizontal temperature gradients recorded in this transition are particularly extreme, changes of about 2.5°C taking place in less than 100 m of track. Gradients of this magnitude were also found by the chain, at depths down to around 70 m. Another point to note here is the fact that while the top thermistors in the spar are experiencing major gradients, the bottom thermistors are relatively undisturbed, even though only 7.5 m separates them. The warmer water only reaches the lower level some 500 m further along the track (not shown in Figure 5).

The relatively unmixed character of the upper layers in this example compared with the case of Figure 4(b) is associated with the extremely light winds in this case. The example thus represents a different case of infrared image interpretation from that seen earlier, one in which solar radiation, rather than wind, has the dominant effect.

'Warm Intrusion' Western Boundary

In the example considered here, the major boundary of the frontal system, at least as seen in the satellite image, has a pronounced northwards bulge. This can be seen in Figures 1 and 3 situated to the north-east of feature B. This bulge is often called the 'warm intrusion', and it is frequently (but by no means always) seen in infra-red images of the region. It would appear to be associated with the shallower topography at this Western end of the Iceland–Faeroes Ridge. As well as being shallower, the ridge tends here to extend further north, which may both inhibit

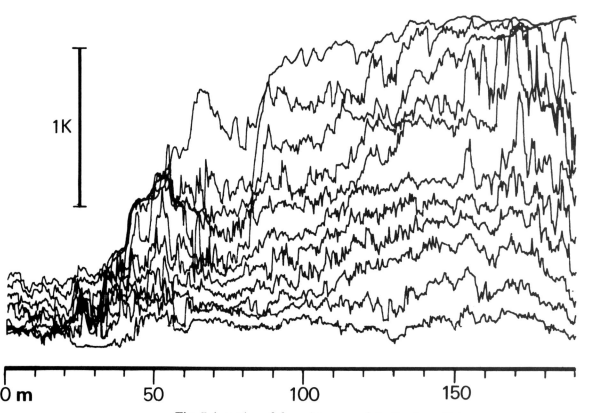

Fig. 5 A section of thermistor spar data for the cold-to-warm transition of the hook feature seen at 'B' in Figure 3. All 11 thermistor outputs are shown here, with arbitrary vertical offsets.

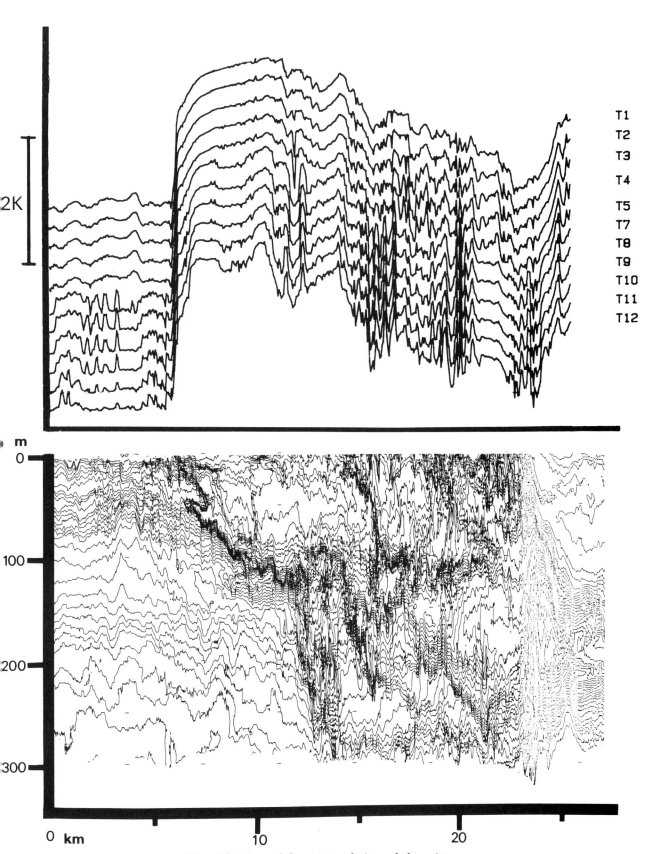

Fig. 6 Sections of thermistor chain and thermistor spar data for a west–east section of the Western boundary of the 'warm intrusion', marked as 'C' in Figure 3 (the scales are the same as in Figure 4): (*bottom*) 0.2°C chain isotherms down to about 300 m; (*top*) the temperature outputs of all 11 of the spar thermistors.

T1
T2
T3
T4
T5
T7
T8
T9
T10
T11
T12

2K

m
0

100

200

300

0   km   10   20

cold water from overflowing southwards and favour northwards flow of the warmer Atlantic upper layers.

In June 1986 the warm intrusion was made up of two distinct parts: that further north being a slightly modified, warmer-surface version of a fairly typical cold Norwegian Sea structure; and the southern part being essentially Atlantic water throughout the upper 200 m. In Figure 3 the two northern boundaries are approximately marked in this area by the 7.5 and 9.0°C contours.

The third feature considered in this paper comes from the point marked C in Figure 3. This is the Western boundary of the warm intrusion, in a location where the boundary separates the cold water outside the intrusion from the Southern, more Atlantic water within.

One of the west–east thermistor chain sections of this boundary is shown in Figure 6(a), and the corresponding data from the thermistor spar are presented in Figure 6(b).

Both in the spar data and in the thermistor chain data the boundary is seen to be heavily layered, and adjacent sections indicated that this layering was approximately perpendicular to the sections. This layering is consistent (at least approximately) with the streaky, or filamentary, nature of the boundary here, which may be observed in the infrared image of Figure 1.

By this time, the in-water data are more than four days later than the image, and more than general agreement would not be expected; however, the filaments seen in the image seem to offer a good interpretation of the in-water variation encountered.

In this third example, the value of cooperative in-water and remotely sensed data for inferring the three-dimensional ocean structure may be clearly seen.

The western boundary of the warm intrusion was also found to be clearly imaged by an airborne X-band real-aperture radar which was overflying the region at the precise time of the data shown in Figure 6. This clearly suggests the potential value of imaging radars for identifying and precisely locating frontal features, although even less work appears to have been done on the physics of the imaging process in this case than in the infrared case.

Surface slicks (presumably of biogenic origin) of similar orientation to the images, both radar and infrared, were also observed, visually, from both the aircraft and the ship. In all cases, the orientation in this region was approximately north–south, consistent with the above interpretation of the infrared image.

ACKNOWLEDGEMENTS

The data reported here could not have been collected without the dedicated efforts of many people, including Claire Burt and Adrian Brown of ARE and a team of engineers from BAJ Ltd. The skills and persistence of the captain and crew of MV *Sea Searcher* are also acknowledged. Special thanks are due to Alan Hall of IOS Wormley, for his collaborative work with the thermistor spar, and to Chris Brownsword, of RAE Farnborough, for his provision of air support during the trial.

REFERENCES

Fedorov, K. N. (1986). *The Physical Nature and Structure of Oceanic Fronts*. Springer-Verlag, Berlin.

Lane, N. M. and Scott, J. C. (1986). The ARE Digital Thermistor Chain. Proceedings, Marine Data Symposium, New Orleans, April 28 to May 1, pp. 127–132.

Shutko, A. M. (1985). The status of passive microwave sensing of the waters—lakes, seas and oceans—under the variation of their state, temperature and mineralization (salinity): Models, experiments, examples of application *IEEE Journal of Oceanic Engineering* 10: 418–437.

Thorpe, S. A. and Hall, A. J. (1980). The mixing layer of Loch Ness. *Journal of Fluid Mechanics* 101: 687–703.

Thorpe, S. A. and Hall, A. J. (1987). Bubble clouds and temperature anomalies in the upper ocean. *Nature* 328: 48–51.

8

A Long-Term Intercomparison of Wave Sensors in the Northern North Sea

N. M. C. Dacunha and *T. J. Angevaare,* Shell International Petroleum
Mij, The Netherlands

Wind, wave and meteorological data have been measured by Shell UK Exploration and Production (a Shell/Esso joint operating company) in the Northern North Sea since 1975, initially at the Brent Bravo Production Platform and latterly, since September 1983, at the North Cormorant production platform located in the Northern North Sea north-east of the Shetland Isles.

The objectives of the data collection programme have been to:

- obtain reliable data for the prediction of more reliable design criteria for new structures in the Northern North Sea;
- provide real-time data for the safe management of daily operations around the platform;
- provide information to weather forecasting services; and
- evaluate new instrumentation for wave measurement in order to improve the reliability and associated costs of wave data collection.

Because the largest component in the costs of data collection has been the measurement of wave data, there is a clear need to move towards platform-based systems which offer the potential for reducing

maintenance costs. The wave sensors that have been deployed at North Cormorant are:

- a Datawell waverider buoy since September 1983;
- an EMI Infra-Red Laser Wave Height Monitor since September 1983 (platform-based);
- a Datawell WAVEC buoy (ref. 1) since April 1986, operational after July 1986;
- a MIROS microwave radar (ref. 2) since March 1986 (platform-based) but operational after December 1986.

The first two are omnidirectional wave sensors while the last two can measure wave height and direction. The need to measure wave direction arises out of the need to establish with confidence the directional wave climate and to use this in reducing the costs of new structures in the area. This chapter describes the operating experiences with these sensors and then proceeds to detailed comparisons of the measured parameters.

SENSOR AND SYSTEM PERFORMANCE

The waverider and EMI laser form part of the North Cormorant Metocean system. The two sensors are

Advances in Underwater Technology, Ocean Science and Offshore Engineering, Volume 16: Oceanology '88

TABLE I

Details of calculation of wave parameters

	Spectra			Formulae for wave parameters		
Sensor	*Frequency resolution (Hz)*	$f_1(Hz)$	$f_2(Hz)$	H_s	T_s	T_p
Waverider	0.0078	0.04	0.06	$4\sqrt{m_0}$	$\sqrt{(m_0/m_2)}$	$(m_{-2}m_1)/m_0^2$
EMI laser	0.0078	0.04	0.6	$4\sqrt{m_0}$	$\sqrt{(m_0/m_2)}$	$(m_{-2}m_1)/m_0^2$
WAVEC	0.005	0.03	0.45	$4\sqrt{m_0}$	$\sqrt{(m_0/m_2)}$	$1/f_p[*]$ $(m_{-2}m_1)/m_0^2[†]$
MIROS	0.011	0.03	0.45	$4\sqrt{m_0}$	$\sqrt{(m_0/m_2)}$	$1/f_p$

$$M_n = \int_{f_1}^{f_2} f^n S(f)\, df, \text{ where } S(f) = \text{spectral estimate at frequency } f$$

[*] In comparisons with MIROS wave direction
[†] In comparisons with EMI laser

sampled simultaneously by a data acquisition unit and are then processed identically. Spectral analysis is carried out offshore, and the processed parameters are stored, together with quality control flags. The data are quality controlled onshore and stored on a computer databank.

The WAVEC buoy is a pitch-roll buoy (see ref. 1) where the data are processed as described in ref. 3. The MIROS is a microwave transmitter and receiver (see ref. 2) which measures water particle velocity and uses a first order wave theory to convert this to wave amplitude. The sensor has a beam width of 30° and hence samples wave energy from a restricted direction. By rotating the sensor head it is possible to provide information for all directions. The sensor at that stage was unable to distinguish between waves moving towards it and those moving away (referred to as the '180-degree ambiguity').

The data from the four sensors as deployed on North Cormorant all undergo spectral analysis, with differing frequency resolutions and limits. The analysis details are summarized in Table I.

Operational experiences with the Waverider and the EMI laser

The North Cormorant Metocean System was commissioned on 2 September 1983. The system runs off the platform power supply, which can be unstable, and failures in the power supply resulted in a data loss of approximately 2–3% on average during the period to the end of December 1986. A summary of percentage data return for the two instruments and the reasons for data loss for each year are shown in Table II.

The waverider went adrift six times, was re-deployed twice, and replaced three times. In addition, it had to be removed from its location and re-deployed once because of the need to clear the area for survey operations. There was a 19% loss of data due to the buoy being off-site, and the overall data return was 74.3%. The buoy finally went adrift in January 1987 and has not been replaced as there now is a WAVEC buoy permanently at the site.

The EMI laser developed faults soon after commissioning which resulted in substantial data losses until the faults were finally rectified in May 1984. The sensor then performed satisfactorily until December 1985 when a major fault occurred in the sensor's power supply unit. The sensor was replaced by a spare and returned to the manufacturer for repair and re-calibration.

The sensor was subsequently returned to be kept in reserve as a spare. There were no further sensor failures in 1986. The overall data return for the EMI laser from September 1983 to December 1986 was 72.4% and from May 1984 to December 1986 was 89.7%. The waverider data return over the latter period was 73.4%.

Performance of the WAVEC pitch-roll buoy

The buoy was deployed in April 1986, and the processed data were sent to the Rijkswaterstaat receiving station in the Hook of Holland. Soon after

TABLE II

Data returns for the EMI laser and the Waverider at North Cormorant (%)

	September–December 1983	1984	1985	1986	Total
Waverider					
Valid data return	87.7	62.2	83.9	72.4	74.3
Data losses					
Platform power failure	8.3	4.1	1.4	0.6	2.7
Adrift	—	29.3	9.1	23.8	18.7
Receiver/filter malfunction	—	1.8	0.8	—	0.8
Other	4.0	2.6	4.8	3.2	3.6
EMI laser					
Valid data return	6.0	64.1	87.6	87.8	72.4
Data losses					
Platform power failure	—	2.3	1.4	0.9	1.4
Sensor malfunction	92.7	30.9	7.3	6.3	22.6
Other	1.3	2.7	3.7	4.9	3.5

deployment it became evident that there was a problem with the buoy, which after investigation was traced to an oscillation on one of the channels. The reason for the oscillation could not be identified so the buoy was replaced with a new one in July 1986. It performed satisfactorily thereafter. Data for the period December 1986 and January 1987 were obtained from the Rijkswaterstaat. During this period the overall data returned by the buoy was 85% of the maximum possible.

Performance of the MIROS microwave radar

The radar was installed in March 1986 but underwent one software correction and a hardware fault was rectified in December 1986.

An intercomparison study between the MIROS, the WAVEC and the EMI laser was initiated and was intended to cover the period December 1986 and January 1987. During the study it was discovered that for seastates where a substantial proportion of the wave energy lay in frequencies less than 0.1 Hz, the MIROS underestimated the significant wave height relative to that measured by the WAVEC or the EMI laser. On further investigation by MIROS Norway, this problem was attributed to the fact that the MIROS principally measures water particle velocity which is then translated into a wave amplitude via a frequency-dependent transfer function. The (deep-water) approximation in use for this transfer function was found to be inadequate for long period waves and for intermediate waterdepths.

MIROS Norway therefore implemented a more rigorous calculation of the wave amplitude and proceeded to re-analyse the North Cormorant recorded measurements. This was carried out for the period 6 to 18 December inclusive and the results for this period are reported below.

COMPARISON OF RESULTS FROM THE EMI LASER AND THE WAVERIDER BUOY

General comparisons of the two sensors

As described earlier, both sensors were deployed on the platform for more than three years. During that period there were 16 897 simultaneous hourly measurements, affording an almost unique opportunity to compare statistics as well as examine through regression techniques sensor performance. The comparisons presented in this chapter will consist of the results of regression analyses, a comparison of statistics obtained from both sensors, and a detailed examination of measurements during November 1985, the month when the Norwegian-sponsored WADIC experiment was taking place elsewhere in the North Sea.

Regression analyses on the measured wave parameters

Linear regression on the measured significant wave height, H_s, mean zero crossing period, T_z, and spectral peak period, T_p, was performed on all the wave data and also on only those cases when the wave height was greater than 5 m. The technique was to define in all cases the waverider measurement to be the independent X parameter and the EMI laser to be the dependent Y parameter. The empirical coefficients a (called the slope) and b (the intercept) for the relationship:

$$Y = aX + b$$

were then obtained by minimizing the least square deviation of the measured X, Y pairs from the best fit line.

The results are listed in Table III, where it can be seen that in the case where all wave heights are considered the best fit expressions obtained for H_s, T_z and T_p all have a slope that is very close to 1 and a small intercept. This indicates that on average the two sensors produce results that are very nearly equivalent to each other. The quality of the comparison deteriorates for H_s when only high wave heights are considered, but continues to produce very good results for T_z and T_p.

The results for the H_s comparisons, when H_s was restricted to heights above 5 m, show that the EMI laser has a tendency to slightly underestimate the high wave heights when compared with the waverider buoy, but there are too few data points to draw any firm conclusions.

Statistics

Comparisons of the frequency and cumulative frequency distributions for H_s and T_z from simultaneous measurements are shown in Figure 1. The results show that the statistics for all three parameters are in very good agreement. The H_s cumulative frequency distribution shows that the EMI laser has a small tendency to underestimate the high wave heights when compared with the waverider, but there are too few data points to draw any firm conclusions.

The wave scatter diagram equal probability contours as measured by the two sensors are shown in Figure 2. The agreement shown therein is very good. The statistics compared thus far are those normally provided to engineers for use in structural design and indicate that the parameters as derived from the two sensor signals are very nearly equivalent.

Detailed study for November 1985

Data coverage for the month was 96.5% for the waverider and 94.3% for the laser. The most notable gaps occurred immediately before two severe storms, during which H_s values of approximately 10 m were measured.

A comparison of the average spectra computed from simultaneous measurements for the month is shown in Figure 3(a). The average spectrum here is simply the sum of measured spectral estimates at each frequency divided by the number of measured spectra contributing to the sum. Approximately 200 spectra from each sensor were available. The results show good agreement with a very small shift towards

TABLE III
Results of regression analysis between measurements from the Waverider buoy and the EMI laser

Measurement parameter compared	Number of pairs	Correlation coefficient	Regression results[a] Slope	Regression results[a] Intercept	Standard error normal to best fit line
All wave data	16 897				
H_s		0.984	1.00	−0.07	0.17
T_z		0.966	0.97	0.33	0.22
T_p		0.970	0.98	0.32	0.36
$H_s > 5$ m	931				
H_s		0.854	0.97	0.19	0.36
T_z		0.892	1.06	−0.56	0.24
T_p		0.966	1.03	−0.26	0.33

[a] Regression equation is $Y = \text{slope} \times X + \text{intercept}$
where: Y = EMI laser, X = waverider

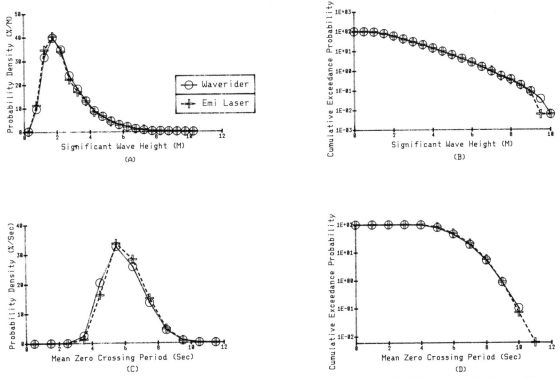

Fig. 1 Comparisons of probability distributions measured by the EMI laser and the Waverider buoy at North Cormorant (total number of simultaneous observations = 16 897).

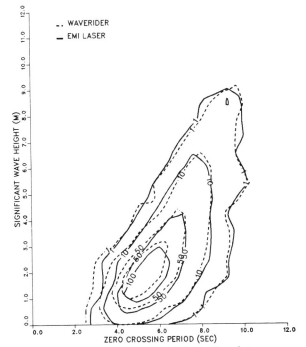

Fig. 2 Simultaneous EMI and Waverider data wave scatter diagram (equal probability density contours in parts per thousand).

low frequencies in the average spectrum from the EMI laser relating to that from the wavrerider.

COMPARISONS BETWEEN THE EMI LASER, THE WAVEC PITCH-ROLL BUOY AND THE MIROS MICROWAVE RADAR

These comparisons cover the period December 1986 to February 1987. WAVEC buoy and EMI laser data were available for the whole of this period but, as discussed above, data from the MIROS were only available for the period 6 to 18 December inclusive. These data had been re-analysed as described above so that the low frequency components of the spectrum were calculated more accurately than had been done previously.

Figure 4 shows the measured H_s time series for the month of December, where it can be seen that for the most part the three sensors agree very well. The MIROS shows a tendency to measure marginally higher wave heights than the other two sensors.

In interpreting the results, it is important to note, as described above, the processing used for the three

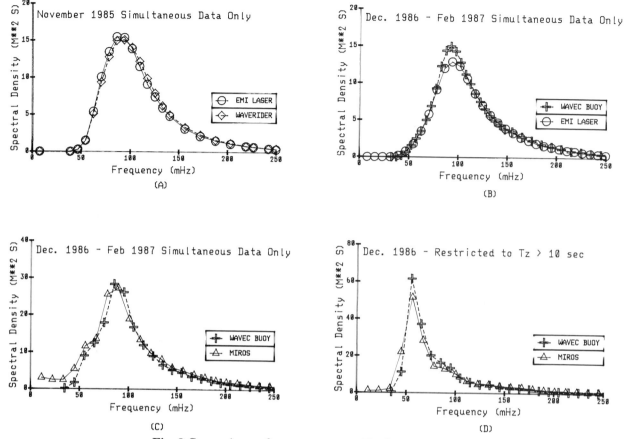

Fig. 3 Comparisons of wave spectra at North Cormorant.

sensors results in spectra at differing frequency resolutions, and different frequency intervals are used in the calculation of spectral moments. Also the sampling times and duration were not simultaneous.

Nevertheless, the results for December show good agreement and will now be discussed in more detail.

Comparisons between the WAVEC buoy and the EMI laser

There were 1230 simultaneous measurements available for December 1986 and January 1987 out of a possible maximum of 1488. The results for regression analyses on H_s and T_z values are shown in Table IV. The best-fit line for H_s shows a slope very close to 1 and a very small intercept. The results for T_z show that in the range where there are measured values, the WAVEC consistently reports higher values of T_z. This is probably due to the different frequency ranges used in the calculation of T_z.

The average spectra measured over the two

months are shown in Figure 3(b). The results show that the WAVEC on average measured higher spectral estimates in the frequency range 70 to 100 mHz than did the EMI laser.

Comparisons between the WAVEC buoy and the MIROS microwave radar

As discussed earlier, re-analysed measurements for the MIROS radar were only available for the period 6 to 18 December inclusive. On 15 December a severe storm occurred with a peak H_s value of approximately 8 m. After the storm there was substantial wave energy at low frequency so that the measured values of T_z were around 10 s or longer. It was the presence of this low frequency energy which identified that the algorithm used in the MIROS wave data processing was not accurate for long period waves.

The average spectra measured by the MIROS and the WAVEC are compared in Figure 3(c). The spectra displayed therein are in close agreement, the only

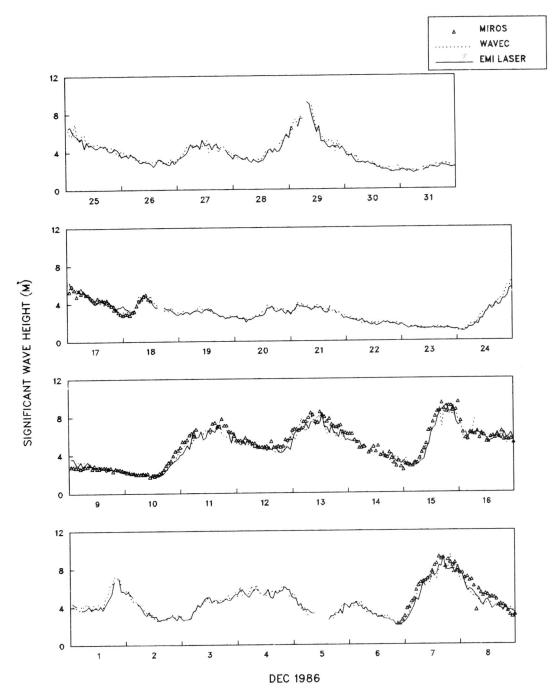

Fig. 4 Comparison of significant wave height.

differences being that the MIROS measures some energy at very low frequencies (less than 0.05 Hz) while the WAVEC measures no energy at these frequencies and the MIROS spectrum is very slightly shifted to lower frequencies relative to the WAVEC spectrum. The MIROS low frequency energy at less than 0.03 Hz does not contribute to the spectral moments (see Table I).

TABLE IV

Results of regression analysis between measurements from the
WAVEC buoy and the EMI laser

Measurement parameter compared	Number of pairs	Correlation coefficient	Regression results[a]		Standard error normal to best fit line
			Slope	Intercept	
All wave data	1 230				
H_s		0.98	1.01	0.06	0.24
T_z		0.96	1.08	−0.35	0.23
T_p		0.97	1.02	−0.32	0.39
$H_s > 5\,m$	223				
H_s		0.85	1.02	−0.07	0.40
T_z		0.91	0.98	−0.51	0.24
T_p		0.98	1.03	−0.28	0.39

[a] Regression equation is $Y = \text{slope} \times X + \text{intercept}$
where: Y = WAVEC buoy, X = EMI laser

In view of the changes carried out on the MIROS processing software, a further spectral comparison was considered desirable. This comparison was restricted to seastates where the WAVEC buoy measured T_z to be greater than 10 s, thus ensuring that the bulk of the wave spectral energy was at frequencies less than 0.1 Hz.

Average spectra for seastates measured simultaneously by the two sensors are shown in Figure 3(d). A total of 15 spectra for each sensor was averaged over and the spectral peak frequencies for the average spectra were approximately 0.055 Hz. The results show that the MIROS and WAVEC produce spectra that are in close agreement on average even when most of the wave energy is at low frequency.

A comparison of the mean direction of the spectral peak frequency is shown in Figure 5 as a time series plot. Since the MIROS had a 180-degree ambiguity, the WAVEC data have been plotted on the same basis. It can be seen therein that these directions agree well except on 10 December, when the wave direction kept changing, and at times on 14 December when the spectral peak periods also differed.

The comparisons between the WAVEC and the MIROS shown here are encouraging but cover only a short measurement period. Further work is necessary to establish unequivocally the reliability and accuracy of the MIROS system and whether the WAVEC and the MIROS do produce equivalent wave height and direction statistics.

SUMMARY AND CONCLUSIONS

A Metocean system containing a number of wave sensors has been collecting measurements on the North Cormorant platform since September 1983. The wave sensors deployed at various times have consisted of a waverider buoy, an EMI infrared laser wave height monitor, a WAVEC pitch-roll buoy and a MIROS microwave radar. Operating experience has shown that:

(a) Over the period September 1983 to December 1986 the EMI laser and the waverider recorded almost the same percentage of good data. Most of the data loss from the EMI laser occurred in the first eight months, and in the period May 1984 to December 1986, the data return from the laser was significantly greater than that from the waverider;

(b) The wave parameters measured by the waverider and the EMI laser are in close agreement, but there is evidence to suggest that the laser may marginally underestimate the high wave heights relative to the values registered by the waverider;

(c) The omnidirectional wave parameters measured by the EMI laser and the WAVEC buoy are also in close agreement. Once again there is a slight tendency for the EMI laser to underestimate the high wave heights relative to the WAVEC;

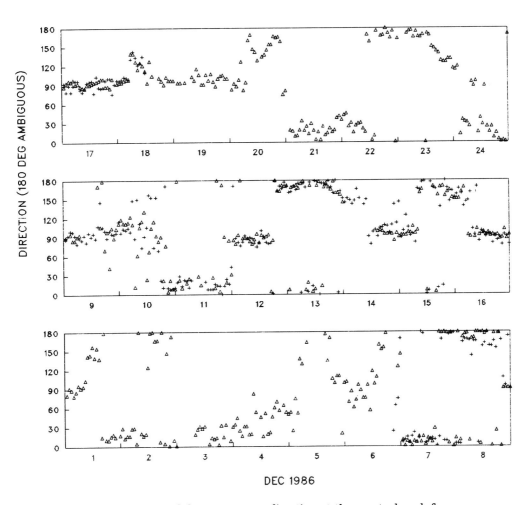

Fig. 5 Time series of the mean wave direction at the spectral peak frequency.

(d) The EMI sensor and data-processing configuration has been demonstrated to be acceptable as an alternative to the waverider and similar data-processing methods. As a consequence, the EMI sensor will continue to be used as a standard means of monitoring wave parameters, together with the waverider buoy.

(e) Software changes and hardware problems with the MIROS meant that only two weeks of good data were available. During this period the omnidirectional parameters measured were in good agreement with those measured by the WAVEC. The wave direction at the spectral peak was also in reasonable agreement.

(f) The advantages of having dual wave sensors at a prime Metocean station have been clearly demonstrated in terms of obtaining significantly higher data returns than could have been achieved with either one.

Data collection will continue at North Cormorant with at least two wave sensors; the EMI laser and the WAVEC pitch-roll buoy. Further investigations of the raw wave data measured by the EMI laser are in progress. It is also intended that a further inter-

comparison study between the WAVEC buoy and the
MIROS microwave radar be carried out at some
future date.

ACKNOWLEDGEMENTS

The authors would like to thank Shell UK Exploration and Production for permission to use the data,
the Dutch Rijkswaterstaat for the WAVEC data and
MIROS for assistance during this study.

REFERENCES

1. Van der Vlugt, A. J. M. (September 1981). The WAVEC Buoy for Routinely Measuring the Direction of Sea Waves.
   Conference on Wave and Wind Directionality Applications to the Design of Structures, Paris, France.
2. Gronlie, O., Brodtkorb, D. C., Woien, J. (June 1984). MIROS—A Microwave Remote Sensor for the Ocean Surface.
   Symposium on Description and Modelling of Directional Seas, Copenhagen, Denmark.
3. Kuik, A. J. and Van Vledderm G. Ph. (June 1984). Proposed Method for the Routine Analysis of Pitch-Roll Buoy
   Data. Symposium on Description and Modelling of Directional Seas, Copenhagen, Denmark.

9

Wave Height Distributions in Shallow Water

C. J. Martin, Marex, *C. K. Grant,* BP International *and R. A. Sproson,*
Marex

The distribution of individual wave heights in shallow water has been shown by many workers to differ from the Rayleigh distribution. This chapter describes the detailed comparison of storm wave data measured at the West Sole Field in the southern North Sea with a variety of distributions and the identification of the most appropriate distribution for design purposes. To achieve this each selected distribution was expressed as a Weibull function of the form:

$$P(H > H_0) = \exp(-x^{\gamma}/\varphi) \tag{1}$$

where P is the probability that the height, x, exceeds the value H_0.

The Gluhovski distribution simplified for the local conditions can be described by equation (1) with $\gamma = 2.2006$ and $\varphi = 9.9724$, for data normalized using the rms sea surface elevation. Of the distributions used this was found to provide the best fit to the normalized cumulative probability data. The consequences of using this distribution for the calculation of the most likely wave in a given period are examined further.

THE WAVE DATA

Time series of sea surface elevation measured using a calibrated wavestaff were recorded on paper charts. The ten most severe storms during the period March 1974 to July 1986 were selected using wave, wind and fetch information. In general, wave records are 17.07 min long and digitized at 2 Hz to give 2048 data points for further analysis. Because the West Sole chart records are only 10 min long, digitization was carried out at a rate of 4 Hz (in terms of the time series) over a period of 8.53 min to produce the required 2048 data points. Sixteen 8.53 minute samples, recorded at three-hourly intervals, centred on the peak of each storm were digitized, giving a total of 160 digital records. The digital data range was ± 15.24 m with a resolution of 0.73 cm. The storm periods used are summarized in Table I. The wavestaff calibrations were then applied to convert the digital samples from 'recorded height' to 'true height'.

Time series and wave spectra were plotted in order to identify records less than 8.53 min long, noisy records, and records containing errors as a result of the

Advances in Underwater Technology, Ocean Science and Offshore Engineering, Volume 16: Oceanology '88

TABLE I
Storm data periods

2100 27:10:74–1800 29:10:74
1800 16:11:75–1800 18:11:75
1800   2:1:76–1200   4:1:76
0300 24:11:77–0000 26:11:77
0900 11:1:78–0900 13:1:78
0000 11:1:79–0000 13:1:79
0000 14:2:79–2100 15:2:79
0600 28:11:80–0300 30:11:80
0600 25:4:81–0600 27:4:81
0000   1:2:83–0300   3:2:83

digitization process. The wave data were also flagged automatically to give an indication of the number of failures of quality control checks on each data sample. The data were checked for deviation from the sample mean and standard deviation, change in steepness, and breaking waves. A total of 16 invalid records were identified and these were removed from the data base, giving a total of 144 valid records.

The time series records were analysed by hand using the Tucker–Draper method (refs 1 and 2). Computerized analyses were also used to calculate the root mean square of the sea surface elevation and other commonly used spectral and deterministic parameters, the analysis of which is not discussed here.

Normalization of wave data

The raw wave data were normalized by dividing each data value by the rms value of the 8.53 minute sample. The normalized data for all 144 valid records were then combined and the heights of the zero-crossing waves were divided into height classes with an interval of 0.5. The data were then arranged as a table of probability of exceedance versus number of waves.

Wave height distributions

Many theoretical and semi-empirical distributions have been proposed for short term wave height statistics. In the present study five distributions were selected following a literature review and these are described below.

(1) Rayleigh distribution

Longuet-Higgins (ref. 3) showed that for a narrow-band frequency spectrum, assuming the sea surface to be the sum of many sine waves, the wave amplitudes are distributed according to a Rayleigh distribution defined by:

$$P(A > A_0) = \exp(-A_0^2/a^2) \tag{2}$$

where a is the rms amplitude of the sea surface elevation, and $P(A > A_0)$ is the probability that the amplitude A of any given wave exceeds the value A_0 (where A_0 is the amplitude for which the probability of exceedence is required).

(2) Longuet-Higgins modified Rayleigh distribution

A modified version of the Rayleigh distribution has been proposed by Longuet-Higgins (ref. 4). This distribution is defined by equation 2 with:

$$a = 0.925\sqrt{(2m_0)} \tag{3}$$

where m_0 = the variance of the sea surface elevation

(3) Two-parameter Weibull distribution

Forristall (ref. 5) fitted hurricane storm wave height data, normalized by the square root of the zero order spectral moment ($\sqrt{m_0}$), to a two-parameter Weibull distribution of the form given in equation (1). The data used in the present study were fitted to a distribution of this form using the method of least squares to fit to all the normalized data and those data where the normalized wave height was greater than 3. Wave height (for a narrow band spectrum) is assumed to be twice the crest elevation from the mean line, i.e. twice the wave amplitude, A.

Given that $E(x)$ is the probability of exceedence for a height x of the data and $R(x)$ is the probability of exceedence calculated using the Rayleigh distribution it is possible to plot $\ln(\ln[E(x)/R(x)])$ versus $\ln(\text{wave height})$, and this is presented in Figure 1. This plot shows that there is little deviation from the Rayleigh distribution for normalized wave heights less than three (natural log approximately equal to 1); however, above this value the deviation is significant and generally increases with increasing wave height.

(4) Gluhovski distribution

A semi-empirical probability density function relating probability of occurrence to water depth has been proposed by Gluhovski (ref. 6). The distribution is defined as:

$$P(H) = \{\pi/(2h[+H^*/(2\pi)^{1/2}][1-H^*])\}$$
$$\times (H/h)^{\{1+H^*\}/\{1-H^*\}}$$
$$\times \exp\{-\pi/(4[1+H^*/(2\pi)^{1/2}])(H/h)^{2/(1-H^*)}\} \tag{4}$$

where $H^* = h/d$

h = mean wave height
d = water depth

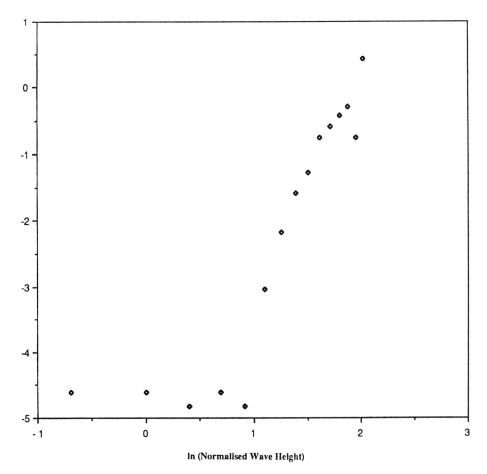

Fig. 1 Error in wave height using the Rayleigh distribution.

In deep water (i.e. $H^* \rightarrow 0$) the Gluhovski distribution assumes the form of the Rayleigh distribution. In order to compare this distribution with the cumulative probability data, and for ease of comparison with the other distributions, a simplified form of the exceedence distribution was sought.

For normalized data

$$h = (2\pi m_0)^{1/2}/m_0^{1/2} = \sqrt{2\pi}$$
$$= 2.507$$

Therefore, the non-dimensional depth is

$$d^* = hm_0^{1/2}/d = H^*\sqrt{m_0}$$

For this location for all storm data (averaged over all storms),

$$m_0^{1/2} = 1.03 \approx 1$$

The sensitivity of this assumption was checked and the difference between using 1.03 and 1 was found to be insignificant. Thus from equation (4) the exceedence probability is given by

$$P(H > H_0) = \exp\{-\pi/4A'(d^*)(H/h)^{K(d^*)}\} \quad (5)$$

where $A'(d^*) = (1+d^*(2\pi)^{-1/2})^{-1} = 0.9649$
$$K(d^*) = 2(1-d^*)^{-1} = 2.2006$$

and $P(H > H_0)$ is the probability that H exceeds H_0

which reduces to equation (1) with $\gamma = 2.2006$ and $\varphi = 9.9724$. Henceforth this will be referred to as the Gluhovski distribution. The parameters for the various distributions expressed in terms of equation (1) are summarized in Table II.

The percentage difference of each of these distributions from the data is shown in Figure 2.

In the calculation of design parameters the extreme wave heights are of interest, and therefore the fit to the upper tail of the data is of primary concern.

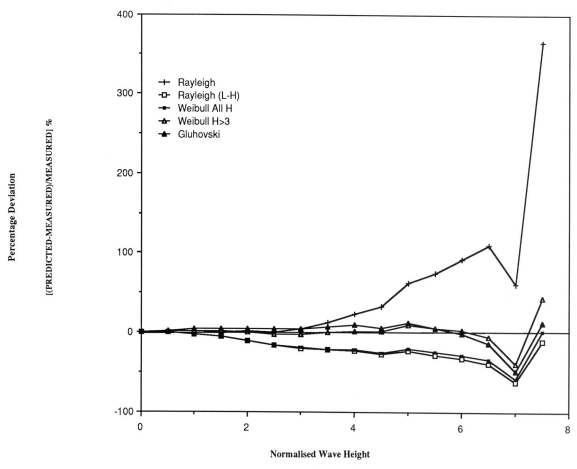

Fig. 2 Percentage deviation of distributions from the
measured data.

Figure 3 compares the fit for the five distributions with the storm data and shows that the Weibull fit to data where $H > 3$ gives the best fit. The Gluhovski distribution, however, does fit the data well, and the improvement in fit when compared with the Rayleigh distribution is significant.

TABLE II
γ and φ parameters for the various distributions

Distribution	γ	φ
Rayleigh	2	8
Longuet-Higgins modified Rayleigh	2	6.48
Weibull fit (normalized data)	1.99	6.44
Weibull fit (normalized data, $H > 3$)	2.105	8.465
Gluhovski (site specific form)	2.2006	9.9724

Note: $P(H > H_0) = \exp(-x^{\gamma}/\varphi)$

The Gluhovski distribution also has the advantage that the variation in γ and φ is parameterized in terms of wave height and water depth, so the relationship between significant wave height and maximum wave height may be determined without having to analyse individual wave data to estimate values of γ and φ and the effects of water depth.

EFFECT OF EMPIRICAL WAVE HEIGHT

Two of the most important wave height parameters of interest to engineers concerned with the design of offshore structures are significant wave height (H_s) and maximum wave height (H_{max}).

Where the individual wave heights follow a Rayleigh distribution,

$$H_s \approx \sqrt{2} H_{rms}$$

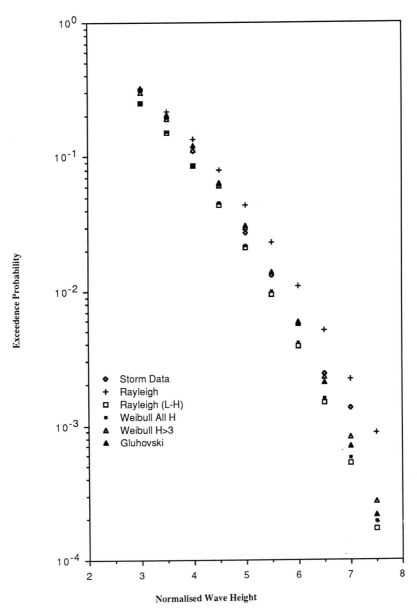

Fig. 3 Cumulative probability diagram for normalized
wave heights > 3.

and

$$H_{max} = H_s (\ln N_z/2)^{1/2}$$

where

H_{rms} = the root mean square zero-crossing wave
height

and

N_z = the number of zero-crossing waves in the
sample

The relationships between H_s and H_{rms} and H_{max} and H_s have been examined further using methods described by Forristall (ref. 5) and Tann (ref. 7) and then extended to the general case where the distribution of individual wave heights is not Rayleigh.

TABLE III
$A_{(\rho)}/a$ for the various distributions

ρ	$A(\rho)/a$				
	Rayleigh $\gamma = 2$ $\varphi = 8$	Rayleigh (L-H) $\gamma = 2.00$ $\varphi = 6.48$	Weibull $H > 3$ $\gamma = 2.105$ $\varphi = 8.465$	Weibull (Forristall) $\gamma = 2.126$ $\varphi = 8.42$	Gluhovski $\gamma = 2.2006$ $\varphi = 9.9724$
0.01	2.3592	2.1235	2.2045	2.1594	2.1822
0.05	1.9855	1.7871	1.8709	1.8356	1.8658
0.10	1.7998	1.6199	1.7040	1.6734	1.7065
0.20	1.5911	1.4321	1.5153	1.4897	1.5255
0.25	1.5172	1.3656	1.4482	1.4244	1.4608
0.30	1.4538	1.3085	1.3904	1.3681	1.4050
0.3333	1.4156	1.2742	1.3556	1.3338	1.3713
0.4	1.3466	1.2121	1.2925	1.2726	1.3106
0.5	1.2561	1.1306	1.2094	1.1914	1.2293
0.6	1.7564	1.0582	1.1352	1.1189	1.5676
0.7	1.1014	0.9914	1.0663	1.0516	1.0892
0.8	1.0307	0.9278	1.0004	0.9871	1.0242
0.9	0.9608	0.8648	0.9348	0.9228	0.9592
1.0	0.8860	0.7976	0.8638	0.8530	0.8879

Estimation of significant wave height from H_{rms}

The form of the generalized two-parameter Weibull distribution that describes the probability that the height of a zero up-crossing wave, H, is less than H_0 is

$$P(H < H_0) = 1 - \exp(-x^\gamma/\varphi)$$

The probability density function is given by Forristall (ref. 5) as

$$f_x(x) = d[P(x)]/dx = \gamma/\varphi \, x^{\gamma-1} \exp(-x^\gamma/\varphi)$$

Following Forristall and the notation in his paper for compatibility, the proportion, $1/\rho$, of heights which exceed a certain value, x_0, is

$$1/\rho = \exp(-x_0^\gamma/\varphi)$$

Note, $x_0 \equiv H_0$
Re-arranging gives,

$$x_0 = (\varphi \ln \rho)^{1/\gamma}$$

The mean value of the heights, $H_{(1/\rho)}$, greater than x_0 is given by

$$\int_{x_0}^{\infty} x f_x(x) dx = x_0 \exp(-x_0^\gamma/\varphi) + \int_{x_0}^{\infty} \exp(-x^\gamma/\varphi) dx$$

from which

$$H_{(1/\rho)} = (\varphi \ln \rho)^{1/\gamma} + \rho\{(1/\gamma)\varphi^{1/\gamma}\Gamma(1/\gamma) - \int_{\infty}^{x_0} \exp(-x^\gamma/\beta) dx\}$$

(6)

Assuming that,

$$H_{(1/\rho)} = (2\sqrt{2})A_{(1/\rho)}/a$$

where $A_{(1/\rho)}$ is the mean value of those amplitudes greater than $H_{(1/\rho)}$ and a is the root mean square amplitude. Solving the integral numerically and evaluating for the five distributions gives the values of $A_{(1/\rho)}/a$ presented in Table III. From this table the relationship between significant wave height ($H_{1/3}$) and H_{rms} can be found since,

$$H_{(1/3)} H_{\mathrm{rms}} = A_{(1/3)}/a$$

The estimation of maximum wave height from H

The number of zero-crossing waves in a recording period of L hours is given by,

$$N_z = L/T_z$$

where T_z is the mean zero-crossing period.

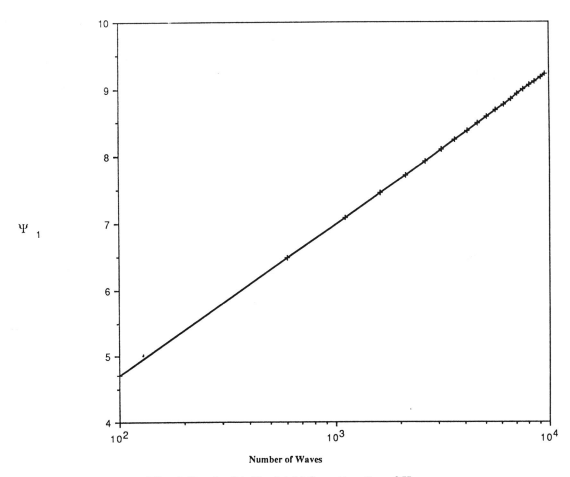

Ψ_1

Fig. 4 Graph of ψ (Rayleigh) for estimation of $H_{\max}$.

Assuming an exceedence probability of the form

$$P(H_1 \leqslant x) = [1-\exp(-x^\gamma/\varphi m_0^{\gamma/2})]^{N_z} \qquad (7)$$

where H_1 is the largest wave in three hours then following Tann (ref. 7) the probability density function is given by

$$2N_z/(\varphi m_0^{\gamma/2})[x \exp(-x^\gamma/\varphi m_0^{\gamma/2})]$$
$$\times[(1-\exp(-x^\gamma/\varphi m_0^{\gamma/2})]^{N_z-1}$$

The turning point of this equation occurs when

$$\psi_1 = \ln N_z - \ln[1-(1/\gamma\psi_1)(1-\exp^{-\psi_1})]$$

where $\psi_1 = x^\gamma/\varphi m_0^{\gamma/2}$

Solving the equation,

$$0 = [\ln N_z - \ln(1/\gamma\psi_1(1-\exp^{-\psi_1})]^{-\psi_1}$$

numerically gives the relationship between ψ_1 and N_z, as shown in Figure 4. There is less than 0.5%

difference in ψ_1 for any value of N_z for the distributions considered in this study and so only one line is presented.

Following Tann's assumption that the height of a zero up-crossing wave is approximately twice the height of the crest measured from the mean line then,

$$H_{\mathrm{rms}} = 2A$$

From equation 6

$$H_{1/3} = KH_{\mathrm{rms}}$$

where $K = A_{(1/3)}/a$

and $a = \sqrt{(2m_0)}$ (refs 8 and 9)

Therefore,

$$H_{1/3} = K2\sqrt{(2m_0)}$$

Now,

$$H_{\mathrm{rms}} = 2\sqrt{(2m_0)}$$

and

$$\psi_1 = x^\gamma/\varphi m_0^{\gamma/2}$$

So

$$x^2/H_{rms}^2 = (\psi_1\varphi/8^{\gamma/2})^{2/\gamma}$$

and

$$H_{max}(L) = 2\sqrt{(2m_0)}(\psi_1(L)\varphi/8^{\gamma/2})^{1/\gamma}$$

However,

$$H_{1/3} = K2\sqrt{(2m_0)}$$

So

$$H_{max(L)} = H_s/K(\psi_1(L)\varphi/8^{\gamma/2})^{1/\gamma}$$
$$= (\sqrt{2}/4)(H_s/K)(\psi_1(L)\varphi)^{1/\gamma}$$

For the West Sole Field the values of normalized 50 year extreme maximum wave height were calculated as 7.5, 7.1 and 7.2 for the Rayleigh, Gluhovski and Weibull distributions respectively.

PROBABILITY OF EXCEEDENCE OF H_{max}

An expression for the probability of exceedence of the most likely maximum of N values from a Rayleigh distribution is given by Carter and Challenor (ref 10). The maxima of the cumulative distribution of N values from a two-parameter Weibull distribution is found from the first derivative of:

$$P(x_{max} < x) = F(x) = [1-\exp(-x^\alpha/\beta)]^N \qquad (8)$$

where

$$\alpha = 2, \quad \beta = 2$$

for the Rayleigh distribution and

$$\alpha = 2.2006, \quad \beta = 2.1695$$

for the Gluhovski distribution using the approximation

$$df/dx = (F_1-F_0)/(x_1-x_0) \qquad (9)$$

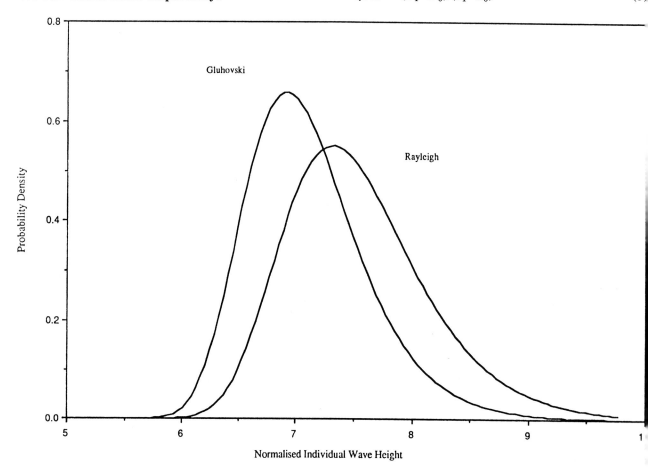

Fig. 5 Rayleigh and Gluhovski distributions for H_{max}. $N = 1059$ waves.

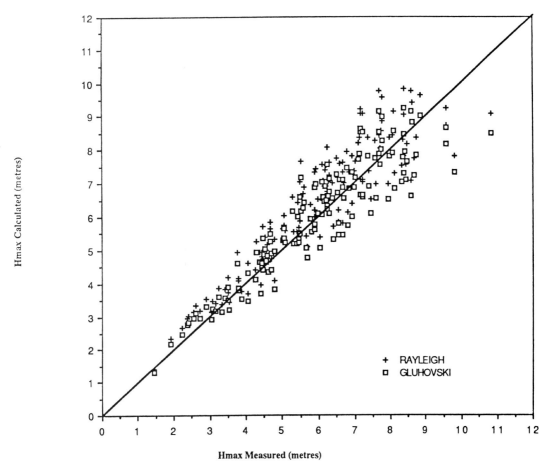

Fig. 6 Regression analysis of H_{max} (measured) against
H_{max} calculated using the Rayleigh and Gluhovski
distributions.

Evaluating equation (9) for $F(x)$ given by equation (8) in the range 0–25 metres (Fig. 5) the maximum of the first derivative of equation (8) was found by inspection, this being the percentage exceedence probability of the distribution. The results were 60.52 and 60.54% for the Rayleigh and Gluhovski distributions respectively for 1059 waves which is a typical number of waves in an extreme three hour sea state.

COMPARISON OF H_{max} (MEASURED) AND H_{max} (CALCULATED)

The maximum wave height for each of the 8.53 minute samples was plotted (Fig. 6) against the values calculated using the Rayleigh and Gluhovski distributions. The results of a regression analysis of these data are given in Table IV.

The difference between the two distributions for a period of 8.53 minutes is small. The Gluhovski distribution does, however, appear to fit the measured data at least as well as the Rayleigh distribution. Within the limits of statistical significance the correlation coefficient between the measured and calculated values are the same for both distributions with the covariance slightly lower for the Gluhovski distribution and the major axis regression line constrained to pass through the origin having a slope closer to one.

CONCLUSIONS

The distribution of indivdual wave heights at The West Sole Field in the southern North Sea has been shown to be represented well by the Gluhovski dis-

TABLE IV

Results of regression analysis of H_{max} (measured) against H_{max} (calculated)

	Rayleigh	Gluhovski
Correlation coefficient	0.916	0.914
Covariance	3.09	2.92
Major axis through origin regression		
Slope	1.0664	1.0000
Standard Error	0.54	0.52

tribution, simplified and expressed as a two-parameter Weibull distribution. The most likely maximum wave height for a given sampling length and return period calculated using this equation is 7% lower than that estimated using the Rayleigh distribution.

The Gluhovski distribution has been reduced to a site-specific form expressed as a two-parameter Weibull distribution and compared with the cumulative percentage exceedence of the most likely highest wave, H_{max}, in 1059 waves has been calculated to be 60.5%. This is equivalent to the value for the Rayleigh distribution for the same number of waves.

ACKNOWLEDGEMENTS

This paper arose from a research project performed by Marex Technology Ltd on behalf of BP International Ltd. The authors gratefully acknowledge BP's permission to publish this paper.

REFERENCES

1. Tucker, M. J. (1963). Analysis of records of sea waves. *Proc. Instn Civ. Engrs* **26**: 304–316.
2. Draper, L. (1963). Derivation of a design wave from instrumental records of sea waves. *Proc. Instn Civ. Engrs* **32**: 291–304.
3. Longuet-Higgins, M. S. (1952). On the statistical distribution of the heights of sea waves. *J. Mar. Res.* **11**: 245–266.
4. Longuet-Higgins, M. S. (1980). On the distribution of the heights of sea waves: some effects of nonlinearity and finite band width. *J. Geophysical Res.* **85**: No. C3, 1519–1523.
5. Forristall, G. Z. (1978). On the statistical distribution of wave heights in a storm. *J. Geophysical Res.* **83**: No. C5, 2353–2358.
6. Ibragimov, A. M. (1973). Investigation of the distribution function of wave parameters in regions of wave transformation. *Oceanology* **13**: 584–589.
7. Tan, H. M. (1976). The Estimation of Wave Parameters for the Design of Offshore Structures. IOS Report No. 23.
8. Goda Y. (1985). *Random Seas and the Design of Maritime Structures*. University of Tokyo Press.
9. Tayfun, M. A. (1984). Non-linear effects of the distribution of amplitudes of sea waves. *Ocean Engng* **11**: 245–264.
10. Carter, D. J. T. and Challenor, P. G. (1981). Estimating Return Values of Wave Height. IOS Report No. 116.

Analysis of Waves from Arrays at Abu Quir and Ras El-Bar, Egypt

M. H. S. Elwany, ECY-Systems Management Associates Inc, Encinitas, CA, USA
A. A. Khafagy, Coastal Research Institute, Alexandria, Egypt
D. L. Inman, Scripps Institution of Oceanography, La Jolla, CA, USA
and A. M. Fanos, Coastal Research Institute, Alexandria, Egypt

This chapter gives wave statistics along the Mediterranean Coast of Egypt. This information can be used to estimate longshore transportation of sand from the direction of propagation and the energy flux of waves at the break point (ref. 1). In addition, the divergence of the longshore transport rate identifies areas of potential erosion/accretion, while the magnitude of the divergence is proportional to the rate of erosion/accretion (refs 2–4).

Reliable data on wave climate prior to this study are not available for the Nile Delta littoral cell. Some wave data are presented in refs 5–7, but do not provide adequate wave direction. Inman *et al.* (ref. 7) give tables of wave climate based on shipboard observations, but this type of data is notably inaccurate. In order to obtain an accurate description of wave climate along the coast of Egypt, two stations for the measurement of wave energy and direction were installed off the Nile Delta at Abu Quir and Ras El-Bar (Fig. 1). Each station consists of four basic units plus cables for interconnections: the recorder unit, acquisition unit, power unit, and sensor array. The sensor arrays of the two stations incorporate three wave pressure sensors. The system is developed and produced by the Center for Coastal Studies, Scripps Institution of Oceanography and is called cassette acquisition system (CAS). This system is well described by Boyd and Lowe (ref. 8).

The pressure sensors at Abu Quir Station are at about 18 km from the shoreline, mounted 6 m below mean water level (MWL) and 12 m above the sea bed. At Ras El-Bar the sensors are at 6.5 m below MWL and 0.6 m above the bottom, approximately 900 m offshore of the shoreline. The sensors are positioned linearly parallel to the Coast. At these two stations, data are recorded for 34 minutes, once every 6 hours, so that there are 4 wave records per day at each station. Figure 2 gives the configurations of the Abu Quir and Ras El-Bar wave arrays.

The measured data have been subjected to statistical and spectral analysis. For each 34-minute data set, the total variance $\langle \eta^2 \rangle$ of the sea surface displacement η is obtained. The significant wave height H_s, the wave period T_p corresponding to the reciprocal

Advances in Underwater Technology, Ocean Science and Offshore Engineering, Volume 16: Oceanology '88

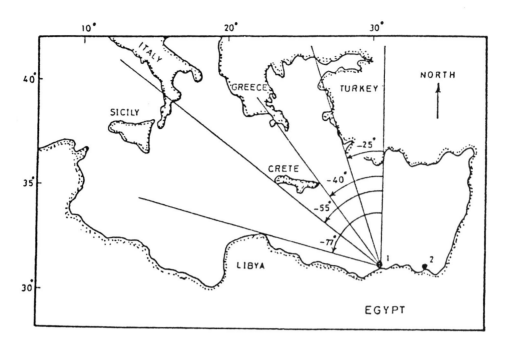

Fig. 1 Location of Abu Quir (1) and Ras El-Bar (2) wave arrays.

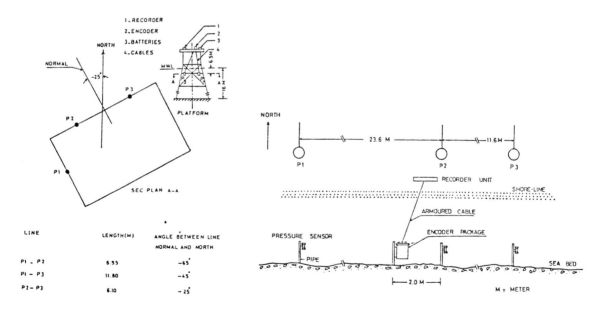

LINE	LENGTH(M)	ANGLE BETWEEN LINE NORMAL AND NORTH
P1 - P2	6.55	-65°
P1 - P3	11.80	-45°
P2 - P3	6.10	-25°

Fig. 2 Arrangement of Abu Quir (*left*) and Ras El-Bar (*right*) wave arrays.

of the frequency of the band with maximum energy density, and a mean direction A_p for peak frequency are computed.

ANALYSIS OF WAVE DATA

The three pressure time-series were subjected to pre-processing treatment designed to detect areas of invalid data, based on expected maximum and minimum recorded values for each time series maximum excursion from the mean and expected slew rates. Once the magnitude and slew rate checks have been applied, a linear or quadratic interpolation is applied across the set of spurious values detected. The processed time-series resulting from the above treatment is used as input for analysis programs.

Each 34.1 minute pressure time-series was Fourier transformed. The Fourier coefficients of sea surface elevation were obtained by applying the frequency dependent depth correction given by linear wave theory. The very large correction required for high frequency waves limits the upper frequency which can be studied to about 0.35 Hz. A significant wave height is then obtained from the mean of the variances of the three pressure sensors through the formula

$$H_s = 4(\langle \eta^2 \rangle)^{1/2} \qquad (1)$$

The analysis of wave direction requires more than one sensor. In the following, we shall discuss the problem of estimation of wave direction from data measured by two pressure sensors.

Assuming that the waves are known to approach within a 180° arc, the direction of a single wave train can be determined from the relative arrival times of a wave crest at the sensors. The phase difference between the signals of two sensors expressed in radians is related to the direction of the wave train $A(f)$ by the equation

$$\phi_i = 2\pi l \sin A(f_i)/L_i \qquad (2)$$

where l is the distance between the two sensors, X is the angle of the wave approach to the normal to a line connecting the two sensors, $L_i = 2\pi/k_i$ is the wave length for a wave of frequency f_i, and k is the wave number at frequency f given by

$$(2\pi f)^2 = gk \tanh kh \qquad (3)$$

where g is gravitational acceleration and h is the water depth. The phase difference of the wave signals of the two sensors as a function of frequency is obtained from the cross-spectrum analysis. From equation (2) $A(f_i)$ is given by

$$A(f_i) = \sin^{-1}(L_i \phi_i / 2\pi l). \qquad (4)$$

There are two conditions for which it is not possible to calculate the wave direction $A(f_i)$. These include

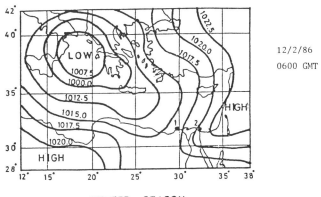

12/2/86
0600 GMT

WINTER SEASON

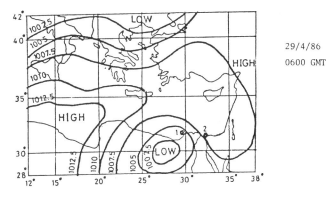

29/4/86
0600 GMT

SPRING SEASON

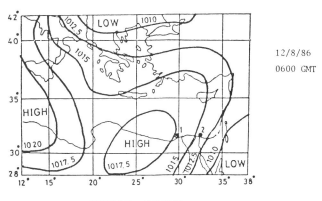

12/8/86
0600 GMT

SUMMER SEASON

Fig. 3 Typical sea level pressure distributions (pressure units in millibars). (After ref. 9.)

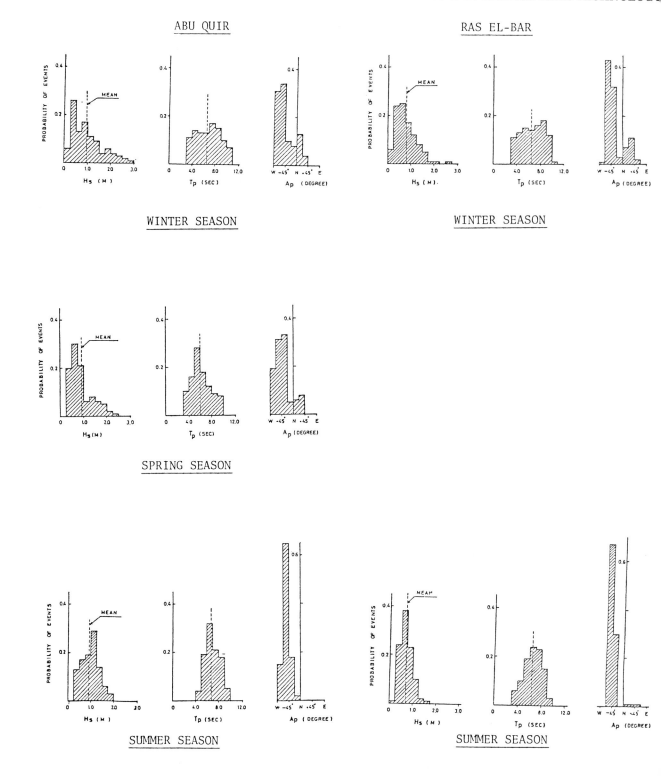

Fig. 4 Histograms of wave parameters H_s, T_p and A_p for various seasons.

poorly conditioned wave data, presumably due to spectral leakage, and spatial aliasing due to large separation distance between the two sensors. If the data are poorly conditioned for determining wave direction, the absolute value of the quantity within the brackets in equation (4) may exceed unity, a physically impossible condition since the extreme values of sine function are ± 1. The second condition is related to spatial aliasing, which necessitates that one-half the wave length be equal to or greater than the projection of the sensor separation distance l in the direction of wave propagation.

In this study the angle of the waves in the frequency band with the maximum energy density is selected as the representative direction for a particular run and we referred to it as A_p.

RESULTS AND DISCUSSIONS

Wave action along the Nile Delta Coast is seasonal in nature and is strongly related to large-scale pressure systems whose limits overstep the boundaries of the Mediterranean area and extend towards the Atlantic, Eurasia and Africa. Inspection of a large number of synoptic weather maps shows that there are three distinct patterns of sea level pressure distributions, each associated with a certain period of the year. The three various patterns (Fig. 3) cover all the year. Further details are given below.

Winter season

The Azores anticyclone often extends over the Libyan desert, and North Atlantic depressions may enter the Eastern Mediterranean area and bring in masses of cold Arctic air. This cold air, meeting the warm and moist air of the Mediterranean area, produces vertical instability which often leads to moving atmospheric perturbations associated with meteorological fronts. The travelling depressions when associated with ridges of high pressure over the north-western Libyan Desert, generate high waves in the Eastern Mediterranean responsible for most damage on the northern Egyptian coast. Winter season occurs from November to March.

Spring season

The desert depression (see Fig. 3) affects the area during April and May which causes strong wind associated with sand storms.

Summer season

The summer season covers the period from June to September and is free from moving depressions with light wind. Sea swells are primarily due to north to north-west wind.

The measured data have been subjected to statistical and spectral analyses. Each wave event of 34.1

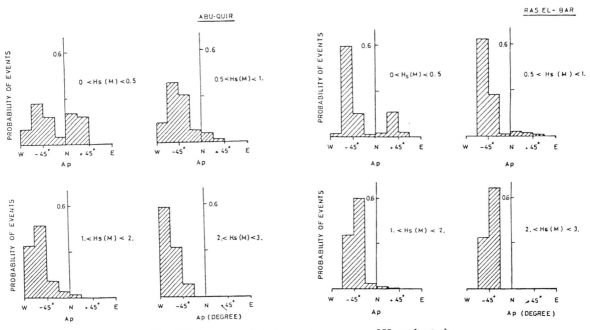

Fig. 5 Histograms of A_p for various ranges of H_s evaluated from all the measured data at wave arrays.

minutes duration is characterized by three parameters obtained from the wave frequency spectra and their phase difference:

(1) The significant wave height, H_s defined in equation (1)
(2) The 'peak frequency', $f_p = 1/T_p$, where f_p is the central frequency of the frequency band containing the maximum energy density in the frequency spectrum and T_p is the peak wave period.
(3) The 'peak direction' A_p defined as the direction of approach of the waves in the peak frequency band.

Histograms of the three parameters H_s, T_p and A_p are shown in Figure 4 for seasonal changes at both stations. Figure 5 shows the predominant wave direction for various ranges of the significant wave height, H_s. Figure 6 is presented as an example of data collected by the CAS. A time series of H_s, A_p and T_p at

Abu Quir is plotted with data collected simultaneously at Ras El-Bar. Figure 7 shows the cumulative percentage of occurrence of H_s and T_p for the available data. The bivariate distributions of H_s and T_p for both stations are shown in Figure 8.

Computation of the directional spectrum of the wave field at Ras El-Bar using maximum likelihood (MLE) and iterative MLE (IMLE) directional estimators (see refs 10 and 11) have been carried out. All the information from the three sensors are utilized in this analysis. In designing Ras El-Bar array, distances between sensors are chosen for maximum directional spectrum resolution obtainable from a three-sensor linear array. Several data sets have been analysed. Figure 9 shows directional spectra for two frequencies on 18 May 1985. The figure indicates that at frequency = 0.18 Hz wave trains from two different directions passed over the array at the same time. Referring to Figure 1, these two directions correspond to the windows between Libya–Crete and Turkey–Crete.

CONCLUSIONS

Accurate directional wave measurements were continuously carried out at two locations in Egypt in front of Abu Quir and Ras El-Bar. As the wave climate varies from year to year, it is planned to continue this study to update the data to acquire a statistically reliable database. These data led to better understanding of coastal problems. A summary of the findings from this study are presented below:

- The CAS system has proved to be a viable method for obtaining data from nearshore waters.
- At the Abu Quir wave array the predominant directions of the waves are west north-west followed by north-west and west, while at the Ras El-Bar wave array the predominant wave direction is north-west.
- At the Ras El-Bar wave array westerly waves are usually absent.
- Generally the phase difference with north-west wave activity between Abu Quir Bay and Ras El-Bar is about 6 hours.
- The wave climate of the winter and spring months is much more severe than the summer, with alternating high storms (say, H_s = 2–3 m) and calm periods. There is also an increase in the occurrence of medium to low energy 4–6-second chop waves.
- The directional information indicates that there are many components to the winter and spring wave climate, while the summer wave climate is

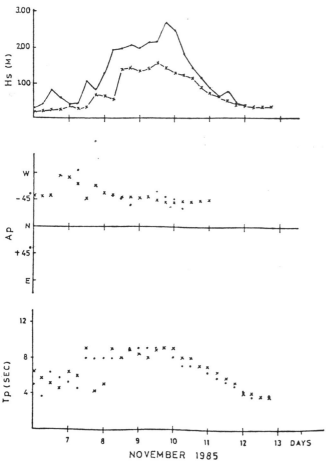

Fig. 6 Comparison of Wave data collected at Abu Quir (●) and Ras El-Bar (×) wave arrays.

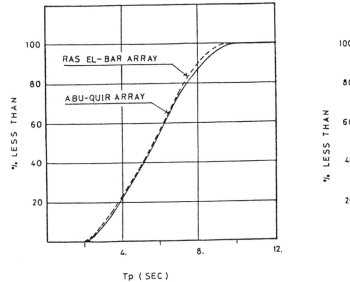
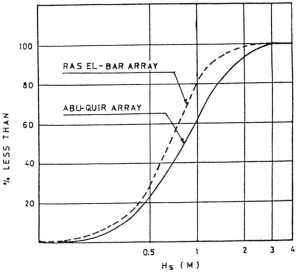

Fig. 7 Cumulative distribution of H_s and T_p.

the most narrowly directed with north-west predominant direction.

ACKNOWLEDGEMENTS

This report presents the results of an AID-sponsored study 'Coastal Management and Shore Processes in the Southern Eastern Mediterranean'. Many people aided in this study. We express our gratitude and indebtedness to Mr B. Abel, Mr B. Lowe and Mr W. Boyd of the USA and to Mr Hossam El-Morsy of Egypt.

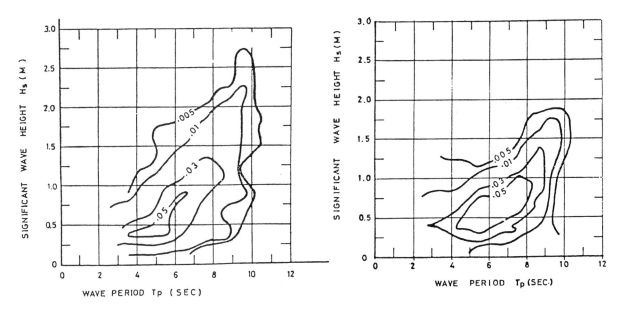

Fig. 8 Joint probability of H_s and T_p: (*left*) Abu Quir; (*right*) Ras El-Bar.

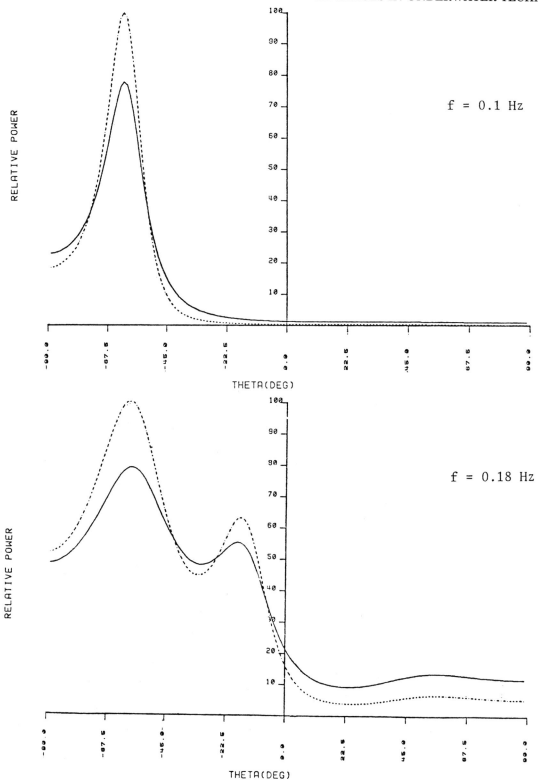

Fig. 9 Directional spectra of waves at Ras El-Bar for two
frequencies on 18 May 1985. Each spectrum is scaled to
have maximum density 100. (———— MLE; – – – – IMLE.)

REFERENCES

1. Komar, P. D. and Inman, D. L. (1970) Longshore sand transport on beaches. *J. Geophys. Res.* **75**: 5914–5927.
2. Inman, D. L., Aubrey, D. G. and Pawka, S. S. (1976). Application of nearshore process to the nile Delta. UNDP/UNESCO Proc. of Seminar on Nile Delta Sedimentology, Academy of Scientific Res. & Tech., pp. 205–255.
3. Monhar, M. (1981). Coastal processes at the Nile delta coast. *Shore and Beach* **49**: 8–15.
4. Inman, D. L. and Jenkins, S. A. (1984). The Nile Littoral Cell and Man's Impact on the Coastal Zone of the Southern Mediterranean. 19th Coastal Eng. Conf. Proc., ASCE, pp. 1600–1617.
5. Monhar, M., Mobarek, I. E. and Fanos, A. M. (1974). Longshore Currents and Waves at Burallus Coast. Proc. Coastal Eng. Conf., 14th Amer. Soc. Civil Engin., Vol. 2, pp. 685–689.
6. Monhar, M., Mobarek, I. E., Fanos, A. M. and Rahal, H. (1974). Wave Statistics Along the Northern Coast of Egypt. Proc. Coastal Eng. Conf., 14th Amer. Soc. Civil Engin., Vol. 1, pp. 132–147.
7. Monhar, M., Mobarek, I. E. and El-Sharaky, N. A. (1976). Characteristics of Wave Period. Proc. Coastal Eng. Conf., 15th Amer. Soc. Civil Engin., Vol. 1, pp. 273–288.
8. Boyd, W. and Lowe, R. L. (1985). A High Density Cassette Data Acquisition System. Ocean 85: Ocean Engineering and the Environment, Marine Technological Society and IEEE, Vol. 1.
9. Hamed, A. A. (1983). Atmospheric Circulation Features over the Southeastern Part of the Mediterranean Sea in Relation to Weather Conditions and Wind Waves along the Egyptian Coast. PhD Thesis, Faculty of Science. Alexandria University, Alexandria, Egypt.
10. Inman, D. L. and Guza, R. T. (1984). Island Sheltering of Surface Gravity Waves: Model and Experiment. *Continental Shelf Res.* **3**: 35–53.
11. Oltman-Shay, J. and Guza, R. T. (1984). A data-adaptive ocean wave directional-spectrum estimator for pitch and roll type measurements. *J. Phys. Oceanogr.* **14**: 100–110.

11

The Influence of Changing Wind Conditions on Sea Surface Residual Currents Measured by HF Radar

J. P. Matthews, School of Ocean Sciences, University College of North Wales, Menai Bridge, Gwynedd LL59 5EY, UK

Early theoretical investigations of the effect of wind stress on coastal waters were performed by Ekman (1905) and later by Jeffries (1923). It was recognized that the well-known Ekman spiral type of solution obtained for the open ocean, in which surface current is inclined to the wind at 45° towards the high pressure side, must be modified by the presence of a coastline which prevents any resultant drift developing across vertical sections parallel to the coast. For example, according to the deep coastal water solutions obtained by Jeffries, hurricane force winds blowing alongshore produce a surface drift which is inclined at an angle of 18° to the wind—a smaller inclination than that produced in the open ocean. However, for hurricanes meeting the coast at right angles, the surface drift and wind are again inclined at 45°, since the presence of the coastline does not affect the depth averaged drift. For more commonly encountered wind speeds (of, say, ~ 10 m s^{-1}), the mathematics indicate a surface response which follows the direction of the coastline for a wide range of wind vectors. This result is of particular importance for the present chapter. The theory also indicates that the local response time under changing wind conditions has a predicted dependence on the square of the distance from the shoreline. For distances from the shoreline of relevance here (2–40 km), a steady state can develop within time scales of $\leqslant 1$ hour, which is much shorter than typical response times for the open ocean.

Computer modelling of the dynamics of selected oceanic regions, such as the Irish Sea (e.g. Proctor, 1981), has greatly improved our insight into wind-driven surface currents, but relatively few detailed attempts to compare theory or modelling predictions with experiment have been made, mainly because true surface currents are difficult to measure. An interesting study along these lines was made by Murray (1975), who focused on a comparison between predictions for near-shore wind-driven currents and flows determined by tracking drogues (using a theodolite technique) to within a distance of 800 m of a straight coast. This work indicated that currents generated by local winds are directed parallel to the shoreline to within a few degrees, nearly independent of wind direction or speed, in agreement with the

Advances in Underwater Technology, Ocean Science and Offshore Engineering, Volume 16: Oceanology '88

results of analytic theory. However, current speed was found to be strongly controlled by the wind angle to the shoreline and, to a lesser degree, by wind speed.

The drogue tracking technique employed by Murray (1975) is best suited to operations in restricted areas of water close to the coast. It is difficult, however, to get a truly synoptic view of currents in more extensive regions using drogues, since strong biasing of the data in time and/or space is normally produced. Ideally, a constant throughput of drogues, evenly distributed across the region under study, is required, but this approach poses severe logistic constraints in all but the narrowest strips of coastal water. Current meters give useful data at selected points, but are less reliable in providing information near the surface—a region of crucial importance in wind-related studies. One answer to the problem of determining surface currents is to employ the HF radar technique to provide current maps free from

spatial or temporal aliasing. Large areas of ocean (of order 500 km^2) can be surveyed by this method, since HF signals are able to propagate beyond the normal optical and microwave horizon.

The HF Doppler radar technique relies on the fact that the ocean surface is able to backscatter signals transmitted from a shore-based radar, in a way analogous to the diffraction of X-rays by a crystal. In both cases, for strong backscatter, a matching condition must be met by the wavelength of the incident radiation and the spatial separation of scatterers. This requirement (called the Bragg resonance condition) implies that, for the ocean, radar signals are strongly backscattered by ocean waves of half the radar wavelength.

The radar echo from the sea surface ideally contains two peaks, which, in the absence of oceanic surface currents, are symmetrically spaced about the transmitted frequency. With an underlying surface

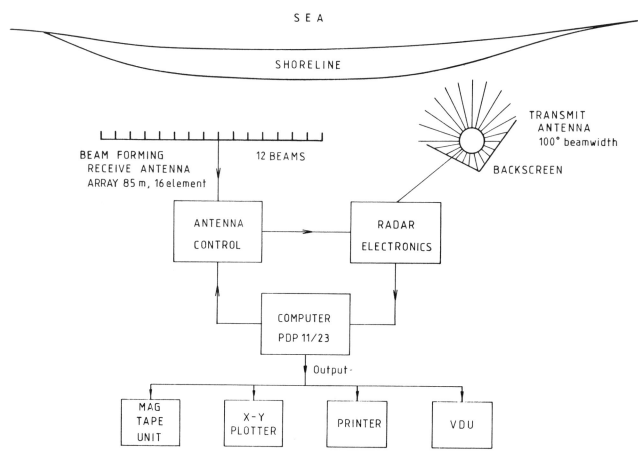

Fig. 1 The OSCR experimental configuration. Note that two such systems are required to define the vector surface current.

current present, however, these peaks are displaced in frequency by an amount proportional to the component of current along the line of sight of the radar. The shift is positive (negative) for currents travelling towards (away) from the radar. Measurement of this shift provides one radial component of the surface current, and so two radars are required to measure the vector current.

EXPERIMENTAL DETAILS

The ocean surface current radar (OSCR) (King *et al.*, 1984) HF radar system employed in the present study (Fig. 1) uses a transmitter working at 27 MHz. This frequency implies that resonance with sea waves of wavelength near 5.5 m takes place for Bragg scattering and that the radar-derived surface currents represent depth averaged values for the top 30 cm (= $\lambda/12\pi$; λ is the radar wavelength) of the water column (Stuart and Joy, 1974). The radar transmissions are

of 8 µs duration and peak power of ~ 2 kW, and the backscattered sea echoes are detected by a 16-element beam forming antenna array, which produces a series of narrow beams each of width 5°.

Data collection is performed through an interleaving procedure in which backscattered signals from one radar site are recorded for 2 minutes and then analysed during the following 3 minutes. Within this 3-minute interval, backscattered signals are recorded at the other site, so that a cycle for 12 beams at each site takes one hour (i.e. data points are obtained once per hour). Range values along each of the beams are calculated from the time delay between transmitted and received signals, and the echo data are grouped into bins of length 1.2 km.

In May 1985 an OSCR experiment was performed from the coast of North Wales, UK, by the University College of North Wales, IOS Bidston, and the University of Cambridge. Figure 2 shows the locations of the two radar sites (Point Lynas and Great Orme's

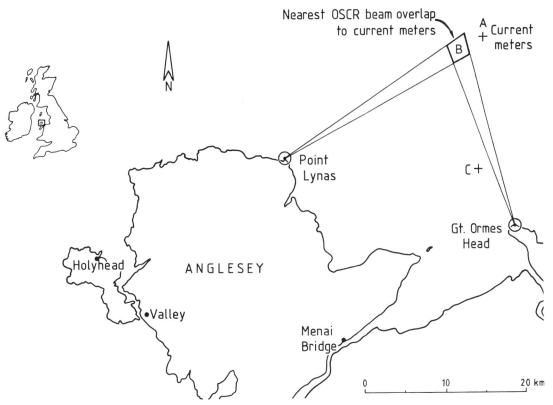

Fig. 2 Map of the May 1985 HF radar Campaign. OSCR units were located at Point Lynas and Great Orme's Head and measured surface currents within the triangle formed by intersection point B and the radar sites. Current meters were located at point A, and an acoustic Doppler current profiler at point C.

Head) during this campaign. Surface currents were derived throughout the triangle formed by Point Lynas, Great Orme's Head, and beam intersection point B, for a period of nearly four weeks. The data set collected during this experiment has proved to be extremely useful. Griffiths (1986) compared results obtained from an acoustic Doppler current profiler (ADCP) located at point C of Figure 2, and surface currents measured in the vicinity of point C by the OSCR radar. Matthews *et al.* (in press) employed a similar approach with data obtained from three current meters moored at point A of Figure 2, and used the analytical model of Prandle (1982) to obtain a surface current prediction for comparison with the radar data. The difference between the radar-derived and predicted surface flow fields was represented by an ellipse. The major axis of this ellipse was directed close to north, indicating the presence of an additional flow component in the radar results of magnitude 0.13 m s^{-1} (0.07 m s^{-1}) in the north–south (east–west) direction.

In confirming the internal consistency of these OSCR data, Prandle (in press) found a standard error of < 0.03 m s^{-1} in the derived surface currents. His investigation into the role of wind forcing across the radar survey region was based on an empirical orthogonal functional (EOF) analysis, which showed that a 'slab-like' surface current was the most likely mean response to wind forcing across the radar survey region. A high correlation coefficient (0.73) was obtained between the dominant east–west mode and the wind-stress time series, although the correlation

was weaker (0.32) in the north–south directions. The analysis of wind-driven currents presented here complements that of Prandle (in press), but is more in the line of a case study, since we are primarily concerned with the details of the structure and development of these currents under changing wind conditions.

WIND OBSERVATIONS AND RADAR-DERIVED SURFACE CURRENTS

The surface currents described here were measured during the period 12–15 May, 1985, when well-defined changes took place in the wind field. To illustrate these changes, wind information from Valley, Anglesey are displayed in Figure 3. These data, together with evidence from other stations (particularly Bidston, Merseyside) and from Daily Weather Summaries, were used to identify the major wind fields influencing the radar survey region. The data period has been divided into five intervals, each representative of a reasonably steady wind field or of a transitional period between wind fields. These are listed as follows:

(1) Largely north-easterly winds at average speed of 7 m s^{-1} from 0000 GMT 12 May to 0200 GMT 14 May.
(2) North-easterly winds at an average speed of 3.5 m s^{-1} from 0200 to 1300 GMT 14 May.
(3) A short transitional period from 1300 to 1500 GMT 14 May.

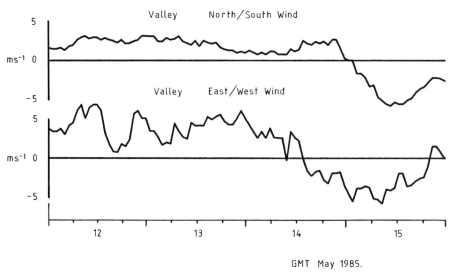

Fig. 3 North–south and east–west components of wind observed at Valley (Fig. 2) during the period under study.

4) North north-westerly winds at an average speed of 5 m s^{-1} from 1500 to 2300 GMT 14 May.

5) A transitional period from 2300 GMT 14 May to 0900 GMT 15 May.

6) South south-westerly winds at an average speed of 9 m s^{-1} from 0900 to 2300 GMT 15 May.

Our aim is to compare the wind conditions during these six periods with the corresponding surface residual currents derived from the radar. These residuals are obtained by tidal analysis of the individual time series obtained for each line of sight component at the radar beam intersection points. This procedure gives a prediction for the tidal flow components which can be subtracted from the original data. The resulting residuals are then smoothed, grouped into time windows, each of 1 hour duration, and then plotted in the form of surface current maps.

(1) 0000 GMT 12 May–0200 GMT 14 May

The data shown in Figure 4(a) are representative of the surface residual distribution resulting from a period of sustained forcing by largely north-easterly winds blowing at a mean speed of 7 m s^{-1}. Near to the coast, the flow is alongshore (the coastline is irregular but in an average sense can be considered to run from west north-west to east north-east). However, in deeper water, at the northern apex of the radar survey region, currents are directed toward the north-west. This northward bias may indicate that further from the coast the large scale circulation pattern set up in Liverpool Bay as a whole is more

important than the alongshore flow. However, it is more likely that this northward bias points to the influence of density gradient currents directed parallel to the roughly north-south aligned isopycnals (Matthews *et al.*, 1988). Typical values of the wind-driven residual currents observed at these times are ~ 0.15 m s^{-1}.

(2) 0200–1300 GMT 14 May

During this period, north-easterly winds continue, but the wind speed drops to a mean value of 3.5 m s^{-1}. The overall pattern (Fig. 4(b)) is similar to the case (1), though typical wind-driven currents are now smaller at ~ 0.07 m s^{-1}. Again, the flow field is divided into two regions. Some channelling of current parallel to the coast seems to develop near the western apex of the radar survey region, while near the northern apex a northward flow is visible.

(3) 1300–1500 GMT 14 May

The data gathered during periods (3) and (4) give some insight into the time scales required for surface currents to respond to a change in wind conditions. These changes began at 1300 GMT and involved a switch from north-easterly to north-westerly winds over a two-hour period. At 1430 GMT (Fig. 5(a)) the residual flow field near the coast still shows no organized motion in response to these changes, except possibly in the northern portion of the radar map. In particular, the earlier pattern (Figs 4(a), (b)) is not present.

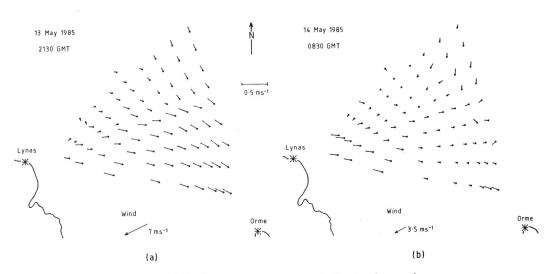

Fig. 4 (*a*) Surface current response to forcing by north-easterly winds blowing at a mean speed of 7 m s^{-1}; (*b*) surface current response to forcing by weak north-easterly winds blowing at a mean speed of 3.5 m s^{-1}.

(4) 1500–2300 GMT 14 May

By 1630 GMT (Fig. 5(b)), however, there is clear evidence of an alongshore flow component close to the coast. Note that the direction of this flow is opposite to that shown in Figure 4(a) and (b), consistent with the change in wind direction from north-easterly to north north-westerly (i.e. alongshore flow in both cases). In fact, a weak response to these wind changes is first visible at 1530 GMT (data not shown) some 30 minutes after the vector wind field has stabilized.

(5) 2300 GMT 14 May–0900 GMT 15 May

During this period the wind veers steadily to the south, and the vectors in the northern apex of the map form a strong north-eastward current (see Fig. 6(a)), while those closer to the coast continue in roughly alongshore flow (cf. Figs 5(b) and 6(a)).

(6) 0900–2300 GMT 15 May

Figure 6(b) is representative of the flow fields measured during this period of forcing by strong south south-westerly winds. There is a well-defined alongshore component close to the coast, whereas flows in the north are directed towards the north-east. Typical residual current values are ~ 0.22 m s^{-1}.

DISCUSSION

A feature of the surface flows examined in this short period is the predominantly alongshore response exhibited for a variety of wind directions by current vectors close to the coast. As the direction of the coastline has a strong east–west component, this sheds some light on why Prandle (in press) found a high correlation between the dominant east–west EOF mode and the wind-stress time series. Another important aspect is the northward bias of surface residuals in the northern apex of the radar survey region, which, as mentioned earlier, is most likely related to the presence of density gradient currents (Matthews *et al.*, in press), or to the large scale circulation patterns set up in Liverpool Bay. A third aspect related to the time-scales on which surface residuals respond to a change in the wind direction.

The alongshore response to wind forcing can be anticipated on the grounds of analytic theory, given that certain simplifying assumptions (such as a long straight coastline) are valid. The speeds of the wind driven currents observed here are in reasonable agreement with the well-known '2% law' (i.e. that surface current equals 2% of the wind speed), as is easily deduced from the data of Figures 4, 5(b) and 6(b). These results provide a working guide to the prediction of surface currents for forthcoming HF radar experiments and are useful if future drogue intercomparison experiments are to be optimized within HF radar survey regions. In the data presented here, the wind changed direction after 1300 GMT on 14 May over a two-hour interval. During this period, surface currents were in confused state (see Fig. 5(a)), and it was not until roughly 30 minutes after the wind had stabilized that a weak response to the new wind conditions was first detected. By 1630 (Figure 5(b)) the anticipated flow was well defined. Therefore the time required for a reversal

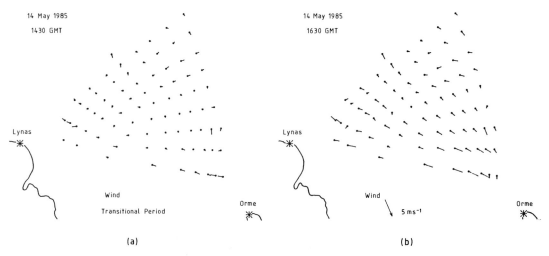

Fig. 5 (*a*) Surface currents measured during a period of changing wind. (*b*) Surface current response to forcing by north north-westerly winds at a mean speed of 5 m s^{-1}.

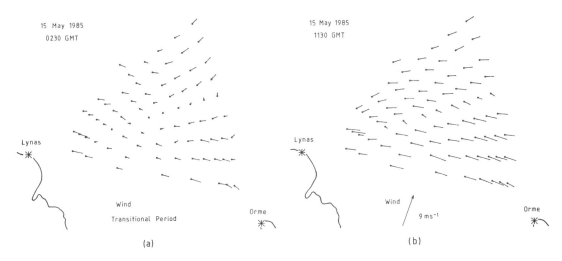

Fig. 6 (*a*) Surface currents measured during a period of changing wind; (*b*) surface current response to forcing by strong south–south-westerly winds at mean speed of $9\,\text{ms}^{-1}$.

in the direction of the flow field (cf. Figs 4(*a*) and 5(*b*)) is ~ 1 hour in this region. Time scale predictions for the establishment of a steady state based on analytic theory (e.g. Jeffries, 1923) are of order 10 minutes for a distance of 10 km from the shoreline, somewhat lower than that observed here.

In conclusion, we underline the usefulness of HF-radar-derived surface current maps to the study of wind-driven circulation. These data, gathered across a wide area of the ocean, and unbiased in space or time, would take an immense effort in time and money to collect by conventional current measuring techniques.

ACKNOWLEDGEMENTS

I acknowledge useful discussions with my colleagues in the School of Ocean Sciences, University College of North Wales. I thank D. Prandle for helpful advice on data processing and interpretation and acknowledge the work of Marex engineers in obtaining the radar data set. Thanks are also due to D. Roberts of the Computing Centre at the University College of North Wales and to A. R. Forcer-Evans for typing the manuscript.

REFERENCES

Ekman, E. (1905). On the influence of the earth's rotation on ocean currents. *Ark. Mat. Astron. Fys.* **2:** 1–53.

Griffiths, G. (1986). Intercomparison of an Acoustic Doppler Current Profiler with conventional instruments and a tidal flow model. Proc. Third IEEE Conference on Current Measurement, Airlie, Virginia, USA.

Jeffries, H. (1923). The effect of a steady wind on the sea-level near a straight shore. *Philosophical Magazine* Vol. xlvi.

King, J. W., Bennet, F. D. G., Blake, R., Eccles, D., Gibson, A. J., Howes, G. M. and Slater, K. (May 1984). OSCR (Ocean Surface Current Radar) observations of currents off the coast of Northern Ireland, England, Wales and Scotland. Paper presented at the Society for Underwater Technology Conference on Current Measurements Offshore, London.

Matthews, J. P., Simpson, J. H. and Brown, J. (in press). Remote sensing of shelf sea currents using the OSCR HF radar system. *J. Geophys. Res.*

Prandle, D. (1982). The vertical structure of tidal currents and other oscillatory flows. *Continent. Shelf Res.* **1:** 191–207.

Prandle, D. (in press). The fine structure of near-shore tidal and residual circulations revealed by HF radar surface current measurements. *J. Phys. Oceanogr.*

Proctor, R. (1981). Tides and residual circulation in the Irish Sea: A numerical modelling approach. PhD thesis, Liverpool University.

Stuart, R. H. and Joy, J. W. (1974). HF radio measurements of surface currents. *Deep Sea Res.* **21:** 1039.

Part III

Techniques and Instrumentation

12

Parametric Investigations into Dynamics of Tethered Subsea Units during Launch/ Recovery Operations using Model Experiments

D. Vassalos, D. Dutta and *P. Macgregor,* Ship and Marine Technology,
University of Strathclyde, Glasgow G4 0LZ, UK

As the development of offshore fields moves into deeper waters and the technology of subsea production systems improves, a large variety of manned and unmanned subsea units (SSU), such as diving/ observation bells, ROVs, habitats and modules, will be launched and recovered. This calls for safe and cost-effective methods of deployment using conventional marine cranes from platforms, rigs and readily available non-specialist support vessels. The launch/ recovery operation is influenced by:

- Station-keeping ability and motions of the mother vessel
- Impulsive wave loading on the SSU and its motions
- Large dynamic loads (snap loading) in the handling cable.

This chapter addresses the problem of snap loading in the handling cable which reduces the fatigue life and could lead to failure of the cable, with serious consequences. Goeller and Laura (Refs 3 and 4) have shown that the snap condition is easily initiated in steel cables, and they measured peak loads of nine times the static load. Such large dynamic forces in the handling cable can severely limit the operational seastates. It is therefore important that the snap loading phenomenon is thoroughly understood and the necessary steps are taken to avoid it and/or to minimize the adverse effects resulting from it.

This chapter contributes to this overall aim by investigating the effects of key parameters on snap loading in the handling cable and proposing possible solutions and guidelines to enhance the safety and efficiency of the operation. The parameters investigated are:

- Amplitude and frequency of excitation
- Mode of excitation
- Subsea unit (SSU) buoyancy
- Cable elastic property

THE EXPERIMENTAL SET-UP

The experimental set-up used a spherical subsea unit, 20.32 cm in diameter weighing 46.32 N in air and

Advances in Underwater Technology, Ocean Science and Offshore Engineering, Volume 16: Oceanology '88

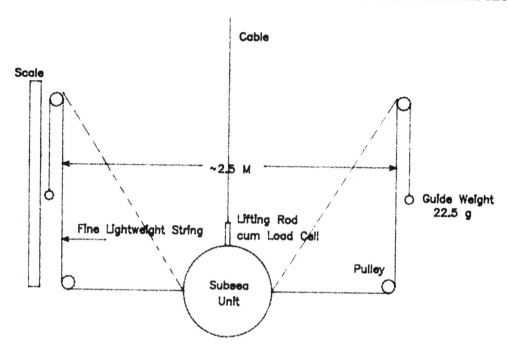

Fig. 1 Guide weight arrangement to restrain rotation of subsea unit.

6.01 N when fully submerged in fresh water. It was attached to a synthetic rope suspended from a platform rigged at a height of about 6.0 m. The upper end of the rope was excited at the point of attachment by horizontal, vertical, and circular motions with varying amplitudes and frequencies. Figures 1 to 4 show various aspects of the experimental set-up. The heave and sway accelerations were measured by accelerometers encased in the model SSU, while the tension was measured by a strain-gauged bar joining the cable to the model. Particulars of the model SSU and the cables used are given in the Appendix.

To obtain meaningful sway accelerations of the SSU, its rotation was restrained without significantly affecting the linear motions. The arrangement adopted is shown in Figure 1. A minimum of 22.5 g was required as guide weight to keep the restraining string in tension. The movement of the guide weight along a measure scale was noted, and this provided a fair indication of the magnitude of SSU motion. For vertical oscillations of the cable upper end, the SSU motion was observed to be predominantly heave, and in this case the restraining string was routed directly over the upper pulley, as indicated by the broken lines in Figure 1.

The driving mechanism for exciting the cable was essentially a crank-type device (an electric motor with a flywheel) attached to a sliding arm (Fig. 2). A balancing weight was necessary to achieve uniform motor speed. The instrumentation used is briefly described below:

Measurement instrumentation

Upper end motions   The amplitude and frequency of the upper end motion were measured using a simple potentiometer mounted on the sliding arm of the driving mechanism.

Cable tension was measured using a round bar-shaped load cell, which makes use of strain gauges. The output signal from the load cell was fed to a bridge amplifier and then on to the recording device.

Data-recording instrumentation

The measured signals were recorded on a LINSEIS TYP 2065 six-channel strip chart recorder, and the data were also stored on magnetic tapes using a Tektronix 4052 desktop computer with software for automatic data acquisition.

THE MEASUREMENTS

The measurements were performed with the model SSU suspended in air (high static tension) and with the SSU submerged in water (low static tension). The

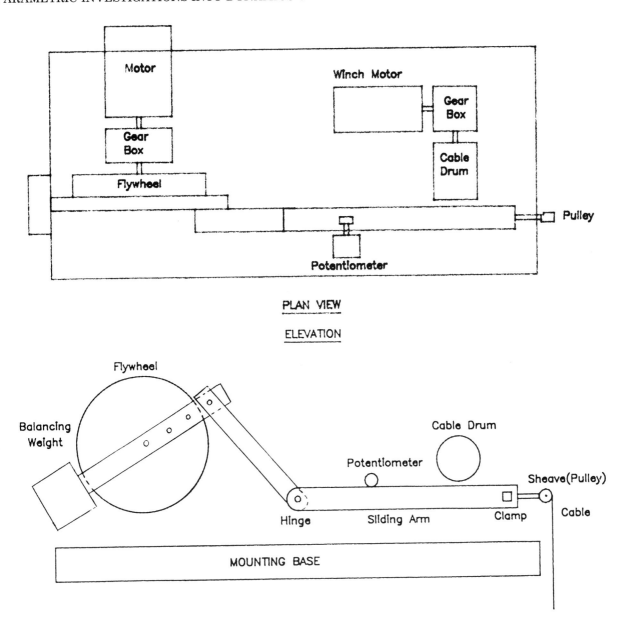

Fig. 2 Upper end excitation system mechanism.

cable lengths with the SSU in air and water were 4.573 m and 5.234 m respectively. For both positions of the SSU, the point of suspension of the cable was excited with horizontal, vertical and circular motions with amplitudes of 50 mm, 100 mm and 150 mm. The frequency was varied from 0.51 to 6.7 rad s^{-1}. The natural frequencies of in-line oscillations and pendulum motion were measured with the SSU in air as well as fully submerged in water and are given in the Appendix.

Measurements with horizontal excitation

With horizontal excitation, the motion of the SSU when not submerged in water was almost purely sway (pendulum motion) at low frequencies. As the

Fig. 3 The experimental rig.

frequency was increased beyond the natural frequency, the sway motion reduced significantly and, more importantly, the SSU began to oscillate vertically (heave). At frequencies greater than $4.0 \, \mathrm{rad \, s^{-1}}$ (frequency ratio of 2.86) the sway motion almost disappeared and only high frequency vertical oscillations of very small amplitudes occurred. Owing to physical restrictions caused by the tank boundaries and the scaffolding of the erected structure, it was not possible to excite the pendulum motion at resonant frequency. With the SSU submerged (the cable almost entirely out of water), the natural frequency of pendulum motion was very low, owing to hydrodynamic effects. In this case all the measurements were made above this natural frequency and the variation in cable tension was minimal.

Measurements with vertical excitation

All the natural frequencies for in-line oscillations, except one, were far beyond the maximum motor speed. This resonant condition could only be excited with the smallest amplitude of 50 mm, once the water depth in the tank was only 0.7 m. At larger amplitudes the SSU hit the tank bottom or emerged out of water. When the SSU was located in air, the maximum excitation amplitude was limited to 100 mm, as the driving mechanism motor was sluggish. With this type of excitation it was observed that at frequencies below $1.0 \, \mathrm{rad \, s^{-1}}$ there was slight sway,

Fig. 4 The SSU model with load-cell, accelerometers, and amplifier units.

although the predominant motion was heave. As the frequency was increased, the motion was purely heave.

Measurements with circular excitation

Circular excitation induced both sway and heave of the SSU. However, at frequencies just above the natural frequency of pendulum motion, it was observed to be predominantly heave. Upon increasing the frequency further, the sway motion disappeared altogether at about 2.8 rad s^{-1} and the motion was purely heave. In this case, when both sway and heave exist, the SSU moved in a somewhat elliptical path, and this made the interpretation of the guide weight movement difficult.

RESULTS OF PARAMETRIC INVESTIGATIONS

The effects of varying any one parameter at a time are examined below.

Effect of amplitude and frequency of excitation

An increase in amplitude causes an increase in the cable tension. Excitation frequency has a similar effect. This is shown in Figure 5. The highest frequency that could be achieved—6.66 rad s^{-1}—corresponds to a frequency ratio of 0.20.

Effect of mode of excitation

Three types of excitations were considered, and their effects are shown in Figure 6. The cable used was specimen no. 1, and the subsea unit was fully submerged in water. The amplitude of top end excitation was 100 mm. It shows that horizontal excitations have practically no effect on cable tension, which remains nearly constant. In contrast, both vertical and circular excitations cause sharp increase in tension. Snap condition in the cable is initiated at a frequency of about 2.0 rad s^{-1}, which corresponds to a frequency ratio of approximately 0.31.

Effect of subsea unit buoyancy

The buoyancy of the subsea unit, which determines the static tension, plays a vital role in causing or preventing snap condition which gives rise to large dynamic loads. For snap to occur, one condition is that the tension must drop to zero (slack cable). Therefore, for positively, neutrally, and slightly negatively buoyant units, the probability of occurrence of snap is much higher than that for a unit with large negative buoyancy. This effect is shown in Figure 7, where the tension variation is only marginal when the subsea unit is in air (high static tension) as compared to the variation of tension with the unit submerged in water.

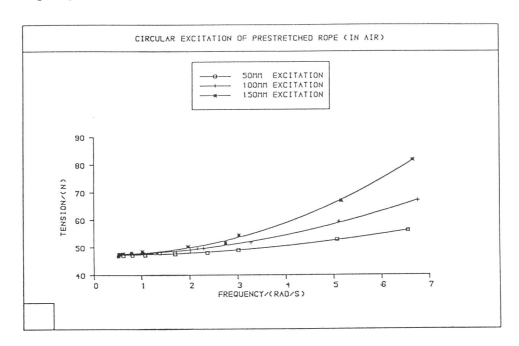

Fig. 5 Effect of amplitude and frequency of excitation on tension.

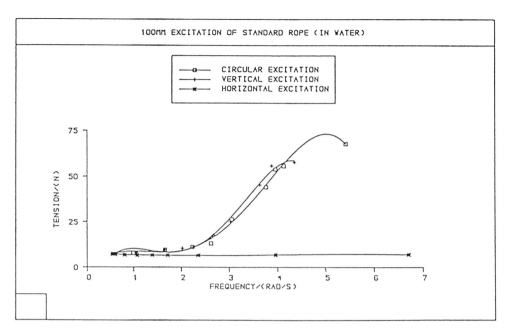

Fig. 6 Effect of mode of excitation on tension.

Effect of cable elastic property

Tension measurements were performed with two specimens of polyester ropes with non-linear elastic behaviour. The spring constants, calculated from measured natural frequencies were found to be 1390 N m^{-1} for specimen no. 1 and 5020 N m^{-1} for specimen no. 2. This difference in stiffness is primarily due to the difference in their diameters. These correspond to elastic moduli of 2.445×10^9 N m^{-1} and 3.38×10^9 N m^{-1} for specimens no. 1 and 2 respec-

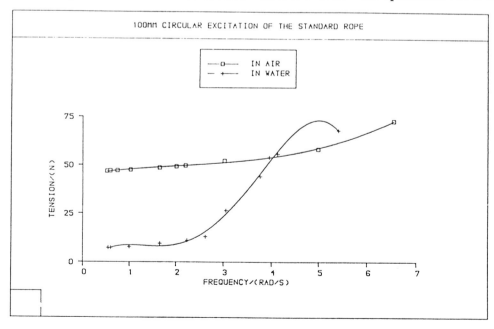

Fig. 7 Effect of SSU buoyancy on tension.

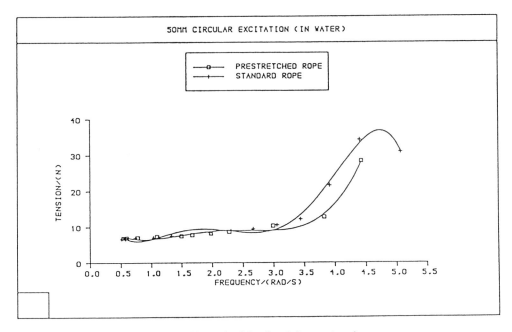

Fig. 8 Effect of cable elasticity on tension.

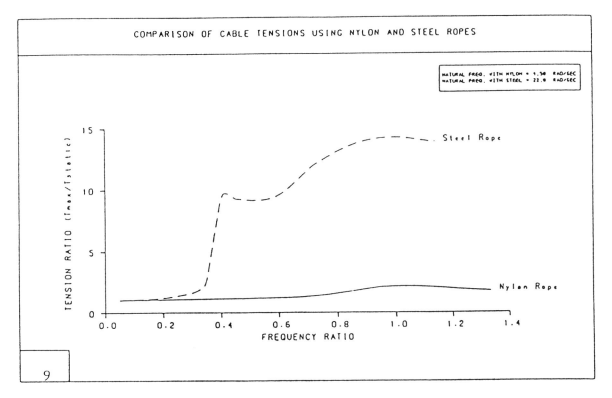

Fig. 9 Effect of cable elasticity on snap loading.

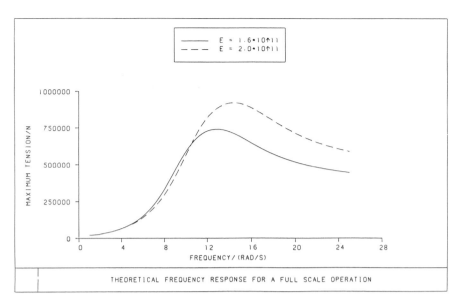

Fig. 10 Effect of cable elasticity on maximum tension.

tively. These values are given for comparison only. For computation of tension, the force extension diagram is used. It can be seen in Figure 8 that, contrary to expectations, the stiffer rope gives smaller tension than the more compliant rope. This behaviour can be attributed to the fact that because of the higher natural frequency of the stiffer (prestretched) rope, it is operating at a lower frequency ratio than the more compliant one, which has a significantly lower natural frequency. These investigations also confirm that the snap condition can occur at a frequency much lower than the resonant frequency for the cable–body system.

DISCUSSION

The results and trends of the experiment validate the key parameters established through theoretical parametric investigations using mathematical models (Ref. 2). Some of the interesting features of the theoretical results are:

(1) The dynamic tension increases with increasing frequency until resonance is reached. An increase in excitation amplitude causes onset of snap to occur at a lower frequency. Before the snap condition occurs, the increase in tension due to increase in amplitude or frequency is marginal. Once the snap condition is initiated, a small increase in frequency results in sharp rise in tension up to a certain point beyond which there is first a slight drop in tension and then a gradual increase until resonant frequency (Fig. 9).

This rise and drop in tension beyond the 'snap initiation frequency' (SIF) are believed to be due to either subharmonic resonance or transient phenomena caused by the sudden rise in tension (Refs 1 and 4).

(2) Critical length increases with the elastic modulus and decreases or increases with an increase in the excitation frequency depending on whether the frequency is above or below the resonant frequency.

(3) For a given excitation frequency, dynamic tension decreases with increase in elastic modulus, as long as it is less than the natural frequency for the smallest elastic modulus (Fig. 10).

Based on the findings of the experimental investigations, the following guidelines, which, it is believed, would improve the safety, efficiency and effectiveness of the operation, are proposed.

(a) To minimize snap loading in the cable

- The operating range should be restricted to less than 25% of the natural frequency of longitudinal oscillations, as in this range the response is stiffness dependent and can be controlled effectively by choice of cable properties.
- Subsea units which do not have adequate negative buoyancy should be made sufficiently heavy using a launch-frame or garage during passage through the air–sea interface. This will ensure that the static tension will nearly always exceed the compressive part of the dynamic tension.

If this is not practicable, two other approaches are

recommended:

- A compliant system where the slack in cable can be efficiently taken up by a spring actuated or pneumatic system should be designed.
- A shock mitigator should be designed into the cable system. An example is to add a small length of nylon cable at the subsea unit end of the steel cable. The nylon cable acts like a shock absorber, the energy being dissipated by elongation of the cable.

Finally,

- By determining the magnitude of the maximum tension, the cable can be chosen such that the loading does not exceed 20 to 25% of the breaking strength. This will greatly improve the fatigue strength of the cable. This is usual practice for design of mooring lines, which are also subjected to dynamic forces due to waves and current.

(b) To minimize pendulum motion of the SSU

To minimize the pendulum motion of the SSU, particularly during the passage through the air gap, the natural frequency of this motion should not be close to the frequency of the excitation.

CONCLUSIONS

(1) The experimental investigations have confirmed that the key parameters with significant influence on the snap loading are the amplitude and frequency of in-line excitation, cable elasticity and the submerged weight (buoyancy) of the SSU.

(2) Snap loading can be several times the weight of the SSU. The magnitude depends on the amplitude and frequency of in-line excitation. The transverse mode of excitation has no appreciable effect on cable tension.

(3) The snap initiation frequency can be much lower than the resonant frequency of the cable–body system.

(4) Snap loads can be mitigated and even eliminated by proper selection of cable properties and submerged weight of the SSU.

ACKNOWLEDGEMENTS

This work was funded by the UK Science and Engineering Research Council. Their financial support is gratefully acknowledged.

REFERENCES

1. Dutta, D. (1986). Dynamics of Tethered Subsea Units During Launch/Recovery Through the Air-sea Interface. PhD Thesis, University of Strathclyde, Glasgow.
2. Dutta, D. Launch and Retrieval of Tethered Submersibles: State of the Art Review Project MASS Task 3. 1B Internal Report, University of Strathclyde, August 1983.
3. Goeller, J.E. and Laura, P.A. (1971). Analytical and Experimental Study of the Dynamic Response of Segmented Cable Systems. *Journal of Sound and Vibration,* **18:** 311–324.
4. Goeller, J.E. and Laura, P.A. A Theoretical and Experimental Investigation of Impact Loads in Stranded Steel Cables During Longitudinal Excitation. Proceedings of the 5th Southeastern Conference on Theoretical and Applied Mechanics (pp. 177–206), 1970, North Carolina and Duke University.
5. Liu, F.C. Snap Loads and Bending Fatigue in Diving Bell Handling Systems OTC 1981, Paper No. 4089.

APPENDIX   DETAILS OF THE MODEL EXPERIMENTS

Natural frequencies of motions (rad s^{-1})

Mode of motion	Position of SSU	Rope	
		Specimen 1	Specimen 2
Pendulum motion	In air	1.40	1.40
Pendulum motion	In water	0.50	0.50
In-line oscillations	In air	17.14	32.58
In-line oscillations	In water	6.44	11.42

The SSU was designed to be only slightly negatively buoyant so that snap condition could be initiated. The particulars of the SSU and the two synthetic ropes used in the experiments are given below.

Particulars of the model SSU used in the laboratory experiment

Geometrical shape	Spherical
Material	Wood
Diameter	20.32 cm
Weight in air (including all instrumentation)	46.32 N
Submerged weight (including all instrumentation)	6.01 N

Particulars of cables used in laboratory experiments

	Specimen 1 (standard)	Specimen 2 (pre-stretched)
Specification	8-plait standard polyester	3-strand prestretched polyester
Manufacturer	H. & T. Marlow Ltd, Hailsham, East Sussex	
Nominal diameter	2.0 mm	3.0 mm
Weight in air	4.155 g/m	8.067 g/m
Ratio of actual to nominal cross-sectional area*	0.827	0.960

*Calculated using a specific gravity of 1.38 for the polyester rope.

13

Impact of Non-linearities on the Dynamic Behaviour of Tension Leg Platforms

Filiberto Galeazzi and *Pierluigi Zingale*, Tecnomare SpA

The tension leg platform is at present the most promising concept for the exploitation of offshore fields in very deep water. One of the major tasks in the design of such structures is a correct evaluation of the wave-induced motions: they represent the major contribution to the forces acting on the tethers, which are the most critical structural components of a TLP.

The TLP dynamic behaviour is characterized by low frequency, large amplitude in plane motions (surge, sway and yaw) induced by the slowly varying drift forces and by the first order responses to the direct wave excitation in all six degrees of freedom.

Although such dynamic behaviour is common to a large variety of moored vessels, TLP differs in that it may experience also a resonant condition in heave, roll and pitch motions. Despite the fact that the tether stiffness is chosen to keep the natural periods of these motions out of the wave frequency range, an excitation is still present, due to the second order, high frequency, wave forces. Consequently, tether slack may be experienced or tether fatigue life may be drastically reduced by high frequency cyclic loads.

The importance of a correct evaluation of the fluid–structure interactions, together with an accurate mathematical modelling of the rigid body dynamics is now evident.

Nowadays a good level of accuracy has been reached in the estimate of the first order wave forces either using programs based on 3-D potential theory or, in some cases, a Morrison type formulation (refs 1–2). Many efforts have been made to evaluate the wave forces up to the second order in both the low and high frequency range (refs 3–4), but several uncertainties are still present. In particular, no reliable tools are available for the estimate of the low and high frequency damping, which is, of course, of paramount importance when dealing with resonant motions. Furthermore all the computer procedures which have been developed are very expensive and cannot be used for parametric studies or in an early stage of a project.

For this reason Tecnomare has developed a computer program, specifically devoted to the TLP dynamic analysis, which can tackle the problem at different levels of accuracy and make possible a substantial

Advances in Underwater Technology, Ocean Science and Offshore Engineering, Volume 16: Oceanology '88

saving of computer resources. A comparison has been performed with experimental results relevant to a TLP in 800 m water depth, which shows how even a simplified approach may be sufficiently accurate.

TLP CONFIGURATION

The TLP analysed in this study has been designed to operate in 800 m water depth. A square deck is supported by four circular columns with bottom footings (Fig. 1), which are interconnected by submerged circular pontoons. The tether system comprises four tubes of 50 cm external diameter placed at the centre of each column.

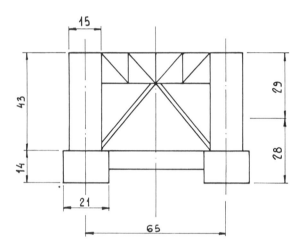

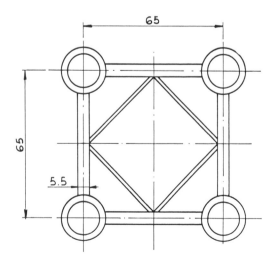

Fig. 1 Front and top view of the TLP (units: m).

The pre-tension (bouyancy−weight) and the tether axial stiffness were chosen to give the following natural periods at the rest position:

Surge and sway	119 s
Heave	3.2 s
Roll and pitch	3.5 s
Yaw	112 s

METHOD OF ANALYSIS

The motion analysis of the TLP has been performed using a computer program which integrates the rigid body equations of motion in the time domain. This approach makes it possible to account for any non-linear contribution such as viscous forces and non-linear mooring stiffness. One of the most interesting features of this procedure lies in the possibility of performing the analysis at different levels of accuracy. This is particularly true with the evaluation of linear and non-linear wave forces, as briefly described in the next item. This feature makes possible the use of a reasonably sophisticated procedure also during feasibility or parametric studies, when time and costs do not allow for running programs based on 3-D diffraction potential theory.

Scattering and radiation first order forces

Two alternatives are available:

(1) Wave force transfer functions, added mass and potential damping matrices come from the application of the 3-D potential velocity theory (Tecnomare's DINDIF program);
(2) Scattering and radiation forces acting on the columns are evaluated using a closed form solution of the potential function relevant to a vertical cylinder (ref. 5). Morrison's equation is used for the pontoons.

The latter approach does not account for hydrodynamic interaction among the structure components and this can lead to erroneous results when the column spacing is not more than three or four times the column diameter. For this reason, in the present analysis, we have adopted the first alternative.

The frequency dependence of the added mass and potential damping matrices cannot be tackled directly in a time series solution, and so the method of convolution integrals is used (ref. 6).

The added mass and potential damping contribution to the force becomes:

$$R(t) = \gamma a(t) + \sum_{\tau} \delta(\tau)v(t-\tau) \qquad (1$$

A series of computations confirmed that the almost constant parameter γ could be used throughout as an average added mass. The retardation function δ is used to evaluate the instantaneous potential damping force.

Second order wave forces

First order wave forces are obtained by integrating the hydrodynamic pressure over the mean wetted surface of the body. In addition only the linear term of the Bernoulli's equation:

$$p = -\rho\frac{\partial\phi}{\partial t} - \frac{1}{2}\rho(\nabla\phi)^2 \qquad (2)$$

which gives the relationship between the pressure and the velocity potential function is accounted for.

The integration of the pressure as far as the actual wave profile and the evaluation of the non-linear term in equation (2) represent the largest contribution to the second order wave forces in both low and high frequency range.

Following the Pinkster's formulation (ref. 7) where a random sea state is described as:

$$\eta(t) = \sum_{i=1}^{N}\eta_i\cos(\omega_i t + \varphi_i) \qquad (3)$$

the second order wave forces may be written as:

$$F^{(2)-}(t) = \sum_{i=1}^{N}\sum_{J=1}^{N}\eta_i\eta_J P_{iJ}^{-}\cos[(\omega_i - \omega_J)t + (\varphi_i - \varphi_J)]$$

$$+ \sum_{i=1}^{N}\sum_{J=1}^{N}\eta_i\eta_J Q_{iJ}^{-}\sin[(\omega_i - \omega_J)t + (\varphi_i - \varphi_J)] \qquad (4)$$

$$F^{(2)+}(t) = \sum_{i=1}^{N}\sum_{J=1}^{N}\eta_i\eta_J P_{iJ}^{+}\cos[(\omega_i + \omega_J)t + (\varphi_i + \varphi_J)]$$

$$+ \sum_{i=1}^{N}\sum_{J=1}^{N}\eta_i\eta_J Q_{iJ}^{+}\sin[(\omega_i + \omega_J)t + (\varphi_i + \varphi_J)] \qquad (5)$$

where the coefficients P_{iJ}^{-}, Q_{iJ}^{-}, P_{iJ}^{+} and Q_{iJ}^{+} are quadratic transfer functions. They are evaluated for each couple of frequencies which contribute to the low and high frequency force spectrum. Because a great computational effort is required to accomplish this task, this formulation is suitable to be used only during the engineering phase of a project. An approximated formulation of the low frequency forces due to Newman (ref. 12) became very popular and has

been used also in this investigation. Following Newman's approximation the equation (4) may be written as:

$$F^{(2)-}(t) = \sum_{i=1}^{N}\sum_{J=1}^{N}\eta_i\eta_J (P_{ii} + P_{JJ})/2\cos[(\omega_i - \omega_J)t + (\varphi_i + \varphi_J)] \qquad (6)$$

where P_{ii} and P_{JJ} are the mean drift force coefficients at frequencies ω_i and ω_J, whose evaluation is straightforward when using both the methods for the estimate of the first order forces described in the previous item. Unfortunately Newman's approximation is not valid for the high frequency forces, so we have also implemented in our procedures a simplified formulation of both low and high frequency wave forces, which will be referred to as 'waterline approximation'. This simply consists of a direct pressure integration as far as the instantaneous wave profile, disregarding the diffraction effects. Neglecting the quadratic term of Bernoulli's equation seems acceptable for cylinders as demonstrated in ref. 8.

Current and wind forces

Current and wind force coefficients can be obtained from specific model tests as a function of the heading. The general formulation of the current force is:

$$F_c = \frac{1}{2}C_c(\alpha)\rho_{wa}S_c(v_c - v)|v_c - v| \qquad (7)$$

The fluid–body relative velocity $(v^c - v)$ is considered so as to account for the damping contribution due to the current velocity.

The wind force can be written as:

$$F_w = \frac{1}{2}C_w(\alpha)\rho_a S_w v_w^2(t) \qquad (8)$$

The wind velocity is represented by a mean component and fluctuating component, where the fluctuating one (e.g. gust) may be described by a power spectrum of wind velocities (refs 9–10).

MODEL TESTS

Model tests have been performed at the Danish Hydraulic Institute with a model at a scale of 1 to 75. The experimental set-up is shown in Figure 2. About thirty tests were carried out in regular waves with different combinations of wave height, period and heading and current and wind velocity. Similarly, 34 tests in irregular waves were carried out,

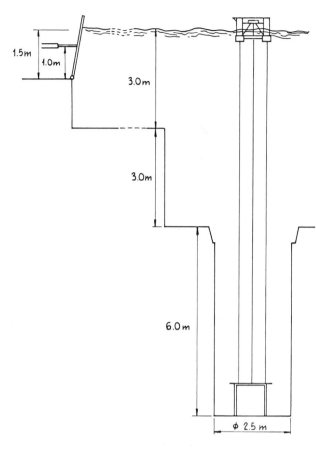

Fig. 2 Experimental set-up.

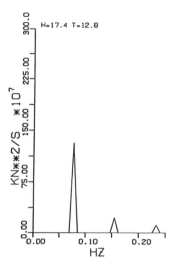

Fig. 3 Spectral density of the tether tension (experimental, model test) (test no. 2460).

using the Pierson-Moskowitz energy distribution of the sea state (ref. 11).

During the tests the following quantities were recorded: TLP displacements and rotations; tether forces; and wave elevation at three locations.

The signals have been processed to obtain the power spectra and the statistical averages.

THEORETICAL–EXPERIMENTAL COMPARISON

In the theoretical–experimental comparison both motions and tether forces have been considered.

The time series relevant to these quantities were processed using a standard Fourier analysis to obtain the spectra and the response amplitude operators. Particular attention was paid to the investigation of the non-linear responses in both low and high frequency range.

The tests were subdivided into two major categories: regular wave tests and random wave

tests. The most significant results relevant to both these categories are reported below.

Regular waves

The non-linear wave forces being proportional to the squared wave height, it is expected, for this category of tests, to find out a response at a frequency twice the wave frequency.

This behaviour was clearly shown by both theory and experiment, as demonstrated by the tether force spectra of Figures 3 and 4 (test no. 2460). The theoretical results were obtained using the 'waterline approximation'. A significant non-linear response was

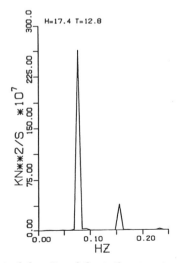

Fig. 4 Spectral density of the tether tension (theoretical, waterline method) (test no. 2460).

present only for the extreme wave conditions, as those relevant to the above cited test ($H = 17.4$ m). For smaller wave heights the response is almost linear and some results are presented in Table I, which still demonstrate the fairly good agreement between the theory and the experiment.

TABLE I
First and second order tether load amplitude (units: kN)

Test no.	H	T	Experimental		Theoretical	
			f	2f	f	2f
2460	17.4	12.8	4517	1865	6391	2464
5150	4.	10.	2271	128	2168	43
5160	4.	12.	2017	59	2061	50

Random waves

According to the second order wave force theory previously described, both low and high frequency non-linear responses are expected inside this category of tests. In the theoretical analysis both Newman and 'waterline' approximations were used. Obviously only the latter can account for high frequency non-linearities, so to compare the two theoretical approaches we first focused our attention on the surge motion, where a great low frequency component is usually present.

From Figures 5 to 7, where theoretical and experimental surge spectra are presented, it appears quite evident how Newman's approximation greatly overestimates the surge response. On the contrary, the waterline approximation shows a good agreement with the experiment. This trend has been found true also for the other tests which were analysed.

For the high frequency behaviour we refer to the tether force, whose value is affected by the contribution of all the motions. The spectra of Figures 8 and 9 show that some energy is present around 3 s, outside the wave frequency range. This is due to the excitation of the TLP high frequency motions by the

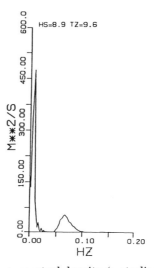

Fig. 6 Surge spectral density (waterline method) (test no. 2480).

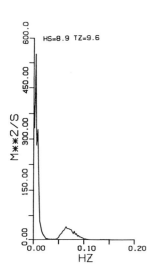

Fig. 5 Surge spectral density (experimental, model test) (test no. 2480).

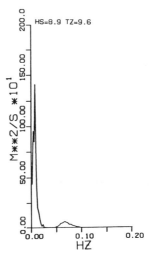

Fig. 7 Surge spectral density (Newman approximation) (test no. 2480).

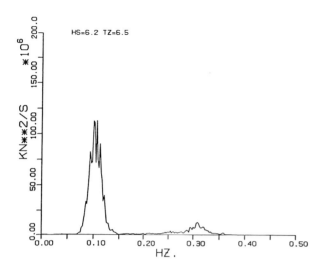

Fig. 8 Spectral density of the tether tension (experimental, model test) (test no. 2790).

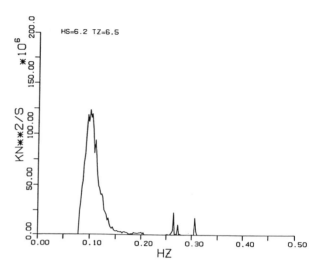

Fig. 9 Spectral density of the tether tension (theoretical, waterline method) (test no. 2790).

second order forces. In particular, in the theoretical analysis (Fig. 9) the contribution of the heave and pitch natural motions is clearly visible.

A more quantitative and comprehensive comparison is given in Tables II and III. In Table III statistical averages are split into three frequency bands: wave band (B) and low (A) and high (C) frequency band.

CONCLUSIONS

The theoretical–experimental comparison has shown that the proposed method for the evaluation of the second order wave forces, even if simplified, can be applied in a preliminary design stage.

Furthermore a more expensive and widely used

TABLE II
Comparison between Newman and waterline formulations

| | Test no. 2480 | | | | | |
| | Model test | | Newman | | Waterline | |
	Variance	Mean	Variance	Mean	Variance	Mean
Surge	5.45	38.35	11.11	39.75	4.93	36.66
Heave	.013	−.83	.028	−.965	.011	−.818
Pitch	5.4e−03	.07	7.6e−03	.061	8.2e−03	.057
Tether 1	4.3e+06	24710	5.7e+06	24870	5.9e+06	24680
Tether 2	2.6e+06	27565	3.8e+06	27560	4.1e+06	27210

TABLE III
Tether tension variances (kN^2) (Test no. 2790)

| Tether no. | A band | | B band | | C band | |
	Theoretical	Experimental	Theoretical	Experimental	Theoretical	Experimental
1	.4e3	9e3	3815e3	3048e3	149e3	537e3
2	1.e3	13e3	248e3	214e3	143e3	73e3
3	4.e3	21e3	3633e3	2143e3	146e3	588e3
4	1.e3	11e3	250e3	195e3	143e3	100e3

pproach such as the Newman approximation does not necessarily give better results. This is a further demonstration that many efforts must still be made to improve the understanding of the second order fluid–structure interactions.

ACKNOWLEDGEMENTS

The authors wish to thank their colleague Silvano Piola for collaborating in the statistical analysis of the experimental data.

NOMENCLATURE

a	acceleration
C	drag coefficient
H	wave height
p	pressure
R	radiation force
S	reference surface
t	time
v	velocity
α	heading angle
γ	constant added mass
δ	retardation function
φ	random phase
η	wave amplitude
ρ	mass density
τ	time
ω	circular wave frequency
ϕ	velocity potential function

Subscripts

a	air
c	current
w	wind
wa	water

REFERENCES

1. Faltinsen, O. M. and Michelsen, F. C. (1975). Motion of Large Structures in Waves at Zero Froude Numbers. DNV Publication No. 90, 1975.
2. Garrison, C. J. and Staccy, R. (1977). Wave Loads on North Sea Gravity Platforms: a Comparison of Theory and Experiment. OTC 2794.
3. Pinkster, J. A. (1980). Low frequency Second Order Wave Exciting Forces on Floating Structures. NSMB Publication No. 650.
4. Eatock Taylor, R. and Zietsman, J. (1982). Hydrodynamic loading on multi-component bodies. Proc. Conf. Boss 82, MIT.
5. Black, J. L., Mei, C. C. and Bray, M. C. G. (1971). Radiation and scattering of water waves by rigid bodies. *J. Fluid Mech.* **46**.
6. Van Oortmerssen, G. The Motion of a Moored Ship in Waves. NSMB Publication No. 510.
7. De Boom, W. C., Pinkster, J. A. and Tan, S. G. (1983). Motion and Tether Force Prediction for a Deepwater Tension Leg Platform. OTC 4487.
8. Petrauskas, C. and Liu, S. V. (1987). Springing Force Response of a Tension Leg Platform. OTC 5458.
9. Harris, R. I. (1971). The Nature of the Wind. Construction Industry Research and Information Association, London.
10. Davenport, A. G. (1961). The application of statistical concepts to the wind loading of structures. *Proc. Inst. Civil Eng.* **19**.
11. Pierson, W. J. and Moskowitz, L. (1964). A proposed form for fully developed wind seas based on the similarity theory of S. A. Kitaigorodskii. *J. Geophys. Res.* **69**.
12. Newman, J. N. (1967). The Drift Force and Moment on Ships in Waves. *J. Ship Res.* **11:** March.

14

Fairings for Oceanographic Cables and Instruments: Have We Yet Solved all the Problems?

*Alan R. Packwood,** Institute of Oceanographic Sciences, Deacon Laboratory, Wormley, Godalming, Surrey, England, GU8 5UB, UK

Cable fairings are routinely used on towed instruments where tow speeds are in excess of 3 knots and where good depth performance is desirable. In order to minimize the cable fairing drag the correct choice of fairing for the application is essential. Typical oceanographic or survey applications utilize wires of 8 to 20 mm diameter at tow speeds of 3 to 10 knots. This corresponds to a Reynolds number range of 10000 to 100000. In terms of aircraft aerodynamics these Reynolds numbers are low. Most of our knowledge of foil section performance comes from model studies of aircraft wings at Reynolds numbers of 1 to 10 million. At these higher Re values boundary layer transition occurs close to the leading edge, and consequently most of the boundary layer on the surface is turbulent. At the Reynolds numbers typical of oceanographic fairings the laminar boundary may persist longer, and laminar separation bubbles are more likely to occur. Furthermore early trailing edge flow separation may be precipitated on poor designs which significantly increases the drag.

The laminar boundary layer is actually very beneficial in that it exerts much less skin friction drag than the turbulent boundary layer. It is possible to design section profiles which make use of this by encouraging the laminar boundary-layer to remain attached and delaying transition. The secret lies in the form of the pressure distribution on the surface. Figure 1 shows the inviscid theoretical pressure distribution calculated for one of the best commercially available cable fairings. The flow visualization sketch for a Reynolds number of 210000 is after Henderson (ref. 1) who reported the wind-tunnel test results. The rounded nose positions the suction peak close to the leading edge, and the remainder of the boundary layer is subjected to a destablizing adverse pressure gradient. The laminar boundary layer separates near the suction peak, forming a separation bubble.

Following reattachment the turbulent boundary layer does not survive the long run to the trailing edge and separates early. The resulting drag co-efficient is 0.04—approximately four times the drag

*Present address: Department of Mechanical Engineering, University of Surrey, Guildford.

Advances in Underwater Technology, Ocean Science and Offshore Engineering, Volume 16: Oceanology '88

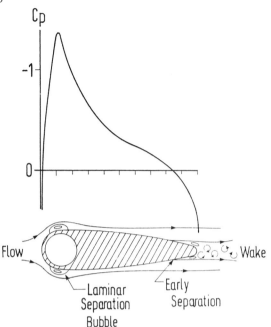

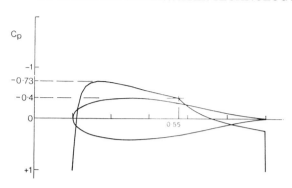

Fig.2 Design pressure distribution and derived foil section to operate at Re = 250 000.

Fig. 1 Inviscid pressure distribution for the commercial fairing section and flow visualization sketch at Re = 210 000 (after Henderson (ref. 1)).

that might be expected for a section like this at Reynolds numbers in excess of 1 million. Henderson also demonstrated that the fairing stalled at an incidence of 4°, and he calculated that the section aerodynamic centre was located at 15.4% chord. The importance of the aerodynamic centre location is in determining the weathercock stability of the freely pivoted fairing, i.e. how well it aligns itself with the flow. The distance between the wire centre and aerodynamic centre on a fairing like this is a direct measure of the weathercock stability. In this case Henderson determined the separation to be only about 0.03 c where c is the chord length of the section. The poor performance of this section at RE = 210 000 is entirely due to the behaviour of the boundary layer. Correct control of the boundary layer, through modification of the section thickness distribution, can potentially yield large performance benefits.

THEORETICAL SECTION DESIGN

At these Reynolds numbers the laminar boundary layer can be encouraged to remain attached for 50% of the chord or more by maintaining a favourable, negative, pressure gradient, or one with a broad level suction peak, often referred to as a 'roof-top' pressure distribution. This can be accomplished by moving the maximum thickness location aft. Here,

however, there is a conflict, for the designer would like the fairing to be as thin as possible, to reduce storage problems on the winch, and yet the cable centre must be well forward on the section in order to maintain good weathercock stability. The result would mean that the fairing would have considerable thickness, perhaps three times the diameter of the cable. Where long lengths of faired cable are employed this would be undesirable. Clearly some compromise is necessary which is dependent on the desired size of fairing and the application. For minimum thickness a 'roof-top' type pressure distribution is most likely to give the desired results.

Thus far we have considered the form of the pressure distribution up to transition. Post transition the turbulent boundary layer can similarly be encouraged to remain attached and exert low skin friction on the section by prescribing the right form of pressure distribution. In this case a Stratford-type pressure recovery (ref. 2) from the roof-top to the trailing edge pressure will yield very low skin friction. The features of a Stratford pressure recovery curve are that the pressure rise is initially very rapid and reduces toward the trailing edge, hence its concave appearance. The turbulent boundary layer is maintained at the point of incipient separation, thus theoretically the skin friction is close to zero. Figure 2 shows an example of a design pressure distribution. This exhibits a roof-top pressure distribution ahead of the estimated transition point at 55% chord, followed by a Stratford-like pressure recovery that is actually somewhat less severe than that described in ref. 2. The section shape shown in Fig. 2 was determined by using an inverse design technique due to Fiddes and Hogan (ref. 3) which is also briefly described in ref. 4. In simple terms this calculates the thickness distribution from the pressure distribution assuming inviscid flow.

A proper design scheme would thus

(1) derive from an understanding of the likely boundary layer behaviour on a foil at a given Reynolds number sketch a design pressure distribution;
(2) using an inverse method determine the section shape; and
(3) check the boundary layer performance on the predicted shape.

To accomplish step 3 we were fortunate in having access to the aerofoil analysis code developed at the Royal Aircraft Establishment by Williams (ref. 5). Not all boundary layer codes work well at these low Reynolds numbers. Williams's code, for instance, cannot be expected to work well if long laminar separation bubbles are likely to occur, but it does predict the laminar separation, reattachment and turbulent separation points. In this case, at a Reynolds number of 250 000, the analysis predicted laminar separation at 46% chord, reattachment at 52% with no early trailing edge separation. The predicted drag coefficient was 0.012 and the aerodynamic centre was located at 25% chord. Thus if the wire centre were at about 10% chord the aero centre to wire centre separation would be 15%. Assuming a wire diameter of 12% chord this means the section thickness is 1.85 D and the chord length is 8.2 D where D is the wire diameter. This is a slightly larger section than that shown in Figure 1 but the aero to wire centre separation is increased by a factor of 5 and the drag reduced by a factor of 3.3. This was seemingly very encouraging but since boundary layer analysis codes have yet to be shown to be completely reliable in this Reynolds number range it seemed prudent to test a model section in a wind-tunnel.

WIND-TUNNEL TESTS

The tests were carried out in the RAE $3' \times 4'$ low turbulence tunnel and are reported in greater detail in ref. 4. For practical reasons the section shown in Figure 2 was modified slightly to thicken the trailing edge 3% chord. Lift, drag and pitching moment measurements, and flow visualization experiments were carried out at Re = 250 000. The results for lift and drag are shown in Figure 3. The minimum drag was 0.021, the aero centre was at 25.4% chord, and the section showed no tendency to stall up to the maximum incidence investigated which was 10°. The flow visualization tests showed that the laminar separation bubble extended from 30 to 70% chord— about twice that predicted, which probably explains

the difference between the measured and predicted drag. No premature tailing edge separation was evident. The long bubble would be unlikely to occur in the ocean environment where the turbulence levels are much higher than in the wind-tunnel used. It is possible therefore that even better performance may be found in service.

For comparison the results obtained by Henderson (ref. 1) for the commercial section shown in Figure 1 are superimposed on Figure 3. Despite the long bubble on the new section there is still a 50% reduction in drag compared to the commercial section and the weathercock stability is increased fivefold. Furthermore the early stall behaviour and early trailing edge separation, evident on the commercial section, have been eliminated. Basically this exercise highlights the sort of improvements that can be made to fairing performance where a proper aerodynamically enlightened design procedure is followed, despite the difficulties in flow prediction in the Reynolds number range 10 000 to 1 million.

TOWED SPAR PERFORMANCE

Another application of a hydrodynamic fairing is shown in Figure 4. Here a rigid streamlined extruded aluminium section is used to mount an array of thermistors and a forward-looking sonar to investigate near surface ocean processes away from the influence of the ship. A full system description and specification is given in (ref. 6). The spar is attached to the supporting catamaran and weight through universal joints and is thus free to pendulum in any direction and turn to align with the oncoming flow. The turning hinge line effectively runs down the spar leading edge so as to give good weathercock stability and thus prevent the spar from kiting or taking on large lateral angles. However, in the first sea trials a speed-dependent instability was encountered which almost wrecked the system. The spar carries angle sensors and a pressure sensor near the bottom. Figure 5 records one such event as monitored by these sensors. As the speed was increased from 2 to 4.5 knots, the mean trail angle increased slowly. The variability on this trace is due to the spar swinging in response to the wave action on the catamaran. The lateral angle increased sharply going off scale at 53° at a ship speed approaching 4 knots. The spar continued to lift with the pressure sensor coming to within 3 m of the surface. The lifting spar contacted the outboard catamaran hull lifting the hull and turning the catamaran on to a collision course with the towing ship. Fortunately a rapid reduction in ship speed

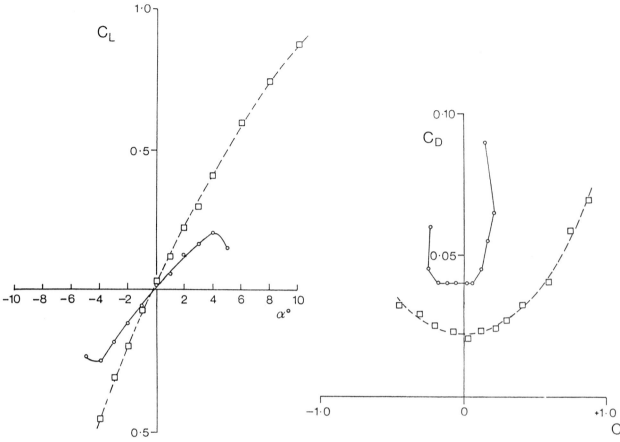

Fig. 3 Wind tunnel results showing lift coefficient
variation with incidence and the drag polar diagram.
o commercial section Re = 210 000
(after Henderson (ref. 1))
□ new section Re = 250 000

recovered the situation before the two collided, and no damage was done.

In further controlled tests the phenomenon was found to be reproducible every time the speed exceeded 4 knots. At lower speeds the spar would take on stable lateral angles of up to 25° at 3.5 knots but the spar rose inexorably at higher speeds precipitating incidents like that described in Figure 5. Clearly the spar was developing a large amount of lift and therefore must have an angle of incidence to the flow. A number of possible causes were investigated and quickly ruled out. These included stiction in the universal joints and asymmetries in the spar or streamlined weight. Only a very careful analysis of the hydrodynamic and mechanical turning moments about the leading edge hinge line resolved the problem.

The hydrodynamic lifting moment generated by the spar itself tends to be stabilizing, i.e. it acts in a sense tending to reduce the spar incidence to the flow. However, the hydrodynamic forces on the sensors protruding ahead of the hinge line and the moment due to the weight of the spar section, which acts behind and above the hinge line when the spar is trailing, both act to increasingly destabilize the spar as the towing speed is increased. When the spar self-stabilizing moment becomes insufficient to balance the other forces with only small angles of incidence (< 1°), the spar begins to lift and the destabilizing weight moment becomes even larger. Furthermore the through-flow thermistor housings probably significantly disturb the flow over the faired section, substantially reducing the lift force giving the correcting moment.

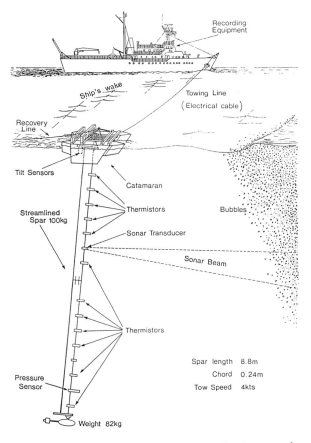

Fig. 4 Sketch of the thermistor spar and catamaran in operation.

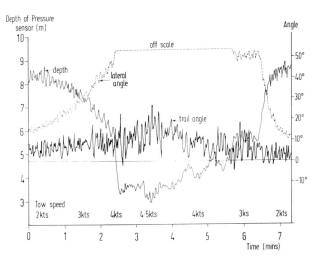

Fig. 5 Traces of the spar lateral and trail angles together with the pressure sensor record as the ship first accelerates then decelerates.

A simple numerical model was devised using Wingham's analysis (ref. 7) to determine the section weight moment and the hydrodynamic lift moment with crude guesses for the effects of the sensors. This estimated the equilibrium lateral angle of the spar, once perturbed from zero, using an iterative procedure. A range of flow velocities and spar trail angles were examined balancing the spar lift against its weight and that of the streamlined bottom weight. What resulted was the stability curve shown in Figure 6. In this context stable means spar lateral angles < 1° and unstable means lateral angles > 30°. The spar performance in trail follows a velocity squared curve like that shown. Where two curves intersect, the spar becomes unstable.

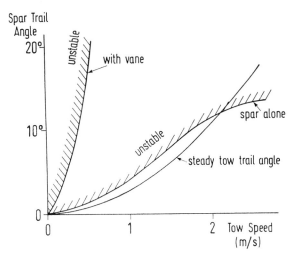

Fig. 6 Spar stability boundaries and trail angle performance assuming steady flow.

Once the mechanism of the instability was clearly understood the remedy was obvious and simple. To the trailing edge of the spar was added a 200 mm long by 10 mm thick extension to the chord length. This was made from polypropylene sheet which helped to reduce the section weight in water; more importantly it doubled the hydrodynamic turning moment tending to stabilize the spar. A sectional view of the spar and vane is shown in Figure 7. The new stability curve for the spar with the vane added is also shown in Figure 6 and is seen to be well clear of the steady spar performance curve. The spar with vane has been tested at sea at tow speeds up to 5.5 knots and has been found to behave extremely well even in gale force 8 conditions.

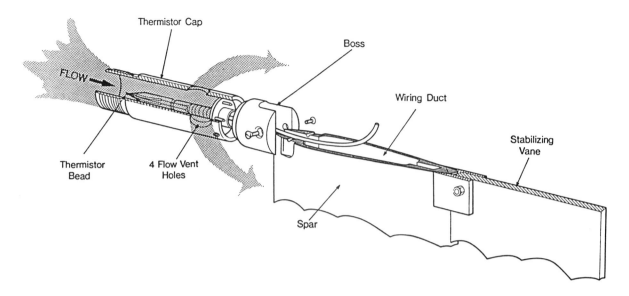

Fig. 7 Exploded sectional view of the thermistor assembly,
spar and stabilizing vane.

CONCLUSIONS: SOME LESSONS LEARNT

The choice of a fairing for a given application is not a trivial problem and has been shown to be strongly dependent on the operational Reynolds number. It should not be assumed that a fairing that works well in one application will be good for another, or that simply scaling a good design will give similar performance figures. Any fairing designed for high performance efficiencies must take account of the boundary layer behaviour. The design scheme outlined here, using current aerofoil design analysis computer codes, has shown that drag reductions of a factor of 2 and stability improvements of a factor of 5 are possible over some of the 'best' commercially available products.

The experience with the towed spar has shown the advantages of carrying out a simple stability analysis where potentially destabilizing forces occur. The exercise makes the designer think much more carefully about the effects of weight distribution and protrusions disrupting the flow. The results assist in setting the operational 'avoid' curves for a particular instrument or fairing.

ACKNOWLEDGEMENTS

The author gratefully acknowledges the assistance of members of the aerodynamics and low speed wind tunnel groups of the Royal Aircraft Establishment, Farnborough for their help in the experimental cable fairing design and model testing.

REFERENCES

1. Henderson, J. F. (1978). Some towing problems with faired cables. *Ocean Engng* 5: 105–125.
2. Stratford, B. S. (1959). The prediction of separation of the turbulent boundary layer. *J. Fluid Mech.* **5**: 1–16.
3. Fiddes, S. P. and Hogan, M. J. (1986). A low-speed aerofoil design method—description and application. Royal Aircraft Establishment (unpublished).
4. Packwood, A. R. and Williams, B. R. (1986). Development of an efficient oceanographic fairing for operation at Re = 250 000. Proc. Conf. on Aerodynamics at Low Reynolds Number, Oct 1986 Roy. Aero. Soc. London.
5. Williams, B. R. (1985). The prediction of separated flow using a viscous–inviscid interaction method. *Aero. J. R. Ae. Soc.* **89**: 185–197.
6. Hall, A. J. and Packwood, A. R. (1987). The towed thermistor spar. 5th Int. Conf. on Electronics for Ocean Tech. Mar. 1987 Edinburgh. IERE Pub. 72 pp 13–19.
7. Wingham, P. J. (1984). Some causes and effects of tow-off. *Ocean Engng* **11**: 281–313.

15

A New Self-contained, Multipath Protected Acoustic Transmission System for the Offshore Environment

Gérard Ayela, Serge Le Reste, Institut Français de Recherche pour l'Exploitation de la Mer, Ifremer, France and *Bernard Bisso,* Syminex, Marseille, France

A transmission system usually includes a transmitter, a receiver, and a transmission channel. The main characteristics of a 'shallow' water transmission channel are 'multipath' effect and noise. Existing acoustic data transmission systems cannot be operated within this environment or they have considerably reduced data rate.

However, the indirect paths of an acoustic wave from the transmitter to the receiver experiences a number of reflections and loose energy; fortunately, the longer the way, the more attenuated it is. Then after a certain time delay, T, the strength on the different signals arriving at the receiver can be selected by using a power criterion.

Starting from this point, Ifremer has developed an original underwater data-transmitting system called the self-contained transmitting system; it is based on an original modulation scheme which fights against the above described problems and enables the user to transmit data at high data rate.

The self-contained transmitting system consists of two main parts: a transmitter subsystem and a receiver subsystem. These two parts are symmetrical: the 'transmitter' achieves the coding of the numerical data and the modulation of the carrier, and the 'receiver' demodulates the acoustic wave and decodes the transmitted information. This equipment can be considered as a sophisticated modulator/demodulator (modem) using an underwater acoustic link.

Protection against multipaths is ensured by an original modulation scheme which can be compared to the spread spectrum method. If the reliability of the transmission must be increased, coding of the data can be introduced but the data rate is reduced.

PRINCIPLE OF OPERATION

General principle

An acoustic wave is modulated using a linear law. The characteristics of this modulation are described in Figure 1 where

F_s is a frequency used to synchronize the system,
γ_F is the frequency deviation,
T is the period.

At the receiver, an internal frequency modulation

Advances in Underwater Technology, Ocean Science and Offshore Engineering, Volume 16: Oceanology '88

is initiated as soon as the synchronization frequency F_s is detected. This modulation has the same characteristics γ_F and T_0 as the transmitted one but the frequency is shifted (see Fig. 1).

A 'beat frequency' can then be created by using the received signal and the internally made reference.

The direct path will always be centred on the frequency ΔF and the other paths, which are delayed because of longer paths, will be centred on frequencies ΔF_i different from ΔF. It will be therefore possible to separate the direct path from the reflected ones by a narrow band filtering around ΔF.

FSK linear modulation

Starting from the above described principle, a special modulation has been developed; it has been called FSK linear modulation. Its characteristics are shown in Figure 2. The frequency of the transmitted acoustic carrier is switched from a slope, which represents the '0' of the binary data, to another slope, which is the '1' of the numerical data, depending on the value of the binary data to be transmitted.

After beating with the local reference, matched filters, centred on ΔF_0 and ΔF_1, are used to determine whether a '0' or a '1' has been transmitted.

EQUIPMENT
Electronics

The transmitter and the receiver subsystems are based on a set of cards called Ifremer 800 which use the well-known CMOS NSC 800 microprocessor. Several parameters can be chosen by the user through an interactive user-friendly software: frequency deviation between two successive bits; the period T; the coding type; and the characteristics of the message to be transmitted. A great flexibility is thus possible.

Communication between each subsystem and the external environment is made through a standard RS232C serial link. A menu enables the user to modify the different basic parameters of the transmission and to generate signals which are used to tune the receiver.

Mechanics

The transmitter and the receiver containers have the following characteristics:

Housing
cylinder body   length   900 mm
                   external diameter   169 mm
                   internal diameter   152 mm
                   weight   20 kg
   maximum pressure 10 bars

Watertight connectors and cables are used to link the hydrophone to the transmitter and to the receiver as well as the sensors and the battery pack to the transmitter. Inside the cylinder, an internal battery

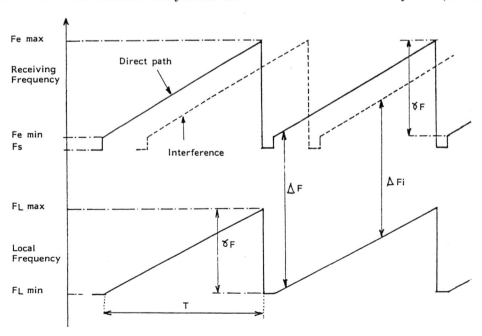

Fig. 1 Acoustic wave modulation.

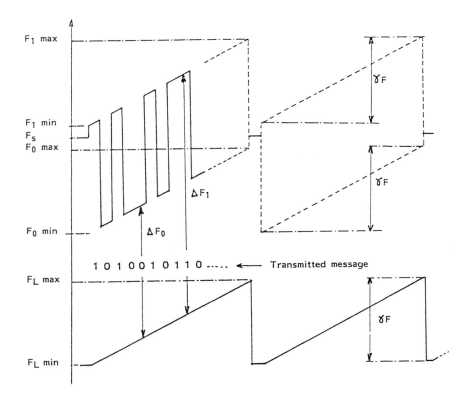

Fig. 2 FSK linear modulation.

back can be installed (100 D-size batteries can be mounted).

Hydrophone

To have a directional hydrophone, the transmitter and the receiver have been equipped with acoustic reflectors which give spacial filtering by eliminating the reflected waves which are not inside their beams.

A specific study and tests in basin have determined their geometry and the materials.

EQUIPMENT CHARACTERISTICS

Operating frequency	from 100 to 150 kHz
Transmission power	5 W maximum
Minimum range (1)	100 m
Data rate	300 bauds
Coding (2)	either no coding or majority or VITTERBI coding

BASIN AND AT-SEA TESTS

Figures 3 and 4 show the material during the tests. The tests of the system have been performed in December 1986 in the Ifremer basin (Brest, France). The geometry of the basin and the nature of its walls are a guarantee of having many multiple paths.

Length	50 m
Width	12.5 m
Depth on ⅓ of the length	20 m
Depth on ⅔ of the length	10 m

At-sea tests have also been conducted.

(1) Range can be increased by increasing the radiated power, but the autonomy is then decreased.
(2) Coding makes the transmission more reliable but decreases the quantity of useful information to be transmitted. For example, a 'majority' coding brings the data rate down to 100 bauds.

Test conditions
Basin

Transmitter depth	3 m
Distance from the basin sides	4 m
Distance between transmitter and receiver	40 m
Radiated power	2 W

Fig. 3 The receiver on its boom during the at-sea tests.

Fig. 4 The transmitter on its platform just before deployment during the at-sea tests.

At sea

Water depth	10 m
Transmitter depth	4 m
Receiver depth	4 m
Range	100 m

Results

To test the acoustic link error rate, a known standard message has been implemented and can be chosen in the menu; the error rate is then automatically computed by comparing the received message with the reference message. Two error rates can be computed: one gives the instantaneous error rate on the N last messages (N can be chosen by the user through the menu) and the other represents the average error rate since the first transmission.

Basin tests

Tests have been carried out continuously for a week. The instantaneous error rate of the uncoded message varied from zero to a few per cent according to the respective positions of the transmitter and receiver. This error is essentially due to the change of the receiving conditions (combination of direct path and interferences) with respect to the geometry of the system. The average error rate being of about one thousandth at the end of these tests.

The screen copy of the terminal linked to the receiver is shown on Figure 5 for two configurations of received messages. The first one corresponds to the 'not coded' standard message and the other one to the majority coded standard message. As can be seen from these examples, the screen shows an active field between the dotted line where the received data

are displayed and a fixed field where the various parameters of the transmission are represented.

At-sea tests

The system has been used at sea over two days; the transmitter was fixed on a tower at a depth of 4 m, and the receiver was fitted on a vertical pool aboard a small boat at a depth of about 4 m. The distance between the receiver and the transmitter was slowly increased and the error rate has been measured.

A typical error rate of 10^{-3} has been obtained at a 100 m range.

FUTURE WORKS

To prove the equipment's ability to fight against industrial noises and multipaths, the self-contained transmitter will be mounted on an offshore rig for a minimum period of one month during autumn 1987. The quality of the transmission will be estimated by transmitting the standard message and by daily recording, at the receiving site, of the average error ratio and the standard deviation of the instantaneous error rate.

APPLICATIONS

The need to telemeter data reliably over long ranges at a high data rate has recently increased as many applications emerged successfully:

- acoustic cathodic protection monitoring systems,
- cathodic potential on subsea line surveys.
- mobile tracking,

- structure deployment and levelling: it is often specified that no umbilical cables for any monitoring equipment should be connected to the templates during installation phase,
- permanently installed underwater engineering measurements,
- data recovering from tidal gauges or other oceanographic sensors that can be left in place to store more data,
- down-hole pressure monitoring systems, along the same systems as above.

CONCLUSIONS

The existing acoustic data transmission systems cannot be operated when strong multipath effects occur. The present, actual results of the Ifremer–Syminex Self-contained Acoustic System have proved its effectiveness and reliability in severe acoustic conditions. Any type of data can be transmitted since the equipment can be considered as a standard serial data link. Only a few minor changes would have to be made to interface existing sensors and data acquisition sub-

```
        IFREMER    Version : V1-9/86   /  Syminex
                                          TRANSMETTEUR AUTONOME
                                          STATION RECEPTION

    DUREE RAMPE :1: 100ms                 CALCUL TAUX ERREUR : 2 : OUI
    TYPE DE RAMPE :1: Normale             MESSAGE PREVU :1:Message type
    TYPE DECODAGE :1: Message non code    NBR MESSAGES POUR CALCUL :5
    -----------------------------------------------------------------
    tHE QUICK BROWN FOx JuMPS OVER THE LAZY DOG 1234567890
    THE QUICK BROWN FOx JUMPS OVER THE LAZ§ DOG 1234567890
    tHE QUICK BROWN FOX JUMPS OvER THE LAZY DOG 1234567890
    THE QUICK BROWN FOx JUMPS OvER THE LAZ§ DOG 1234567890
    THE QUICK BROWN FOx JuMPS OVER THE LAZ§ DOG 1234567890
    THE QUICK BROWN FOX JuMPS OVER THE LAZY DOG 1234567890
    tHE QUICK BROWN FOx JUMPS OVER THE LAZY DOG 1234567890
    THE QUICK BROWN FOX JUMPS OVER THE LAZY DOG 5234567890
    tHE QUICK BROWN FOx JUMPS OVER THE LAZY DOG 5234567890
    tHE QUICK BROWN FOX JUMPS OvER THE LAZ§ DOG 1234567890
    THE QUICK BROWN FOX JUMPS OVER THE LAZY DOG 5234567890
    THE QUICK BROWN FOX JUMPS OVER THE LAZY DOG 5234567890


    -----------------------------------------------------------------
    TAUX ERREUR SUR N MESSAGES :          4.464E-03
    TAUX ERREUR GLOBAL :                  1.055E-03
        IFREMER    Version : V1-9/86   /  Syminex
                                          TRANSMETTEUR AUTONOME
                                          STATION RECEPTION

    DUREE RAMPE :1: 100ms                 CALCUL TAUX ERREUR : 2 : OUI
    TYPE DE RAMPE :1: Normale             MESSAGE PREVU :1:Message type
    TYPE DECODAGE :2: majoritaire caract. NBR MESSAGES POUR CALCUL :3
    -----------------------------------------------------------------
    THE QUICK BROWN FOX JUMPS OVER THE LAZY DOG 1234567890
    THE QUICK BROWN FOX JUMPS OVER THE LAZY DOG 1234567890
    THE QUICK BROWN FOX JUMPS OVER THE LAZY DOG 1234567890
    THE QUICK BROWN FOX JUMPS OVER THE LAZY DOG 1234567890
    THE QUICK BROWN FOX JUMPS OVER THE LAZY DOG 1234567890
    THE QUICK BROWN FOX JUMPS OVER THE LAZY DOG 1234567890
    THE QUICK BROWN FOX JUMPS OVER THE LAZY DOG 1234567890
    THE QUICK BROWN FOX JUMPS OVER THE LAZY DOG 1234567890
    THE QUICK BROWN FOX JUMPS OVER THE LAZY DOG 1234567890
    THE QUICK BROWN FOX JUMPS OVER THE LAZY DOG 1234567890
    THE QUICK BROWN FOX JUMPS OVER THE LAZY DOG 1234567890
    THE QUICK BROWN FOX JUMPS OVER THE LAZY DOG 1234567890


    -----------------------------------------------------------------
    TAUX ERREUR SUR N MESSAGES :          0.000E+00
    TAUX ERREUR GLOBAL :                  0.000E+00
```

Fig. 5

systems to the transmitting/receiving unit. Different parameters of the modulation scheme can be optimized to match the characteristics of the environment. Then the error rate is still decreased. This equipment is the answer to many problems.

Thanks to the original design of the modulation and through the use of suitable methods for signal processing, the self-contained transmitter enables underwater acoustic links in very disturbed environments (industrial noises, multipaths). The use of acoustic baffles, with adjustable directivity, enables, through a spatial masking, to suppress part of multipaths. It is possible to get a more reliable link, to the prejudice of the data rate, by using an error correcting code. The transmitting and receiving electronics have been based round a CMOS microprocessor. It is then possible to choose the parameters through a user-friendly software and operate the whole system with a great flexibility.

The tests, which have been carried out in the Ifremer basin, have proved that the system is able to fight against strong multipath effects. The mean error rate of a coded message is typically around 10^{-4}.

As far as we know, this equipment has no equivalent on the market.

The cost-effectiveness of such acoustic telemetry is obvious when compared to the alternative solution of diving operations—as soon as it can be considered sufficiently reliable. It is precisely the objective of the present development to reach a very high robustness of this link against perturbations due to the specific underwater structures environment.

It is foreseen that operational as well as depth constraints will join to expand the applications of acoustic telemetry for deep offshore platform installation, assembling and monitoring, and for oceanographic parameter acquisition.

ACKNOWLEDGEMENTS

This development has been funded by the French Comité d'Etudes Pétrolières et Marines (CEPM). It would not have been possible without its support.

REFERENCES

Segui, A. and Philipart, R. (Gretsi 1975). Evaluation expérimentale des performances d'un système de transmission adapté à l'existence de trajets multiples.
De Coulon, F. (1984). Théorie et traitement des signaux. Ed. Dunod.
Van Trees, L. (1968). Detection, estimation and modulation theory. Ed. Wiley.
Viterbi, A. J. and Omuka, J. K. Principles of digital communication and coding. Ed. Masson.

16

Real-time, Long Term Communication System for Deep Ocean Instrumentation

C. N. Murray, M. Weydert, Commission of the European Communities
Joint Research Centre—Ispra Establishment 21020 Ispra (Varese), Italy
and T. J. Freeman, Building Research Establishment, Garston, Watford
WD2 7JR, UK

During the last two years, work has been under way on the use of satellite communications for the transmission of data from instrumentation placed in the ocean bed. The set-up of the transmission link has required the development of several new systems, including an underwater vehicle capable of successfully emplacing instrument packages within deep ocean sediment formations, a transmitter–receiver system able to send acoustic signals through both 50 metres of sediments and a 6 km water column, sensors which will give information on sediment characteristics, and a satellite communications system for the automatic quasi-real-time relay of data from the emplaced instrumentation to the laboratory in Ispra (Europe).

BACKGROUND

With the increasing need to investigate and understand processes controlling the distribution and dispersion of materials in coastal and deep water zones, the development of instrumentation that is capable of working autonomously has occurred over the decade. With the advent of more powerful (and cheaper) electronics using little energy, the range of applications has increased as well as the length of time which is required before recovery of the instrumentation. This last point is particularly important because the need to study long-term mechanisms requires the collection of data over comparable periods of time. The advent of satellite transmission systems such as ARGOS (Argos, 1978; Andunson and Fossum, 1980) and METEOSAT (Houet, 1986) have opened up the possibility of obtaining data from remote areas, in real time. Not only has this allowed the development of networks covering large areas it permitted the control of the correct functioning of the system and the replacement of components as required, thus lessening the loss of valuable data.

Murray *et al.* (1987) described work carried out to demonstrate the feasibility of setting up an acoustic link for data transmission between instrumentation emplaced within deep oceanic sedimentary formations and a mobile shipborne surface receiver–transmitter which can then relay data to a land-based European ground station via the European Space Agency's METEOSAT geostationary satellite.

On the basis of these studies, the Commission of

Advances in Underwater Technology, Ocean Science and Offshore Engineering, Volume 16: Oceanology '88

the European Communities, Joint Research Centre, Ispra Establishment (Italy), decided to develop an experimental operating system as part of its research on long-term environmental processes. The aims of the project are to demonstrate that the investigation of slow environmental processes, either within the deep water column or in sedimentary formations, can be studied using specialized instrument systems with the possibility of controlling the investigation and receiving *in situ* data in real time. The communications link uses a mix of satellite and high-frequency radio techniques to communicate with an autonomous surface platform which incorporates a two-way underwater acoustic link for the control of deployed autonomous instrumentation. A laboratory control and data collection centre have been set up at the Joint Research Centre in order to monitor the relay system and the deployed sub-bottom instrumentation. Each of these components is described below.

The METEOSAT Network was chosen as it allows the transmitted data to be received directly at the

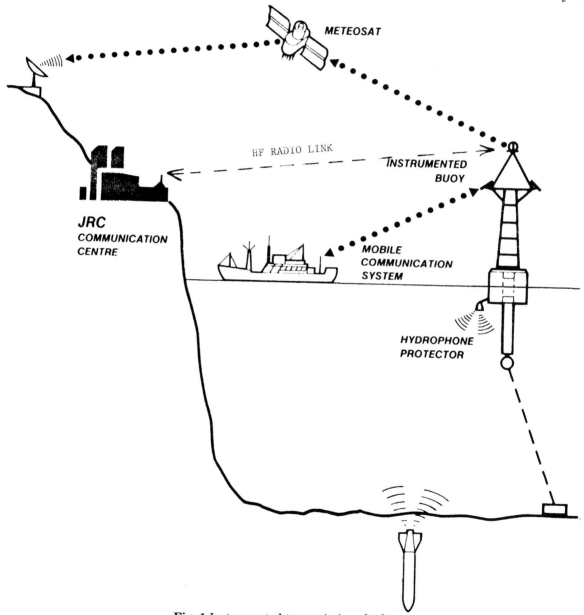

Fig. 1 Instrumented transmission platform.

Institute, so that real-time modelling can be carried out. This is particularly important when rapid decisions have to be made. Other important factors are the amount of data that can be transmitted in a one-minute time slot (~ 5000 bits) and that at present longer time slots can be obtained reasonably easily.

LABORATORY INSTALLATIONS

The installations in the laboratory fall into three parts: the METEOSAT receiver, the support communications (radio), and the data logging and decoding computers. The data transmitted by METEOSAT are received by a dish antenna on the roof of the laboratory building. The carrier frequency is then converted from 1.2 GHz to 137 MHz, preamplified, and sent over an UHF cable to the data collection platform (DCP) receiver in the Laboratory. The built-in computer stores the data on floppy disk. They can be subsequently retrieved and sent to the data logging computer, a PC, where they can be decoded and analysed.

The antenna for the HF/VHF station is at the moment a groundplane (vertical) antenna and dipole antenna to which a loop antenna is being added. The signals (4 and 12 MHz carrier) pass over a 50 m cable to the transceiver, an interface converting the data from the radio format to ASCII, and then into a small 8-bit computer, where they can be stored until requested by the PC. The PC checks the data for errors, omissions and alarms. Then the data are converted to physical units and stored in appropriate files, from where data analysis can proceed directly.

Commands to request data by radio or to run certain control programs can be typed in directly on the keyboard of any one of the computers; these are con-

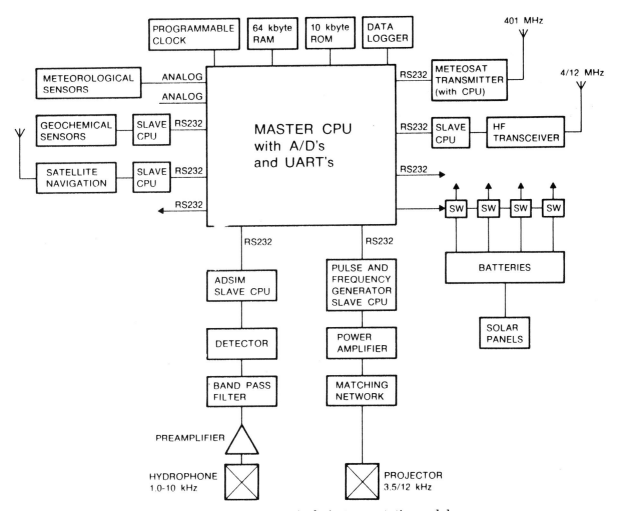

Fig. 2 Electronics systems in the instrumentation module.

verted to the format of the protocol and sent to the buoy. In the text message, the buoy acknowledges the receipt and execution (correct or failed) of the command. Figure 1 shows the overall communication links.

INSTRUMENTED BUOY

The following criteria have been selected for the design and construction of the platform: (i) limited size to allow handling and deployment from standard oceanographic ships; (ii) weight approximately 4000 kg; (iii) anchoring system to keep the buoy within an excursion radius of less than 2.5 km in 6 km of water; (iv) transmissions up to sea state 6; (v) satellite navigation system for easy relocation; (vi) surface instrumentation powered through solar panels; (vii) 3 months of on-board data storage, either in RAM or on cassette; and (viii) cable, sub-surface float and platform retrievable.

The platform contains the master electronics, housekeeping and telemetry systems. The electronics are modular in concept to allow easy exchange and upgrading as necessary. The system is modular, and the buoy consists of a metal shell which will be filled with polyurethane foam. To the single unit, further modules can be added as required by site and load considerations. A 5 m long cylinder 600 mm in diameter slides through the 2 m buoyancy chamber. A 1500 kg weight at its lower end guarantees the stability of the buoy. A 1.5 m long cylindrical container in its upper part houses the batteries and the electronics. The main buoyancy cylinder is topped by a 4 m high pylon supporting solar cells, radar reflector, navigation light, omnidirectional METEOSAT antenna and RF antennas. It should be noted that the present platform is designed to simplify testing of the transmission and power systems; for applications in the open sea, the sensors and electronic modules would be mounted on an ocean-going buoy.

The instrumentation module contains a number of electronics systems (Fig. 2). A microprocessor (INTER 8051) controls the data flow to and from a number of peripheral units such as the METEOSAT data collection platform (DCP), radio transceiver, meteorological station, geochemical sensors, and electro-acoustic transducers. After the data have been collected from the various interfaces or from some of the analog inputs of the master electronics (some of these are used to control the battery voltages), they are formatted and prepared for transmission via the METEOSAT system. This processor is also connected to a data logger (cassette or RAM) so that all received data are recorded for later retrieval in case of incor-

rect functioning of the satellite transmitter and support communications. In addition, it controls a system of relays which allows power to be switched on/off to most of the sensors so that energy can be saved. This is especially important for some of the sensors and those electronics which are not CMOS, when there is insufficient sunlight over longer periods of time or when there is a failure somewhere in the system.

The METEOSAT data collection platform

The METEOSAT data collection platform consists of a transmitter and a omnidirectional antenna (with normally 6 W but also up to 50 W amplifier). Transmissions are completely automatic, the timing being controlled by the high accuracy internal clock.

To receive data from deeply deployed instrumentation an acoustic link has been developed (Flewellen 1986). Low frequency incoming acoustic signals are directed from the hydrophone via a system of filters, preamplifiers, a demodulator and a Schmitt trigger into a receiver and signal convertor where the time-modulated pulse sequence is decoded. Communications with the master CPU are over RS232. A more detailed description of the convertor is given by Weydert and Pelletier (1987).

To interrogate the deployed instrumentation a 3.5 kHz projector is hung below the buoy. A programmable interface transforms commands from the master CPU (sometimes on request via radio) into an acoustic code which can then be transmitted to the transponders on or in the seabed. Through a radio transceiver the master CPU can receive orders or requests for data transmissions at HF or VHF over distances up to 5000 km. Currently, frequencies of 4 and 12 MHz are used.

To control the position of the buoy a passive satellite navigation system is connected via the master CPU to the METEOSAT DCP. The buoy can thus be constantly monitored, allowing the retrieval of the system in case of a loss of its anchoring system.

Power for the onboard instrumentation is provided by four 12 V/105 Ah batteries and one 24 V/210 Ah battery system which store enough energy to run over one month. The batteries are constantly recharged from four 50 W and four 18 W solar panels to ensure longer-term operations.

At present two sets of sea surface measurements and one set of seabed measurements are foreseen; the instrument package for the sea surface measurements (meteorological and geochemical) will be mounted on the platform and the seabed instrumentation (geochemical) in either a penetrator or soft lander vehicle. The two-way acoustic system will

Fig. 3 Free-fall penetrator.

be used for interrogation of the instrumentation deployed within the seabed.

Meteorological sensors

In view of the experimental nature of the present system, the meteorological sensors have been restricted to wind velocity, air temperature, and wave height and direction. Wind velocity and direction are measured using an anenometer, a wind direction indicator, and a compass to monitor the orientation of the buoy. Indication of sea state is provided by a triaxial accelerometer which is programmed to give vector data of wave direction and a mean wave height with standard deviation and range. Temperature is measured at 1 and 5 m above average sea level and is sampled over 1-hour periods.

Geochemical sensors

A geochemical sensor package has been interfaced with the platform data retrieval system for the measurements of subsurface geochemical/physical properties of sea water. These include conductivity, temperature and pressure, oxygen concentration, pH, and the redox potential. In its present form, this unit can be used to a depth of 1.5 km, the data being sent digitally via cable to a dedicated interface on board the buoy.

SEAFLOOR INSTRUMENTATION

As has already been briefly discussed, through the development of the high-frequency radio link (4 or 12 MHz), the buoy is able to receive command signals from the communications centre set up at Ispra (Italy). These signals will be interpreted by the buoy's master central processing unit either as a request for specific data or the running of specific fault finding subroutines or the activation of its acoustic transmitter. In this latter case, the buoy will then interrogate autonomous instrumentation previously deployed in the region of the experimental site.

Two types of instrumentation have been developed for this specific application: free-fall penetrators and a soft lander module. A penetrator is a torpedo-like vehicle (about 2 tons) (Fig. 3) designed to free-fall through the deep water column and to penetrate deep into the underlying oceanic sedimentary formations (Freeman *et al.*, 1984). A parametric study of this vehicle has been reported by Murray and Visitini (1985). The penetrators enter between 30 and 60 m into the bottom sediments at an impact velocity of around 200 km h^{-1}. Through the design of the vehicle and its instrumentation package, sensors have been used to investigate the behaviour of the penetrator during its trajectory in the water and sediment

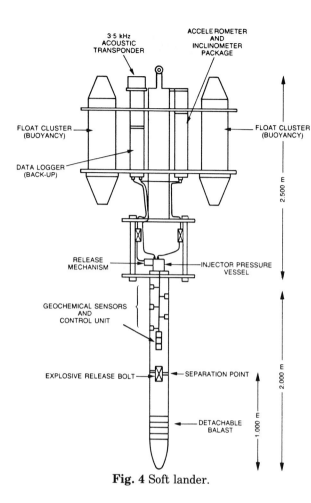

Fig. 4 Soft lander.

columns. The data have been transmitted back to the sea surface using a 3.5 kHz acoustic transmitter (Flewellen, 1986), similar to the one on the buoy.

When instrumentation has to be placed on the seafloor, a specialized soft lander module is used. This module can accept different instrument configurations. It is designed to free-fall slowly to the bottom of the ocean and settle gently on the sediment surface. On command from the surface buoy it will undertake its experimental programme over a period of several months, transmitting the data back to Ispra in real time. On completion of the experiment a command can be sent to the buoy—which in turn interrogates the soft lander and orders it to come back to the surface or to undertake some other (preprogrammed) operation. Figure 4 shows the design of the soft lander which is at present under assembly in the Joint Research Centre workshops.

CONDITIONS

The setting up of the transmission link has required the development of a set of new systems, including underwater vehicles capable of successfully emplacing instrument packages within deep sediment formations, a transceiver to send and receive acoustic signals through both 30 to 60 m of sediments and 5 to 6 km of water, sensors to give information on sediment characteristics, and a satellite communications system for the automatic relay of the data to the Joint Research Centre.

REFERENCES

Argos (1987). User Guide, Data Collection and Positioning by Satellite. Centre Spatial de Toulouse, Toulouse, France, 43 pp.

Andunsen, T., Fossum, B. A. (1980). The use of ARGOS in oceanography. *ARGOS Newsletter* May, No. 8, 1–4.

Flewellen, C. G. (1986). Development of a low-frequency transponder system for penetrator instrumentation. In Joint Research Centre Report Series. Study on the feasibility and safety of the disposal of heat generating wastes into deep oceanic geological formations, Part 4, Deep Ocean Instrument Development, SP. I.07.C2.88.18.

Freeman, T. J., Murray, C. N., Francis, T. J. G., McPhail, S. D. and Schultheiss, J. J. (1984). Modelling radioactive waste disposal by penetrator experiments in the abyssal Atlantic Ocean. *Nature* **310**: 5973, 130–133.

Houet, H. (1986). The Meteosat Data Collection Mission, 1st Meteosat DCP Users Conference, Lisbon, January 8–9, Section Systems Descriptions and Future Plans, 14 pp.

Murray, C. N. and Visintini, L. (1985). Parametric analysis of the performance of free fall penetrators in deep ocean sediments, *J. Ocean Engineering*. **10**: 38–49.

Murray, C. N., Weyder, M. and Freeman, T. (1987). An ocean-satellite link for long-term deep deployed instrumentation: concepts, experimental development and applications. *Deep Sea Research* (in press).

Weydert, M. and Pelletier, H. (1987). Study and construction of a module to interface accoustically transmitted oceanic data to a METEOSAT Data Collection Platform transmitter. In JRC Report Series. Study of the safety of the disposal of heat generating wastes into deep oceanic geological formations, Part 4, Deep Ocean Instrumentation Development (in press).

17

The Hydro-optic Sensor System (HOSS)

S. Svensson, C. Ekström, B. Ericson and J. Lexander, The National Defence Research Institute of Sweden (FOA)

A measurement system for *in situ* studies of optical water quality, salinity and temperature has been designed and built at FOA. Transmission, forward scattering, back scattering and irradiance are measured. The system is unique in using a green HeNe laser (543.5 nm) for this application and in providing an airborne instrument for measuring optical water quality data *in situ*.

This development had two main objects: (1) to map variations in optical water quality as a function of depth, geographical position, season, weather and other hydrological and meteorological parameters and (2) to examine the hydro-optical environment during tests with other optical systems. The system has been tested in the Baltic from a helicopter and from ships.

BACKGROUND

Hydro-optical research has been performed at FOA for about twenty years. The optical properties of Swedish coastal waters and open sea have been measured in different ways during these years. Since the Swedish coastline is very long, the measurement positions form a sparse pattern in time and geography. Hydro-optical parameters have been measured one by one, and correlations to other hydrological and meteorological parameters have not been systematically studied. The need for a multi-sensor system was obvious.

Many new hydro-optical techniques are based on laser technology. In order to predict the performance of this kind of system in Swedish waters, we chose to measure transmission, forward scattering, back scattering, and the irradiance of the downwelling daylight. Temperature and salinity measurements are added to correlate these parameters with the optical ones.

Because FOA budget does not permit extensive exploration of the subsea environment, an instrument should be simple to operate, making it possible for any crew in the Swedish Navy to perform the data collection.

Advances in Underwater Technology, Ocean Science and Offshore Engineering, Volume 16: Oceanology '88

DESIGN

General remarks

The system layout is presented in Figure 1. All the subsea sensors are mounted on the subsea unit according to the system block diagram (Fig. 2). Temperature, salinity (conductivity) and pressure (depth) are measured with commercially available sensors, but the irradiance, scattering and transmission sensors are designed and manufactured by FOA.

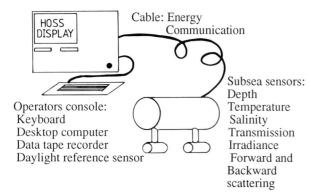

Fig. 1 System layout.

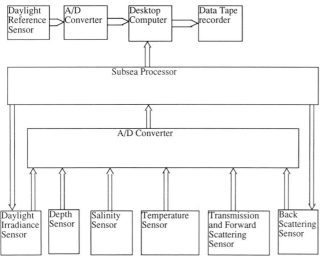

Fig. 2 System block diagram.

All the sensors are designed to give rapid response. The subsea unit has an open layout to secure a rapid exchange of water around the sensors, making acquisition of data possible during continuous up or down movement of the instrument. A depth profile from the surface to 100 m can be recorded within 4 minutes. The data are stored on tape and can later on be fed to a data post processing system for production of water quality maps etc.

The accuracy of the sensors is so far not exactly confirmed, but the ambition has been to keep the errors low without making the instrument too complex or expensive.

Sensor design

Transmission and forward scattering sensors

The transmission and the forward scattering sensor share a light source—a green HeNe laser. This laser was chosen because it is comparatively cheap and it emits in 540 nm which is rather close to a doubled Nd:YAG laser (530 nm). Presently the Nd:YAG laser is the most economic alternative for long-range hydro-optical systems in the Baltic.

Figure 3 presents the principal design of the transmission and forward scattering sensors. (In the actual system layout the lightbeam is folded three times. To reduce the effect of local 'big' particles in the water the lightbeam diameter is expanded to 20 mm. A compromise between the longest appropriate light path of the scattering sensor and the shortest appropriate light path of the transmission sensor resulted in a 0.477 m pathlength.

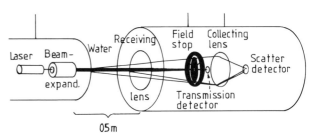

Fig. 3 Transmission and forward scattering sensors.

The optical focusing of the transmitted light is a standard design. Reference (1) shows an example of collecting scattered light onto a photodiode, which is similar to our design. As shown in reference (2) the forward scattering is approximately proportional to the Volume Scattering Coefficient (s).

Back scattering sensor

The layout is small and uncomplicated. Figure 4 shows the mechanical design. A xenon flash lamp is used as lightsource, and the light is not filtered since the scattering properties of seawater scarcely changes with wavelength. Reference (3) describes a

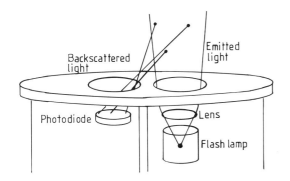

Fig. 4 Back scattering sensor.

similar design in which an infrared diode is the lightsource.

The flash lamp is only fired at the predecided measurement depths. Pulse energy can then be high enough for the detector (a PIN-diode), at the same time as the recovery time of the flash lamp is long enough. The signal from the diode is captured by a peak detector.

Daylight irradiance sensor

A photomultiplier has enough sensitivity for most measurement occasions and was the natural choice for the opto-electrical detector. The vast dynamic range is divided into a number of amplification levels. Each level is associated with a specific high voltage applied over the multiplier. This amplification voltage is automatically set by the subsea processor as a function of the irradiance level.

In front of the photomultiplier there are two filters (short BG 18 and Wratten 58) which limit the response to the blue–green area of the spectrum. Above the filters a diffuser is situated which collects the downwelling daylight and has a cosine-shaped sensitivity regarding angle of incidence.

Daylight reference sensor (not under water)

To eliminate the effect of daylight variations on the assessment of daylight attenuation, a daylight reference sensor was built. This sensor has the same spectral characteristics as the daylight irradiance sensor, but the sensitivity is largely independent of angle of incidence. A PIN-diode is the sensing element, and a logarithmic amplifier gives enough dynamic range for most occasions.

Note that the measurement profiles are recorded rather rapidly, so the elevation of the sun does not change during one profile. In order to be independent of seastate and shadows from the helicopter rotors, the daylight reference sensor only gives relative

values of the daylight. The actual irradiance at the surface is measured by the *daylight irradiance sensor* at the beginning and at the end of each profile.

Other sensors

Temperature, depth (pressure), and salinity (conductivity) are measured using commercial sensors. The depth sensor and A/D-converter together resolve one inch. The salinity sensor uses inductive principles, making it resistant to corrosion and wear.

The temperature sensor is the slowest in response and therefore it limits the profiling speed. The hydrodynamics does to some extent cause the water to move together with the subsea unit giving a noticeable hysteresis on water parameters in depth profiles from lowering to lifting (Fig. 5).

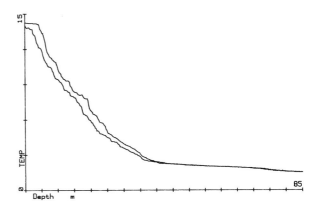

Fig. 5 Hysteresis on temperature versus depth.

Mechanical design

The underwater housing is made out of stainless steel, glass and Delrin plastic. A main cylinder contains the electronics. All the sensors are mounted like spokes from one end of the cylinder. At the other end the underwater cable is connected and the laser light for the transmission and forward scattering sensors are led out into the water. Around the pressure hull there is a cage to make handling easy and to protect the subsea unit from mechanical shocks.

The subsea unit is 0.75 m long, 0.55 m wide, 0.95 m high, and weighs 100 kg.

Electronic design

Amplifiers, power supplies, A/D-converter and microprocessor are mounted in a 19-inch rack inside the pressure hull. 220 V AC is fed through the cable to the rack. A serial current loop transmits data to the surface at a baud rate of 4800. Optocouplers provide galvanic insulation between the underwater system and the main computer at the operators console.

A particular problem was the shielding of the back scattering sensor: the microsecond flash causes a lot of electronic noise if not properly shielded. The salinity sensor also demanded careful installation to avoid disturbing its oscillating frequences.

Software design

The data administration tasks are divided in two parts, one for each microprocessor. Data acquisition from the different submerged sensors is performed by the subsea processor. Its main tasks are to:

- Keep track on the depth from the pressure gauge and initialize the measurements at the predecided depths.
- Adjust the amplification level of the photo-multiplier in the irradiance sensor.
- Administrate the triggering of the flashlamp sensor.
- Control the A/D-conversion and send the data to the surface.

The processor at the operator's console performs the following tasks:

- Storing and downloading software for the subsea computer.
- Passing on commands concerning measurement depths etc. from the operator to the subsea processor.
- Acquiring the daylight reference data from that gauge via the A/D-converter as soon as a new measurement is initialized by the subsea processor.
- Storing the measurement data in RAM during the profiling and on tape after request.
- Performing some real time data processing to help the operator to detect possible mishaps during the profiling.
- Post processing of datafiles to produce tables and diagrams of the profiles (when connected to printer or plotter).

The subsea software is made in assembly language to increase the speed in order to make all measurements at a certain level nearly simultaneously. Speed is not so critical at the operator's console and so the software is written in Basic. To fullfil all our needs the Basic of our system had to be expanded (refs 4 and 5).

TEST EXPERIENCES

The system has been tested from helicopter and from ships in the Baltic. Our experiences so far are very good on the general layout and mechanical ruggedness. No software problems have limited the tests and we have suffered no system break down. All sensors have given promising, repeatable data.

The early helicopter tests revealed some problems. Our tape recorders suffered from this kind of use and one of them stopped working during flight. The pressure gauge had problems finding the correct surface pressure under the helicopter (probably due to rotor effects) and safety regulations caused the cable handling inside the helicopter to be more difficult than expected. System modifications are initiated to overcome these problems.

Results

Currently we are in the testphase of the instrument and there is only a limited amount of data available. Figures 5, 6 and 7, recorded simultaneously at the same site, show the variations in temperature, transmission, and forward scattering as a function of depth. Temperature and optical parameters are highly correlated. This is presumably due to biological activity in the warm layers of the sea producing particles (algae etc.) which influence the optical properties of the water.

Figure 6 shows a comparison between the transmission sensor of HOSS and a Martec transmissometer. Note that the spectral range of the Martec

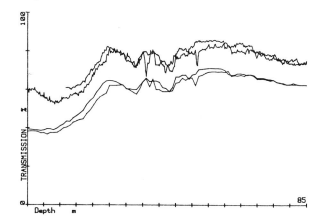

Fig. 6 Comparison between HOSS and a Martec transmissometer. HOSS measures every 0.2 metres while the Martec readings (*lower curve*) are made once a metre. This explains why the HOSS profile (*upper curve*) seems to contain more noise. The readings of the instruments follow one another remarkably well both in the up and down profile.

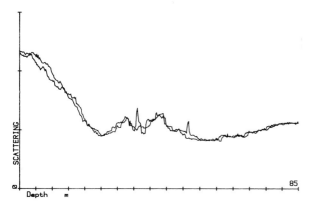

Fig. 7 The forward scattering profile (relative units) recorded simultaneous and at the same site as Figures 5 and 6.

instrument gives maximum sensitivity at 493 nm with the 50% passband 53 nm wide. The HOSS transmission sensor uses laser light of 543.5 nm. This difference should make the Martec reading lower than the HOSS reading, while the Baltic water has a high content of yellow substance shifting the maximum transmission from the blue towards the green part of the spectrum. The difference, however, should not be of this magnitude. One reason for this overestimation of the transmission is probably lack of calibration at this early test.

The high correlation between transmission, back and forward scattering is illustrated in Figure 8. Two obvious shifts in forward scattering are simultaneously matched by shifts in transmission and back scattering. Thirty metres is very close to the bottom at this location and the increase in turbidity close to the bottom is visible in the figure.

ACKNOWLEDGEMENTS

Financial support to the development and good advice concerning airborne equipment has been obtained from the Swedish Defence Materials Administration (FMV). The personnel (of HMS *Urd* and 1st Helicopter Division) of the Swedish Navy has made our field tests possible by their generous cooperation. Mr W. Modig, Mr M. Örbom and Mr S. Gruffman have, among others, contributed at the mechanical construction as well as in practical matters.

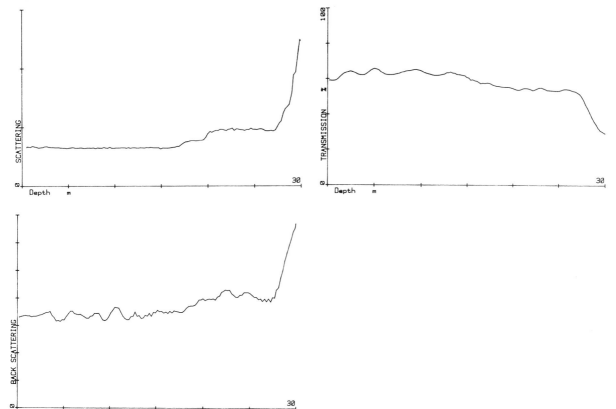

Fig. 8 The forward scattering, transmission, and back scattering profile recorded simultaneously by HOSS at one site.

REFERENCES

1. Harron, J. W., Hollinger, A. B., Jain, S. C., Van Kemenade, C. and Buxton, R. A. H. (1983). An Underwater Optical Transmissometer Incorporating Scattering Correction. The 1983 Annual Meeting of The Optical Society of America.
2. Austin, R. W. (October 1973). Problems in Measuring Turbidity as a Water Quality Parameter. The Environmental Protection Agency Seminar on Methodology for Monitoring the Marine Environment, Seattle, Washington.
3. Moore, C. A., Honey, R. C., Hancock, D. M., Damron, S., Hilbers, R. and Tucker, S. P. (June 1984). Instrumentation for Measuring In-situ Sea Truth for Laser Radar Applications. SPIE Proc. Vol. 489, Ocean Optics VII, Monterey, California.
4. Granlund, J. O. and Miklas, K. (September 1985). Ett filhanteringssystem för Digitals TU-58 implementerat på Luxors ABC-802. FOA Report D 30392, Linköping, Sweden.
5. Koppari, K. (September 1986). Expansion of the Basicinterpretator on Luxors ABC-802 and a Plottsystem for Hitachis Plotter Modell 671-20. FOA Report D 30431, Linköping, Sweden.

18

Determination of Salinity by an Optical Wave Guide Technique

D. Green and *D. Bigelow,* Seastar Instruments Limited, Canada

This chapter describes the results of tests on an optical wave guide sensor for measuring refractive index of sea water and hence salinity and density. The optical sensor used to obtain these results is very sensitive to salinity: it produces a 10 V signal for a range of salinity of 0–50 ppt. The resolution of the sensor is 0.005 ppt with an excellent stability. However, the primary disadvantage of the sensor is its sensitivity to contamination of its surface with the sea. This is a problem that has plagued sensors which rely on internal reflection to measure refractive index of a fluid.

The use of the optical sensor as a device for measuring salinity in expendable probes is discussed. Expendable probes require that the sensor pass through the surface only once, thus avoiding most of the surface slick contamination problem. This optical sensor promises to be cheap and easy to manufacture, essential features for an expendable probe. We present the experimental results obtained to date in testing the sensor for use in expendable probes.

DESCRIPTION OF SENSOR

The sensor consists of a light source, a 2 cm long optical wave guide, and a light detector. Light is sent along the optical wave guide and is partially lost, depending on the refractive index of the fluid in which the optical wave guide is immersed. More light is lost as the refractive index of the fluid approaches the refractive index of the optical wave guide; less light is lost if there is more of a mismatch. The amplitude of the detected light is therefore directly related to the refractive index of the fluid. The refractive index of sea water is directly related to its salinity and density.

REFRACTIVE INDEX AS A METHOD OF DETERMINING SALINITY

The determination of salinity and density via the measurement of refractive index has several inherent advantages over the traditional measurement of con-

Advances in Underwater Technology, Ocean Science and Offshore Engineering, Volume 16: Oceanology '88

ductivity to determine salinity: refractive index is less sensitive to temperature than is conductivity; refractive index increases linearly with salinity to 50 ppt, whereas conductivity does not (sensitivity at saturation is one-sixth sensitivity at low salinities for conductivity); refractive index is hardly affected by changes in the ratios of salts whereas conductivity measurement relies on calibration with a standard salt mixture and is inaccurate when the salt composition changes as in estuaries or ice melt conditions; finally, unlike conductivity, the equation of state for calculating density from refractive index shows that 95% of density change is due to a linear relation with refractive index. Accuracies of 5×10^{-5} cm^3 g^{-1} are attainable easily from the linear relationship. The equation for conductivity is non-linear and temperature dependent (Seaver, 1987).

The main disadvantage of measuring refractive index is that it is more sensitive to pressure than is conductivity. The dependence of refractive index on oceanographic parameters, considering the range of each parameter that a probe would encounter is as follows: temperature $-0.002/30°C$, salinity $+0.009/43$ ppt and pressure $+0.003/200$ kg/cm^2 (Austin and Halikas, 1976).

OPTICAL SENSOR PROBLEMS

The optical wave guide technique has been understood and applied to the measurement of refractive index for several years (Harmer, 1980). However, no commercially successful sensor has been developed for salinity measurement for several reasons.

(1) There is a severe mismatch in refractive index between most optical wave guide materials (glass, plastic) and sea water, limiting the sensitivity of the device.
(2) To obtain the required sensitivity with these optical wave guide materials, many coils of optical guide have to be used. This presents a problem of mechanical stability: movement of the coils affects the result.
(3) There is no reference path in an optical wave guide sensor. Variations in ambient light as well as variations in the intensity of the light source affect the result.
(4) The optical wave guide sensor is affected by anything that affects the surface of the optical guide. Interferences are caused by organic films, bubbles and anything else directly affecting the surface of the wave guide.

The optical wave guide sensor used in these experiments solves three of the above four problems. The resolution is 0.005 ppt using an innovative material with refractive index nearly matching that of sea water. The mechanical stability problem is solved since coils of optical guide are not required: the sensor consists only of a short (2 cm) piece of wave guide with a slight bend. The mechanical stability of this short piece is easy to ensure. The problem of not having a reference signal is dealt with by use of a signal-processing technique that eliminates the effect of variable ambient light. Variations in source intensity modulation are reduced by use of a stable light source.

The remaining problem of surface contaminant effects is a serious one. The surface problems prevent the use of the optical waveguide technique in a laboratory-style salinometer, since repeatability is not sufficiently good to give the accuracy required. Surface problems are reduced in an expendable probe configuration, however, and it would appear that sufficient accuracy can be obtained for this application.

REQUIREMENTS FOR AN EXPENDABLE SALINITY PROBE

Expendable bathythermographs (XBT) have been available for many years. However, expendable devices which measure salinity as well as temperature have not been commercially available because of the difficulty of making a conductivity sensor sufficiently cheaply. The optical waveguide approach presents an alternative. Testing was done on this device to determine if it could meet the performance specifications for an expandable probe. Desired

TABLE I
Performance specifications for an expendable probe

Resolution	0.005 ppt salinity
Accuracy	0.05 ppt salinity
Response time	25 ms (15 cm vertical resolution)
Mechanical stability	Withstand impact at sea surface, and up to 5 ms fall rate without losing calibration
Temperature range	Operates over range of -2 to 35°C
Power demand	Less than 750 mW
Space	Fits in standard XBT or XCP probe
Ease of calibration	Single-point calibration of salinity sensor, no calibration of temperature sensor
Depth	Minimum 200 m, desirable 2000 m
Stability on storage	5 years
Cost	US $40 each at an annual volume of approximately 10 000

specifications are given in Table I. The following sections deal with each of these requirements in turn.

Resolution

The optical sensor has excellent resolution of salinity. A range of 0–50 ppt produces a voltage change of 10 V. Resolution of 0.005 ppt is obtained with some attention to electronic design, so the desired specification for an expendable probe can be met. A comparison of a commercial CTD and the optical sensor is shown in Figure 1.

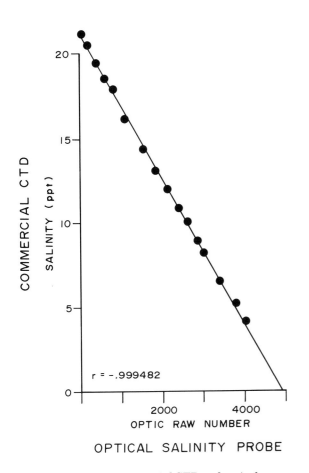

Fig. 1 Comparison of commercial CTD and optical sensor.

Accuracy

Accuracy is much more of a problem. The chief concern is the effect of bubbles and contaminants on the surface of the optical guide.

Contaminants

The results of a test in which the probe was placed repeatedly into a sample showed a mean change between subsequent readings of 0.15±0.05 ppt. With a surface slick present, changes can be in the range of 2–5 ppt between subsequent readings. The instrument is therefore very sensitive to surface contaminants, a feature which makes it very difficult to develop a successful laboratory salinometer. However, the application of an expendable salinity probe is an attractive alternative since it requires only one passage through the surface.

Because only one passage through the surface is required, protection of the probe can be obtained. Once the probe is protected from the surface slick it works well in the water column. The response of the probe to salt slugs injected into water flowing through the probe are shown in Figure 2. Note how the probe returns to baseline after each salt slug passes.

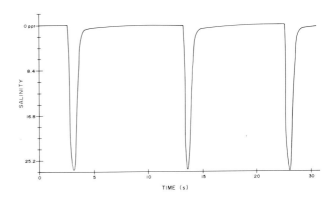

Fig. 2 Fresh water flow with saturated NaCl solution injected through tubing wall. Sensor return to baseline after the salt slugs is within ±0.0053 ppt on average. Rate of flow 2.25 m s^{-1} (0.161 litres s^{-1}).

Bubbles

Air bubbles on the surface of the optical guide cause a change in the detected light amplitude and therefore affect the sensor's accuracy. However, in a flowthrough or profiling mode, bubbles do not adhere to the optical guide and therefore do not present a problem. Further, increasing pressure during profiling will collapse any bubbles on the sensor. Figure 3 shows the effect of injecting air slugs into the flow past the sensor.

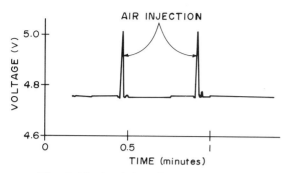

Fig. 3 Air slug injected into flow stream.

Long-term stability

The long term stability of the sensor with tap water flowing through at a rate of 2.25 m s^{-1} (0.16 litres s^{-1}) is as follows:

Short term stability (20 s)	0.003 ppt
Medium term stability (5 min)	0.006 ppt
Long term stability (30 min)	0.008 ppt

Leaving the sensor in the flow over an 8-hour period produces a chart recording indistinguishable from a straight line.

Dissolved organics

Dissolved organics in the water column can be expected to cause problems to the sensor, although much less so than surface films. To check on the effect of dissolved organics, we injected a mixture of organics dissolved in methanol into the water flow. The concentrations of dissolved organics in the water flow were 20 mg litre^{-1}, about 20 000 times the concentration in natural waters. The result of the addition of these organics was to decrease the voltage by 0.19 V or the equivalent of 2.6 ppt. The change was stable, but could be reversed by injection of soap into the waterflow. This is illustrated in Figure 4.

In water with normal dissolved organic concentrations, no drift in the probe could be detected over

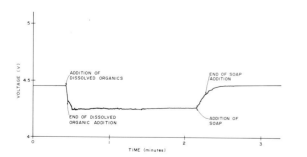

Fig. 4 Effect of injecting dissolved organics into flow, followed by the injection of soap.

an 8 hour pump period in which 0.13 litres s^{-1} o 3744 litres of sea water pumped past the probe.

In conclusion, if the sensor can be carried through the sea surface without becoming contaminated, it will provide a stable, accurate, highly sensitive measurement of salinity.

RESPONSE TIME

The desired response of an expendable probe is that it be able to measure vertical salinity structures of 15 cm while falling at a rate of 5 m s^{-1}. This requires a response time of 25 ms. In the prototype sensor the electronics are set so that a salinity reading anywhere between 0 and 50 ppt can be taken every 22.2 ms. To demonstrate the speed of response of the sensor in practice, it was connected to a tap with plastic tubing. The sensor output was connected to a strip chart recorder. Twenty ml of saturated NaCl solution was injected into the water through the wall of the plastic tubing and the tap was turned on (0.16 litres s^{-1}). When the slug of salt water passed the sensor, the chart recorder was deflected as shown in Figure 2. This graph provides a demonstration of how quickly the sensor can track a sudden change in salinity.

In summary, the optical probe can meet the specification of 25 ms response time and can therefore resolve 15 cm vertical salinity structure at a drop rate of 5 m s^{-1}.

MECHANICAL STABILITY

To be useful the mechanical probe must be able to withstand the impact of up to 10 g from hitting the sea surface and then withstand a flow past the probe of up to 5.0 m s^{-1} as the probe descends. The present configuration of the probe has been tested at flow rates of 2.25 m s^{-1} with a resulting increase in the noise from 13±4 ppm (one minute interval, 0 m s^{-1} flow) to 51±5 ppm (one minute interval, 2.25 m s^{-1}). This corresponds to a noise of 0.003 ppt at 2.25 m s^{-1}, which is not significant for the expendable probe application.

No proper testing of the impact resistance of the probe has been done, although it is not affected by vigorous knocking or banging. No difficulty with mechanical stability is anticipated because the probe can be easily reinforced if the impact resistance is not sufficient to meet the specifications.

TEMPERATURE RESPONSE OF SENSOR

The refractive index of water varies with temperature as well as salinity. If wavelength and pressure remain constant, the refractive index of water decreases with temperature as shown in Figure 5. The approximate variation in refractive index over a temperature range of 30°C is −0.002, which corresponds to a salinity change of about 9.6 ppt. This temperature dependence is far less than the temperature dependence of conductivity: most of the conductivity change in the ocean is due to temperature—not salinity—change.

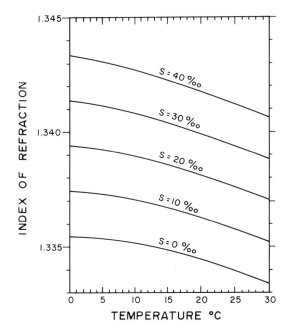

Fig. 5 Relationship between refractive index of sea water and temperature.

The response of the optical sensor to temperature does not follow the expected curve because of the temperature dependence of the probe itself. The actual response of the probe is shown in Figure 6. It exhibits the opposite temperature dependence to refractive index and more than cancels out the expected temperature response. (As temperature increases, the expected response is that the refractive index decreases, causing the sensor output to rise.) The net result is that the sensor exhibits a modest linear decrease in output with increasing temperature (Fig. 6).

Power demand

The current prototype was designed to test the concept, and power consumption was not a design criterion. The power demand of a finalized probe design is estimated to be 52 mA at 12 V or 740 mW, which is within the acceptable limits for an expendable probe.

Space

The sensor optics, electronics and batteries will fit in a cylinder 5 cm in diameter by 10 cm in length with a hole along its axis having a 2 cm diameter. This is similar to the 'form factor' for a standard XBT probe.

Ease of calibration

The linear relationship between salinity and sensor output suggests that a single-point calibration will be sufficient for each probe. This calibration could be done on distilled water. The much lower dependence of the optical sensor on temperature may obviate the need for calibration of the temperature sensor that is required for conductivity sensors.

Depth

The desired depth capability for the probe is 2000 m. At 2000 m depth the refractive index of water will increase by 0.0027 over the refractive index of water at atmospheric pressure. This is equivalent to an increase in salinity of 12.8%. Depth of expendable probes can only be determined to ±2%. The corresponding uncertainty in salinity is 0.25%.

The actual response of the sensor to pressure may be less than is predicted from the refractive index because of cancelling effects on the probe (as with the temperature effect). However, no pressure data are yet available. The design of the probe is suited to working at high pressure and is not expected to

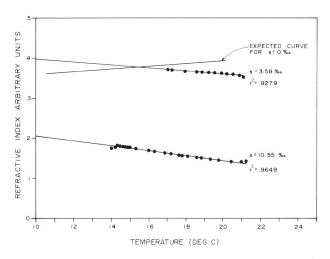

Fig. 6 Expected and actual dependence of output of optical sensor on temperature.

fail. However, careful compensation for pressure will be required in determining salinity.

Stability during long term storage

There have been no long term storage studies done on the optical sensor. The sensor should be stable over storage periods of several years because both the optical and the electronic components are stable over these periods of time. The area of uncertainty is in the surface of the probe, which may tend to become contaminated over long storage periods.

Cost

The electronics required to measure the refractive index is relatively straightforward, and the optical components are quite cheap. The overall price of parts is estimated to be about US $20 in volume quantities of several thousand.

CONCLUSION

The optical sensor described in this chapter meets most of the requirements for an expendable salinity sensor. It is particularly effective in satisfying the requirements for resolution, speed of response, ease of calibration and cost. On the negative side, the sensor is of limited accuracy unless a means can be found to protect it from the surface film when it passes through the sea surface. The effect of pressure on the sensor may be significant and must be studied. Despite these shortcomings, the optical wave guide technique is a promising solution to the problem of how to manufacture a commercially successful expendable salinity temperature–depth probe.

REFERENCES

Austin, R. W. and Halikas, G. (1976). The Index of Refraction of Seawater. Tech. Rep. S10, Ref. No. 76–1. Jan. 1976. Scripps Inst. of Oceanography, La Jolla, CA 92093.

Harmer, A. L. (1980). Device for Producing a Light Signal Corresponding to the Refractive Index of a Fluid. U.S. Patent 4,187,025.

Seaver, G. (1987). The optical determination of temperature, pressure, salinity, and density in physical oceanography. *MTS Journal* **21**: 69–79.

19

Measurement of Current Profiles

D. Flatt, Proudman Oceanographic Laboratory, *G. Griffiths,* IOS Deacon Laboratory *and M. J. Howarth,* Proudman Oceanographic Laboratory

Acoustic doppler current profilers (ACDPs) have the potential to make a significant impact on the measurement of currents in continental shelf seas. Progress made in the development of one ADCP system is described, and results from five different ADCPs deployed in three different ways—mounted in a sea bed frame, on a mooring and on a ship—are presented. Only by deploying the instruments and by comparing their results with other observations will faults in the instrumentation become apparent and the necessary confidence in the technique be gained.

ACOUSTIC DOPPLER CURRENT PROFILERS

Good quality observations of current profiles are essential if optimum benefit is to be obtained from continental shelf seas. The time varying elevation and transport fields of tides and surges, which together with waves dominate shelf sea dynamics, are largely understood, through observations and modelling. It is now feasible, because of both new instrumentation such as ADCPs and new more powerful computers, to develop this understanding and to tackle the real, three-dimensional world of shelf seas.

ADCPs have the potential, in one instrument, to measure the current profile over much of the water column in shelf seas. This chapter will investigate this potential, look at some of the constraints, and discuss ways of realizing the potential. The basic oceanographic design parameters that need specifying (Table I) are the maximum range, the cell size, the accuracy of the measurement, the sample interval, and the duration. To satisfy these the instrument designer has at his disposal the acoustic frequency, f, the beamwidth, φ, the beam angle to the vertical, θ, the speed and memory of the microcomputer, the data logger, and battery capacity. In brief:

$$\text{Maximum vertical range} \propto \frac{\cos \theta}{f}$$

$$dv \cdot dz \propto \frac{1}{F\sqrt{N} \tan \theta}$$

Advances in Underwater Technology, Ocean Science and Offshore Engineering, Volume 16: Oceanology '88

TABLE I
Development specifications POL ADCP

	A. September 85	B. Acceptable	C. Desirable
Frequency (kHz)	XXXX	XXXX	XXXX
Water depth (m)	80–120	10–200	10–200
Max number of cells	10	32[a]	128[a]
Cell size (m)	10	1–5[a]	1 variable
Number of beams	2	4	4
Beam width (degrees)	3	2	2
Deployment duration (months)	1.5	1.5	12
Ping rate (per second)	1	2–10[a]	2–10[a]
Integration period (minutes)	10–15	10–15[a]	1–60[a]
Operating pressure (psi)	500	2000	2000
Maximum currents (m s^{-1})	1	3	5
Measurement accuracy (cm s^{-1})	1	< 1	< 1
Direction accuracy (degrees)	+/−3	+/−1	+/−1
Nearest measurement surface	10% water depth		sea surface
Nearest measurement sea bed	3 m	1 m	sea bed

[a] Programmable parameters. XXXX = Frequency dependent on range

Sample interval	$= N\,dt$
Cell volume	$\approx \dfrac{\pi z^2 \varphi^2 dz}{4\cos^3\theta}$
Cell separation	$\approx 2z\tan\theta$

where

N = number of pings
z = vertical distance
dz = cell depth in the vertical
dv = velocity uncertainty
dt = interval between pings

The minimum value for dt is determined by the time delay between the end of transmission and the return of the backscattered signal from the furthest cell (in shelf seas less than 0.5 s). In practice, however, dt is determined by the time taken for the microcomputer to process the backscattered signal. Given the range determined by the operating frequency, the range over which results can be obtained is constrained both near the instrument and, possibly, at the far field. First, interpretation of the measurement close to the transducers (the first one or two cells) is difficult because of decaying transients. Secondly, measurements near the sea surface for upward-looking sonars and near the sea bed for downward-looking sonars suffer interference between the spectacular return of a (vertical) side lobe and the reverberation return along the main

slant beam. Interpretation of the results within about $(1-\cos\theta)$ water depth of the sea surface/bed is usually difficult. Two further points are worth detailing. First, measurements which cover the whole shelf seas water column are not relevant to small scale processes: the beams can be hundreds of metres apart near the far end of their range where the cell volumes are hundreds of cubic metres. Secondly, the choice of acoustic frequency is fundamental to the design of the instrument: it affects the maximum range (the higher the frequency, the shorter the range), the velocity uncertainty (the higher the frequency, the shorter the time needed to achieve a given accuracy and/or the smaller the cell depth) and, for a given beamwidth, it affects the transducer diameter, $d (d = 1/f\phi)$.

We shall report on five ADCPs, with frequencies between 1 and 150 kHz, deployed in three different ways: in a sea bed frame, on a wire, and from a ship. Four of the systems were designed to cover the majority of the water column in shelf seas and were deployed in sea bed frames or from a ship. Two of these are being developed at Proudman Oceanographic Laboratory (POL), both with a frequency of 250 kHz, and two were made by RD Instruments: a 300 kHz self-contained instrument deployed in a sea bed frame and a 150 kHz ship-mounted system. The fifth system, with a shorter range, operated at 1 MHz and was developed at th IOS Deacon Laboratory. It was deployed on a mooring pointing up to cover the upper 20 m to measure the current profile in the very important wind-driven shear layer not fully covered by lower frequency, longer range systems.

The ADCP development programme at the NERC Deacon and Proudman Oceanographic Laboratories started with the 1 MHz 10 cell Doppler, followed by a 250 kHz 10 cell Doppler and the present 250 kHz 24 cell Doppler, which were designed to the specifications of Table I. Several previous deployments of the 1 MHz instrument in UK coastal waters have been successfully completed and the results reported (refs 1,2). Here the present 250 kHz system is described in more detail, results presented from all five ADCPs obtained during an experiment in the Celtic Sea, January–March 1987, and from a further deployment of the 250 kHz system near Oban in July 1987.

250 kHz ADCP acoustic doppler current profiler

The prototype 250 kHz ADCP was a copy of the 1MHz ADCP with only minimal changes to enable it to operate at 250 kHz and hence an increase in range. The instrument was built and deployed to gain information and experience on ADCPs in measuring the current profile up to a range of 150 m and to determine how realistic our target specifications were.

Initial trials of the instrument were made near Oban, Scotland, using the Scottish Marine Biological Association's (SMBA) facilities at Dunstaffnage. The operating range of the instrument made the choice of a test site difficult. Five criteria were used: water depth between 80 and 150 m, safety of instrument, suitable ship, understandable current flow, and cost. The facilities at SMBA and the waters around Oban meet all these criteria except, because of the presence of islands, understandable current flow.

The first test deployment at Oban (site 1, Fig. 1) gave results that were difficult to interpret because of the unexpected current flow. Although the results showed that the instrument functioned correctly and had the required range, there was a double tidal cycle in the record that could not be explained. This led to some doubt as to the validity of the results which was not expunged until a year later when a conventional Aanderaa current meter was deployed at the site: data from this gave a similar double tidal cycle.

Information gained from the prototype was used in the design of the developed instrument. This instrument was to be of a modular design to allow

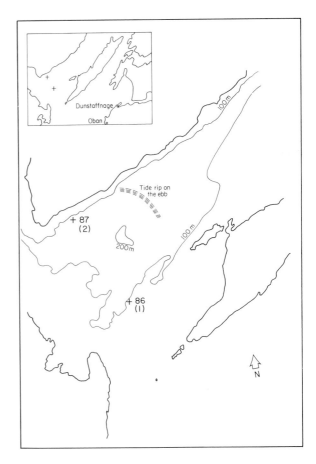

Fig. 1 Trial sites at Oban.

future developments to be incorporated with ease. Operating parameters were to be software controlled in order to make the instrument as flexible to use as possible, and the programme language was changed from Basic to Forth because this promised an increase in processing speed. These changes allowed the instrument to be developed in a short time span that would largely fulfil the initial specification requirements.

The first trials of the developed instrument were held in the Celtic Sea during February and March 1987 and several shortcomings came to light; unfortunately no usable current data were obtained. A later trial at Oban was more successful.

CELTIC SEA EXPERIMENT

Despite problems being encountered with all five ADCPs useful data and information were obtained from each. Each instrument is discussed below.

IOS 1 mHz ADCP mooring

The instrument was deployed at site A (51°0'N 7°0'W). In order to make best use of the 1 mHz ADCP, with its inherent short range, a mooring was designed so that profiles were obtained in the uppermost 20 m of the water column. If conventional instruments are used, this is a difficult region in which to make measurements despite being of great scientific and practical interest, especially in studies of air–sea interaction. A buoyant pressure tube with a stabilizing vane housed the instrument, and the transducers were mounted on a gimballed platform (Fig. 2). The instrument operated from day 71 to 79.

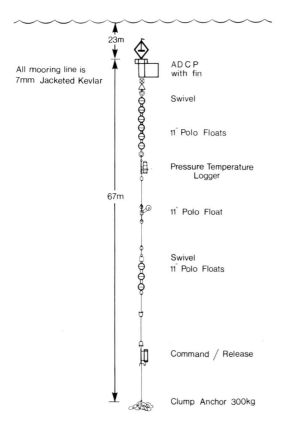

Fig. 2 Mooring encouragement for the IOSDL 1 MHz ADCP, Celtic Sea March 1987.

Signal quality

An important indicator of signal quality is the receiver signal-to-noise ratio (SNR). The instrument is dependent on backscattering from passive tracers of the water velocity. Very high isolated peaks in the SNR can indicate large active targets (e.g. fish), whereas very low SNR warns of poor quality velocity measurements. Figure 3 shows the SNR over the period of the deployment for cells 1 to 10, covering the depth range 19.1 to 4.3 m in 1.65 m increments. Several features can be seen in the diagram:

(a) a low SNR, averaging 25 dB in cell 10, up until day 75;
(b) a slow increase in SNR at all depths from day 75;
(c) high SNR in the near surface cells around days 77.1, 77.5, 78.0 and 78.6 caused by scattering from bubble clouds generated by breaking waves;
(d) from day 76.5 a change in the nature of the fluctuations in SNR especially in cells 1 to 6.

Items (a) and (b) are of particular interest in this chapter in that their effect on velocity measurement can be clearly seen. Valuable information on physical processes in the sea can be obtained from a study of sound scattering. Thorpe for instance (ref. 3) has used measurements similar to (c) above to obtain estimates of turbulence in the upper ocean. We are not yet sure of the mechanism causing item (d).

Figure 3 suggests anomalous behaviour of cells 7–10 in the otherwise smooth decay of SNR with range during the period of overall low SNR up to day 75 (section 1). This deficit is shown in more detail in Figure 4, a log–log plot of mean SNR against range for the two sections, compared to the expected decay curve due to spherical spreading and absorption—corrected for the change of scattering volume with range. It is thought unlikely that a change in scattering characteristics could be the cause, especially near the sea surface. This suggested that there was a problem in the sonar receiver, the cause of which we examine below.

An ADCP is required to operate with a large input dynamic range, to accommodate both range-dependent and site-to-site scattering variations. This requires an adaptive control of the receiver gain over a range of 60 dB. The computed mean signal level within an ensemble is used to control digitally the gain of a post-demodulator amplifier. Because of the low SNR in cells 9 and 10, the receiver was operating with maximum allowed gain at baseband. This had the effect of reducing the bandwidth from that determined by the low pass filters, resulting in truncation of the received spectrum. This spectral clipping also has implications for the accuracy of the velocity measurement, as discussed below. The instrument software has

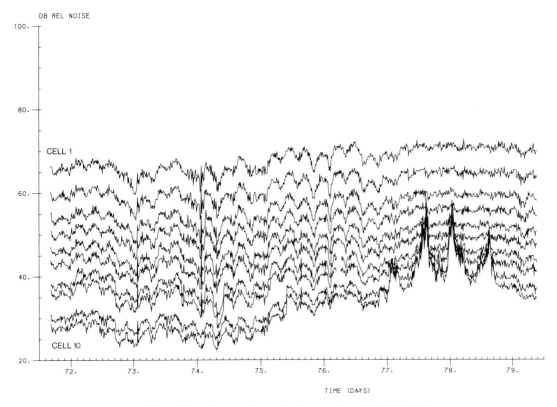

Fig. 3 Signal-to-noise ratio time series, 1 MHz ADCP
Celtic Sea.

now been modified such that the receiver cannot use these gain codes which produce spectral clipping.

Tidal current analysis

The current record obtained from the ADCP was low-pass filtered using a digital filter with a cut-off of 1 hour, to avoid aliasing, prior to running the tidal analysis program TIRA. To examine the effect of changes in SNR and bandwidth on the velocity measurement the record was analysed in two sections: from day 71 19:30 until day 75 18:30 and from day 75 19:30 to day 79 05:30. As these are very short records, care must be taken in interpreting the absolute values of the tidal currents, however, the relative values are more accurate.

The magnitude profiles of the predominant tidal component M2, or lunar semi-diurnal, are shown in Figure 5. Section 1 shows a current maximum at cell 3 (19.1 m depth or 71 m above the sea floor) followed by a decrease towards the surface. However, the currents are suspiciously low in cells 9 and 10. Hansen discussed in detail the effect of non-ideal

ADCP sonar receiver performance (ref. 4). He identified spectral clipping as a major source of bias in the velocity estimates. This mechanism always tends to bias the results towards zero, the bias being inversely proportional to the product of receiver bandwidth and transmission pulse length and proportional to the doppler shift. Using the curves of Hansen for the case of maximum gain (as in cells 9 and 10) for a filter bandwidth of 910 Hz and a transmission pulse of 2.5 ms with a doppler shift of 224 Hz (corresponding to 0.32 m s^{-1}) we obtain a bias of -8%. A smaller correction of 4% is needed for cells 7 and 8 where the effective filter bandwidth was 1.5 kHz. All the other cells had an effective bandwidth of 4 kHz, giving a correction of 1%. As several other factors can cause errors of this magnitude (e.g. SNR, DC offsets, magnitude and phase imbalance in the real and quadrature receiver channels) these cells are not corrected. After correction of the current data, cells 7 to 10 of section 1 are a closer match to section 2. The same cells in section 2 had a much higher SNR, hence a lower

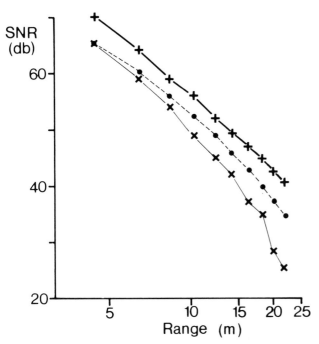

Fig. 4 Signal-to-noise ratio as a function of range: 1 mHz
ADCP Celtic Sea. × = section 1, + = section 2,
• = theoretical decay normalized to cell 1 of section 1.

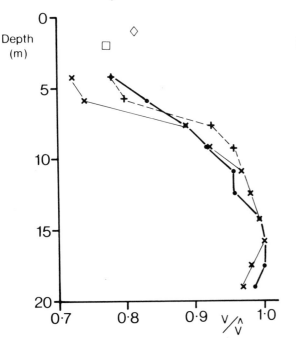

Fig. 5 M2 tidal current profiles, normalized to the peak
magnitude. × = section 1, • = section 2, + = cells
7–10 of section 1 corrected for the effect of spectral
clipping. □ = VAESAT buoy at 2 metres, ◇ = VAESAT
buoy at 1 metre.

gain and minimal spectral clipping. Figure 4 shows
that the mean SNR in cell 10 of section 2 was higher
than cell 7 in section 1.

ADCP noise level

The power spectral density (PSD) of the East
velocity component of cell 1 of the ADCP and that
from a vector-averaging electromagnetic current
meter (VAECM) at a similar depth are shown in
Figure 6. Clearly the ADCP has a substantially
higher noise level than the VAECM, and whilst the
latter generally decreases above 0.3 cph, the ADCP
noise level remains approximately constant. When
the PSD figures are translated into total RMS noise
over the band 0.03 to 2.5 cph, we obtain a noise level
of 0.032 m s^{-1} for the VAECM and 0.085 m s^{-1} for
the ADCP. If we assume that the VAECM noise
level is predominantly due to current fluctuations,
the ADCP has an excess noise level of 0.079 m s^{-1}.
Some of this excess can be attributed to the variance
expected in a pulse doppler system (ref. 5). Field
trials in Loch Ness established the RMS noise level
of the instrument as 0.015 m s^{-1} along the beam (ref.

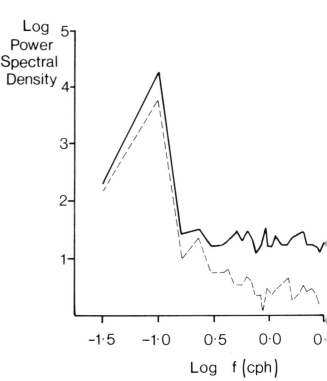

Fig. 6 Power spectral density of cell 1 East component of
the 1 MHz ADCP —— and a VAECM ––––– at a similar
depth.

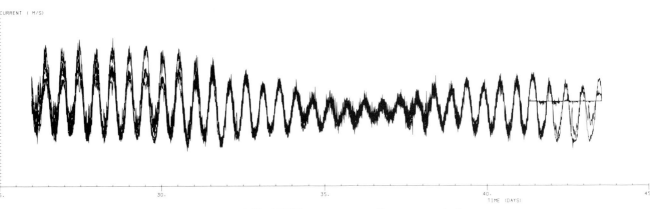

Fig. 7 250 kHz ADCP prototype north component of current, Celtic Sea 1987.

2). In this deployment the beams were inclined at an angle of $\theta = 30°$ to the vertical therefore the equivalent noise level for horizontal currents becomes $Vn/\sin\theta$ or 0.03 m s^{-1}. We postulate that the remainder of the excess noise (0.073 m s^{-1}) is due to aliasing of the mooring and transducer platform motion, caused by the low sampling rate (0.1 Hz) used in the prototype instrument. In support of this we have three observations:

(a) the noise levels found in Loch Ness: a stable low current, hence low motion, environment;

(b) low noise (0.03 m s^{-1}) was also observed during a deployment in the Irish Sea in which the instrument bottom was mounted with the transducers inclined at 30° and rigidly fixed;

(c) the present record shows a high degree of coherence ($c > 0.5$) for frequencies below 1 cph between all cells.

These strongly imply a common source. It is thought unlikely that the cause could be an artefact of the sonar receiver, given the large range of operating conditions covering the ten cells.

POL 250 kHz prototype sea bed mounted

This instrument was deployed, at the same site, for a planned duration of 50 days to measure ten cells of 6.5 m along the beam at 12-minute intervals. The first cell was 15 m from the sea bed.

The returned data showed that there had been an impedance mismatch on one of the transducers. The resultant difference in gain between the two beams and the subsequent vector averaging of them made the data suitable for only limited analysis. The spot compass readings were approximately 270° and so the x,y beams map closely to the N,E axes. The plot of the north component of the current is shown (Fig. 7). At the beginning of the record it can be seen that there is a step in magnitude between the first three cells and the last seven. This is due to one beam only having enough signal to provide a north component for the first three cells. Also shown by the plot is the signal falling off after day 42; this was due to a failure in the transmitter battery.

A plot of signal strength against range along the beams (Fig. 8) gives some indication of the range and attenuation for the Celtic Sea and this instrument. It is difficult to assess the range for an instrument with the transducers correctly matched, but an estimate of signal-to-noise level at the transducers of 76.3 dB has been made using data obtained on a latter deployment, at Oban using the developed instrument. Then, assuming the attenuation is the same, a range along the beam would be 162 m, giving a vertical range of 140 m.

A future version of the software does not include the vector averaging, instead it records the doppler shift of the two beams separately. This will alleviate

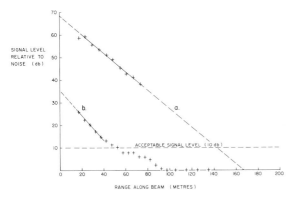

Fig. 8 Signal strength along beams: (a) 250 kHz ADCP prototype, Celtic Sea 1987; (b) 250 kHz ADCP, Oban 1987.

some of the above problems and also speed up the processing by a small amount.

RD instruments ADCP

During the experiment two RD instruments ADCPs were deployed: one ship mounted, the other in a sea bed frame at site A. The ship system was standard, having four transducers in a Janus configuration and operated at 150 kHz. It returned good data during its operation, day 27–40. For the majority of this time 10-minute averages were recorded, each containing about 475 pings. The data were split into 4-m slices and typically in 95 m of water data would be recorded over the range 20–80 m above the sea bed.

The instrument measures two horizontal components of velocity: the vertical velocity and an 'error' velocity; both should of course be close to zero. (The error velocity is the difference between the vertical velocity calculated in two different ways, from the fore and aft transducers and from the port and starboard transducers.) During the experiment it was noticeable that both the vertical velocity and the error velocity were correlated with the ship's speed (Fig. 9). Vertical velocities of these magnitudes would occur if the ship were tilted fore–aft

by between 0.5° and 1°; values of the trim which were indeed measured on departure (0.45°) and on return (1.05°) to port. At sea the trim will change by this order as moorings are deployed and fuel and water consumed. This does not explain the error velocity measurements, which can only arise from imperfections in the head geometry where the angle between beams should be 90° in the horizontal and all the beams should be 30° off the vertical. The measurements imply a deviation of less than 0.5°. Interesting as these observations are, the effect on the horizontal velocity estimates is truly negligible (less than 0.001 m s^{-1} at full speed): it is a stringent test of the ADCP installation, its beam geometry, and the ship's trim.

A self-contained 300 kHz ADCP was deployed in a bottom frame resting on the sea bed twice, days 28–35 and 35–38. Two-minute averages were recorded, each containing 200 pings. The data were split into 4-m slices which covered a range of 11–79 m in water 95 m deep. (Note that although this instrument recorded 2-minute averages as against the ship ADCP's 10-minute averages, its velocity uncertainty was in fact less, because both the frequency was twice as high and also its ping rate was about twice as fast—a ping every 0.58 seconds compard with the ship ADCP's every 1.26 seconds.) No data were recovered from the first deployment because of a malfunction by the data logger, and although data were recovered during the second deployment, a problem was still manifest since half-hour blocks of data were not recorded every 8 hours. During the 60-hour deployment the

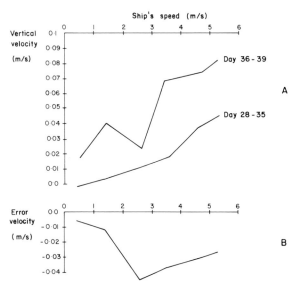

Fig. 9 (*a*) Plot of the average vertical velocity v. ship's speed for two different periods. The data were averaged by splitting the ship's speed into 1 m s^{-1} bins; (*b*) plot of the average 'error' velocity v. ship's speed for the whole period (day 28–39). The data were averaged by splitting the ship's speed into 1 m s^{-1} bins.

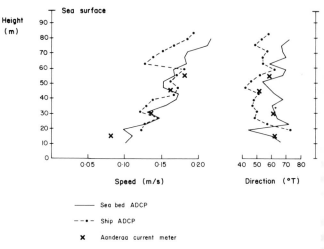

Fig. 10 Current profiles observed at 21:40 day 35 by an ADCP in a sea bed frame, a ship mounted ADCP and a current meter rig containing four Aanderaa RCM4s.

pitch and roll measurements were stable at $-5.8°$ and $-2.9°$, but the compass measurement drifted by $0.8°$. Because the pitch and roll measurements were stable, this is unlikely to have been real.

The measurement by these two ADCPs can be compared for two short periods just after the deployment and just before the recovery of the self-contained instrument. Figure 10 shows the results from a 2-minute sample by the sea bed instrument centred to be at the same time as a 10-minute ship-mounted observation. In general the agreement is good, the most noticeable difference being a direction difference of the order of $10°$.

The measurement from a conventional current meter rig containing four Aanderaa current meters at 15, 30, 45 and 55 m above the sea bed moored at this site are included in Figure 10. Comparisons were also made between the Aanderaa measurements and the results from the second deployment of the sea bed ADCP and between the Aanderaa measurements and the ship ADCP for two earlier 15-hour periods when the ship was in the vicinity of the site. At this stage not much can be said from these comparisons except that the Aanderaa directions were more nearly in agreement with the ship ADCP, that in both cases the Aanderaa measurements indicated greater shear, and that the Aanderaa speeds agreed reasonably with the sea bed ADCP but were up to 15% greater than the corresponding ship ADCP measurements.

OBAN 87 DEPLOYMENT

A trial of the 250 kHz developed ADCP was made at Oban in July 1987, using site 2 (see Fig. 1). An Aanderaa RCM4 was also deployed at a site 200 m from the ADCP, at a height of 45 m above the sea bed. The depth of water was 120 m.

The optimum operating frequency of the transducers fitted was 260 kHz, but the instrument had to be operated at 250 kHz owing to lack of a suitable crystal. It was one of these transducers that had caused the difficulties on the Celtic Sea deployment.

The deployment lasted 36 hours. Both instruments operated correctly and returned data. Again the current flow in this area proved difficult to interpret, with the tidal cycle not being very evident, in fact less so than at site 1.

The two instruments were compared by combining the east and north components from the ADCP into a magnitude and then comparing the plots from the 24 cells with the speed channel of the RCM4. This comparison is shown in Figure 11, where cells 2, 5, 8, 11, 18 and 24 are plotted together with the

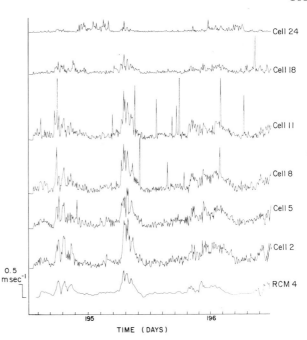

Fig. 11 250 kHz ADCP magnitude and RCM4 speed data, Oban July 1987.

Aanderaa speed data. There is general agreement between the two instruments at the 45 m height, cell 8. The major current features are present in both instruments with the ADCP being somewhat noisier. This is due in part to the low number of pings used in this trial and in part to the nature of the two measuring systems (see 1 mHz deployment in the Celtic Sea above). The data from the ADCP show a reduction in signal strength with range, and this naturally leads to the question; which of the cells are giving usable data? Christensen (ref. 6) shows that for a constant standard deviation in the velocity measurement a minimum SNR of 10 dB is required. The plot of signal level against range (Fig. 8), which takes the noise level to be the signal level found in the quiet part of cell 24, gives a working range of 45 m, the first six range cells. However, Hansen (ref. 4) shows that a significant bias to zero is to be expected with SNRs below 20 dB, reducing the useful range still further.

The correct crystal has now been fitted to this instrument along with an update in the software. Another trial is planned at Oban to test for range and the other changes, non-vector averaging, the addition of tilt sensors, and a cache memory. The cache memory utilizes part of the non-volatile memory to store the last 16 on-the-hour readings. These can be played back on site with a dumb terminal.

CONCLUSIONS

Progress in the development of an instrument to measure current profiles has been charted from level A (see Table I) to level B and towards level C. The principle behind the technique of measuring the doppler frequency shift in a time gated backscatter acoustic signal is simple, and the measurement does not require calibrating. In practice the instrument is complex, involving both acoustic and electronics, and subtle problems are encountered. In addition the measurement is remote, which, whilst offering advantages, can leave the user uncertain as to where the measurements are being made and what makes the interpretation of inter-comparisons difficult. The investigations described here are necessary to give the user confidence in the technique before he can start interpreting the data, a task which will itself be complex, especially in the time- and space-varying case of the ship-mounted doppler.

REFERENCES

1. Griffiths, G. (1986). Intercomparison of an Acoustic Doppler Current Profiler with Conventional Instruments and a Tidal Flow Model, Proc. IEEE Third Working Conference on Current Measurement, Airlie, Virginia, USA January 22–24, 1986.
2. Griffiths, G. and Flatt, D. (1987). A Self-contained Acoustic Doppler Current Profiler: Design and Operation. Fifth Int. Conf. on Electronics for Ocean Technology, Edinburgh, 24–26 March 1987. London: IERE, pp. 41–47. IERE Publication No. 72.
3. Thorpe, S. A. (1982). On the clouds of bubbles formed by breaking wind-waves in deep water, and their role in air–sea gas transfer. *Phil. Trans. Roy. Soc. London* **304:** 155–210.
4. Hansen, D. S. (1985). Receiver and analog homodyning effects on incoherent Doppler velocity estimates. *Journal of Atmospheric and Oceanic Technology* **2** No. 4, December (a).
5. Hansen, D. S. (1985). Asymptotic performance of a pulse-to-pulse incoherent Doppler sonar in an oceanic environment. *IEEE Journal of Oceanic Engineering.* **OE-10** No. 2, April 1985 (b).
6. Christensen, J. L. Ametex DCP-4400 Series Current Profilers. September 1982.

Part IV

Geology, Geophysics, Geotechnics

20

Seismic Interference, Filtering Methods and Revised Noise Limits

J. E. Lie, A/S Geoteam, Norway

Interference between simultaneously shooting seismic vessels is a problem in areas with high exploration activity. If seismic interference exceeds the noise limits set to assure data quality during acquisition, the interfering vessels usually commence time sharing. It is a widely held opinion within the industry that limits based on random noise levels may cause unnecessary downtime when uncritically applied to seismic interference (SI). Revised noise limits for SI have been proposed. These have taken into account both the dip of the noise and the amplitude level. These limits have been based upon the assumption that the noise is to be removed by dip-filtering on shot records. This will limit their application. It will be demonstrated how to decompose the interference noise into elements through which the level of contamination of the stack can be judged. On this basis a revised noise limit for SI that is independent of the filtering methods is proposed.

The noise limits were tested by using 'synthetic' data, where pre-recorded seismic interference noise given the desired characteristics were superimposed on a 'noise-free' testline.

THE INTERFERENCE SIGNAL

Seismic interference noise appears on the shot records as a coherent wave train with characteristic rate of repetition and duration. The interference signal also has a stable RMS level and moveout which only slowly varies along the profile.

Interference noise is due to energy travelling as shallow refractions and mode propagation in the water column (Calvert *et al.* 1984). SI noise may carry over long distances. In the extreme case, vessels at more than 100 km distance have been subject to timesharing operations.

The effect of the interference noise on the data quality depends on the following four parameters.

- Relative moveout of the incoming noise from trace to trace over the shot record. This is decided by the interfering vessel's relative position and movements.
- The recurrent frequency of the SI on the recordings. This reflects the shot interval of the interfering vessel.
- The RMS noise level and duration. These depend on the vessel's separation, size of airguns, water

Advances in Underwater Technology, Ocean Science and Offshore Engineering, Volume 16: Oceanology '88

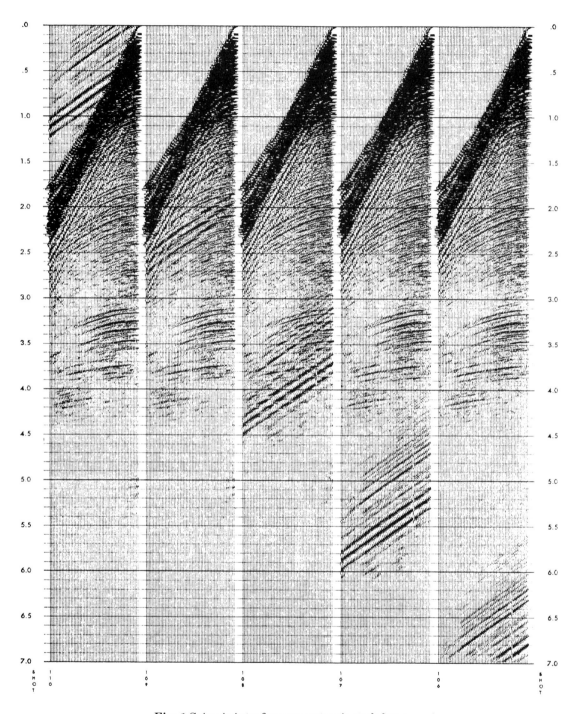

Fig. 1 Seismic interference contaminated shot-records.

depth, seabed geology, seabed topography and sea state.

The effect on the data of these parameters can be studied separately. Their effect can to some extent be quantified, as will be detailed in subsequent setting.

Moveout

On shot records, SI appears as coherent signals with a constant moveout (Fig. 1). On the CDP-gather, the SI is mapped as apparent randomly spaced high amplitude wave trains (Fig. 2). The extent to which SI will be suppressed in the stack is therefore largely

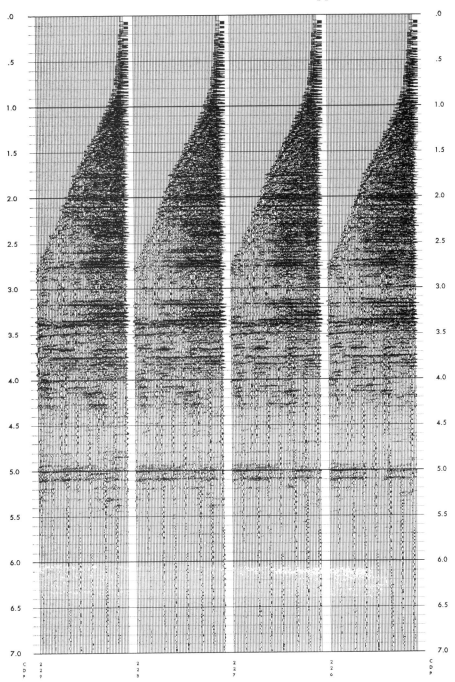

Fig. 2 NMO-corrected CDPs with interference noise.

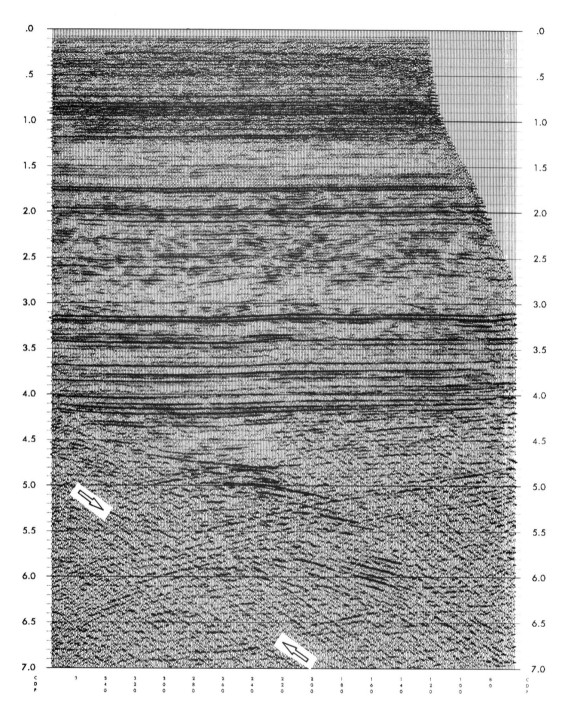

Fig. 3 Stacked section of the testline contaminated with
SI, seen as dipping event on the lower part of section.
25-microbar SI noise level.

independent of its initial moveout. On the stacked section the SI will reappear as stripes with dips depending on the initial moveout plus the stacking velocity function (Fig. 3).

Recurrent frequency and duration

Together the recurrent frequency and the duration of the SI signal will determine the average number of trace samples that will be contaminated across the CDP, that is:

$$n = \frac{2 \times 100 \times D}{RT} \qquad (1)$$

Where n is the number of contaminated sample (%), D = SI signal duration (s), and RT = SI recurrent time (s).

This implies that the level of contamination of the CDP can be estimated in the field. This expression is valid given that the interfering vessels do not shoot with the same cycle or a multiple thereof. The time-shift from record to record of SI should be at least 500 ms to assure that the noise will be evenly distributed on the stacked sections. For example, if the interfering vessel's shot interval is 12 s and the duration of the interference signal is 1.5 s, an average of 25% of the samples across the CDP-gather will be contaminated.

Amplitude level

The RMS amplitude level of SI is of great importance to its effect on the stack. This is difficult to quantify because the signal to SI-noise ration will vary down the record (Fig. 4). Though depending on the number of samples contaminated, SI seems to become visible on the stacked section when its RMS level is comparable to that of the data.

SI REMOVAL STRATEGY

SI noise can be filtered at different stages through the processing sequence. The most obvious might seem to be dip-filtering on the shot records. Such an approach to SI noise removal has a limited application because the vessel's relative positions will decide whether noise and data can be separated in the FK-domain. The relative position of the interfering seismic vessels may also change during the

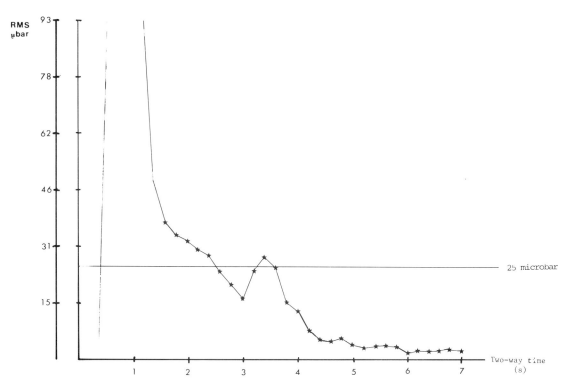

Fig. 4 The RMS amplitude decay in microbars of the recorded data as a function of two way time.

A)   B)

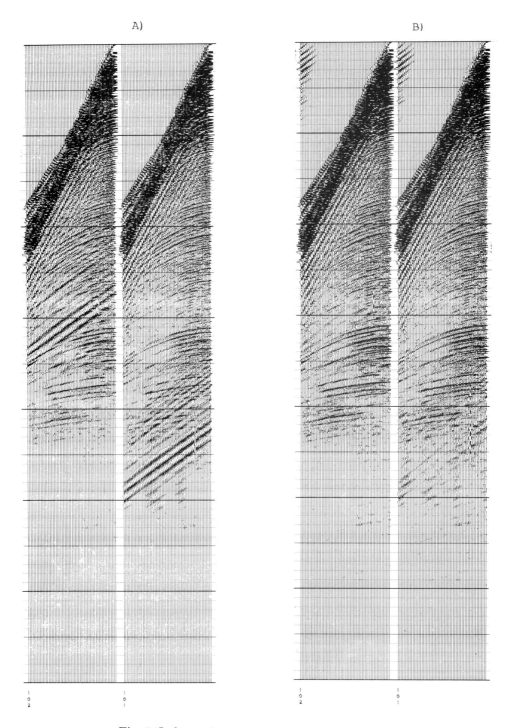

Fig. 5 (*Left*) two SI contaminated shot-records. SI noise
level 25 microbars; (*right*) the same two records after
dip-filtering.

course of a line. This will complicate the filter tuning.

However, when dip-filtering can be applied, it is usually very efficient (Fig. 5). No sign of the noise was seen on the stacked section after filtering.

SI on CDP

On NMO-corrected CDP-gathers SI will be present as apparently random high amplitude wave trains (Fig. 2). The extreme amplitudes can be used to identify SI noise prior to filtering. One approach to filtering such high amplitude noise is to run AGC-scaling on the traces prior to stack. In this way the adverse effect of the high amplitudes are reduced. However, AGC-scaling will also affect the un-contaminated samples in such a way as to reduce the dynamics of the data and thereby deteriorate amplitude information.

A more efficient approach to filtering this type of noise is through the use of some kind of selective stacking. Such methods work by excluding noisy data rather than relying on their adverse effect being averaged out in the stack (Sheriff and Geldart, 1983). Amplitude values that exceed a certain threshold are either excluded from the stack or exposed to forms of inverse weighting. The net effect should be a better SNR on the remaining traces.

Several different selective stacking methods have been developed, but only the most interesting, the trimmed-mean stack, will be considered in detail.

TRIMMED-MEAN STACK (TMS)

This method, also known as alpha trimmed-mean stack (Watt and Bednar, 1983), is conceptionally simple and has been applied to statistical problems for centuries. For each time level, the sample values $ai(t)$ are sorted in ascending order. A certain percentage of the values are then trimmed away at either end of this distribution. The arithmetic mean is calculated from the remaining sample population. For theoretical distributions it can be shown that the trimmed-mean, as a statistical estimator, is very robust against gross error contamination or, in our case, high amplitude noise. In a set of strictly normal distributed observations, the arithmetic mean is the minimum variance estimator, but the arithmetical mean is sensitive to deviations from the normal distribution. The pure median as an estimator is more robust against gross errors, but will perform poorly relative to the arithmetic mean on Gaussian distributions.

The variance of the trimmed-mean as an estimator

can be expressed (Bickel, 1965) as:

$$\sigma_a{}^2 = 2(1-2a)^{-2}\left[\int^{x_{1-\alpha}} t^2 f(t)dt + ax^2{}_{1-\alpha}\right] \quad (2)$$

where $f(t)$ is any symmetrical distributed function, a is the trimming parameter and gives the amount of samples on each flank that is rejected by the trimmed-mean (varies from 0 to 0.5). Symmetrical trimming of a Gaussian distributed dataset will give a very good estimate of the mean even though the fold is reduced. If a total of 20% ($a = 0.1$) of the datapoints are symmetrically removed, the variance will only increase by 5%. Used on a 60-fold dataset a symmetric trimming with $a = 0.1$, will exclude 12 samples but have the same variance as if only 3 samples were excluded from the arithmetic mean. In comparison use of the pure median would give the same variance as if 22 samples were excluded from the arithmetic mean.

It is interesting to study the efficiency of the symmetric trimming operation on general symmetric distributions with heavy tails. This will mimic high amplitude noise contamination. Such a function can be constructed by using two normal distributed datasets $N_1(0,\sigma)$ and $N_2(0,m\sigma)$.

An expression for such a combined distribution is given by Tukey (1960), the so-called gross error model.

$$f(t) = \frac{1-\epsilon}{\sigma}\varphi\left(\frac{t}{\sigma}\right) + \frac{\epsilon}{m\sigma}\varphi\left(\frac{t}{m\sigma}\right) \quad (3)$$

where φ is a normal distributed function, $m\sigma$ is the standard deviation of the broader distribution or gross error (N_2). ϵ gives the amount of gross error contamination. The variance of the arithmetic mean for this kind of distribution is: (Bickel and Docsund, 1977)

$$n\,\text{var}(\bar{x}) = (1-\epsilon)\sigma^2 + \epsilon m^2\sigma^2 \quad (4)$$

(2), (3) and (4) imply that the trimmed mean will be the minimum variance estimator for gross error circumstances relative to the arithmetic mean and the median. This applies on condition that the trimming is tuned to match the level of contamination.

TMS on real data

In a typically bad case of SI, its RMS level would be 25 microbars and 30% of the samples on the CDPs would be contaminated. In such a case, SI-noise would by far exceed the conventional noise limits. Figure 3 showed the testline contaminated with this level of SI conventionally stacked. On the lower half

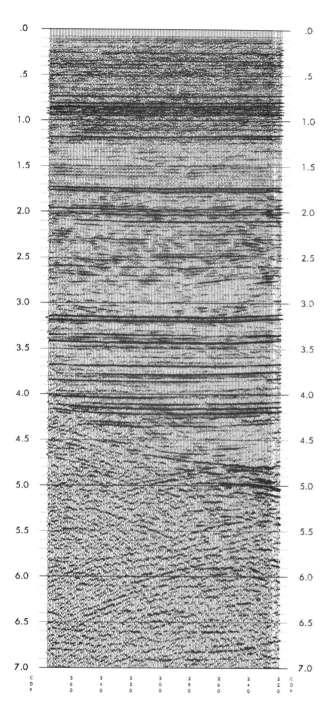

Fig. 6 TMS section of the testline contaminated with SI, RMS-noise level 25 microbar. Trimming parameter, $a = 0.1$.

of the section the interference noise can be identified as dipping stripes. When subjected to TMS the noise can be dramatically reduced or diminished. In Figure 6 the dipping stripes have been efficiently stacked out. This is shown in detail in Figure 7. If the amount of trimming is increased, no further reduction of the noise is observed.

For the trimming parameter set to $a = 0.5$, which gives a pure median stack, the SI noise pattern returns (Fig. 8).

When the contaminated samples are numerous they will influence the position of the median value so much that the noise comes through. With a few and extreme noise samples the relative efficiency of TMS improves.

In Figure 9 the testline is contaminated with an extremely high level of SI (150 microbars) and conventionally stacked. In Figure 10 TMS was applied. The noise on the lowest part of the section has been removed. In the top of the section, where the amplitude level of the data is comparable to the noise, the filtering was less efficient.

The effect of the TMS processor on data without high amplitude noise is also beneficial. When applied with moderate trimming to the non-contaminated testline, increased continuity on weak reflectors was observed. A better result of the TMS may be obtained if the NMO-corrected CDPs are exposed to offset dependent weighting prior to stack. On average the amplitude value from a reflector will decrease with increasing offset (Fig. 11). On 'noise-free' data TMS would have a tendency to trim the nearest and furthest traces. This especially accounts for the shallow parts. Offset dependent weighting would correct this.

Tuning the TMS

When using TMS to suppress SI noise, knowing the number of samples contaminated is a guide for tuning the operator. Trials on the data examples showed that the best result was obtained when the trimming parameter was set to exclude from the stack somewhat fewer trace samples than were contaminated. The contaminated samples will necessarily have values within the range of the distribution of the signals, and therefore cannot be distinguished from these.

Noise limits

Over time the industry has adapted a 3-microbar RMS limit for ambient cable noise measured in the absence of shooting. This limit remains in effect the same, despite data acquisition and processing advancements.

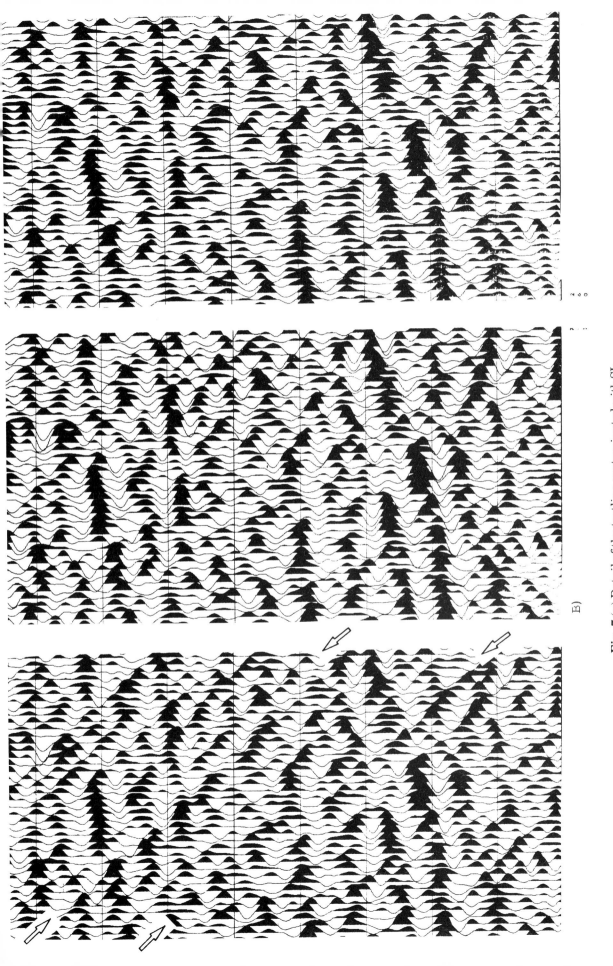

Fig. 7 (*a*) Detail of the testline contaminated with SI; (*b*) the same detail but stacked with TMS; (*c*) the same detail from the non-contaminated testline.

Geophysicists have for years questioned the validity of the 3-microbar limit, arguing that too strict a noise limit will not pay off in terms of better data quality. Hatton (1986) showed that the 3 microbar noise limit is too restrictive when used on weather noise and suggested as an alternative 10 or 15 microbars limits which would not compromise data quality. When the 3-microbar limit was established 25-m group lengths were common and energy sources were of an order of magnitude smaller than today. With the introduction of digital streamers the cables became longer and the group lengths shorter. Shorter hydrophone groups will in general imply fewer hydrophones and thereby less noise suppression. This is because 0.5 m is the

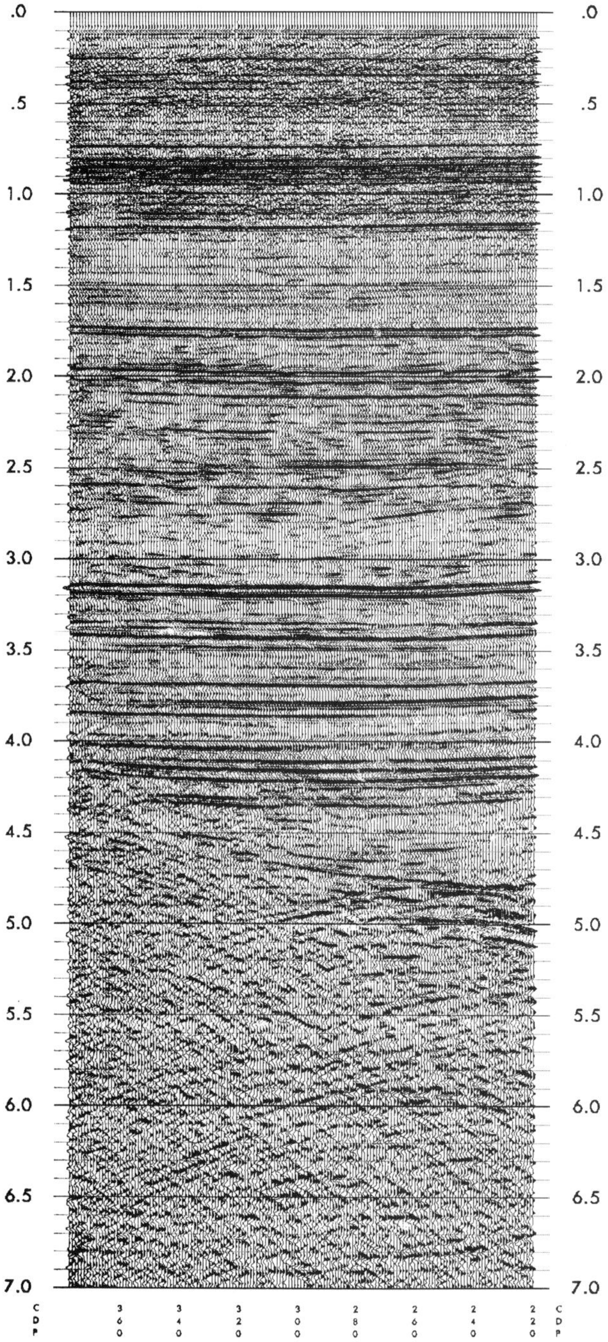

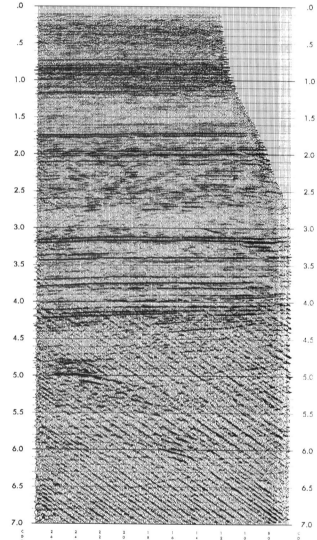

Fig. 8 Median stacked section of SI-contaminated testline. RMS noise level 25 microbars.

Fig. 9 Conventionally stacked section contaminated with SI with an RMS level of 150 microbars.

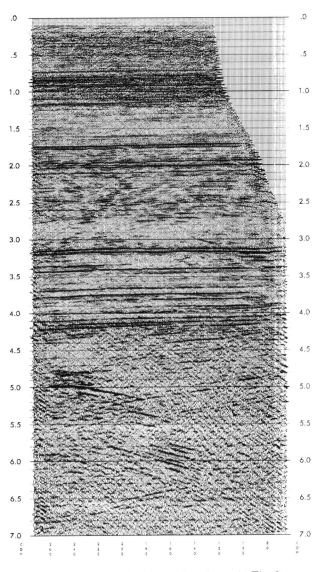

Fig. 10 TMS applied to section shown in Fig. 9.

Noise limit based on recalculation of SI

It is possible to make an estimate of what the SI-noise will be equivalent to as random noise on all traces. This can be done on the basis of the measured RMS amplitude level of SI, its recurrent frequency and duration.

Such an expression is based on two ideal assumptions:

(1) The seismic interference noise is uncorrelated across the NMO-corrected CDP.
(2) For uncorrelated noise:

$$N \, \text{Pre} \, \frac{S}{N_{SI}} = \text{Post} \, \frac{S}{N_{SI}} \tag{5}$$

If n traces across the stack level are contaminated with SI noise, the pre-stack and post-stack SNR will relate as:

$$\text{Pre} \, \frac{S}{N_{SI}} = \frac{\sqrt{n}}{N} \, \text{Post} \, \frac{S}{N_{SI}} \tag{6}$$

Where N is the total number of traces in stack.

If the random noise limit is chosen to be 6 microbars, then (5) and (6) combined will determine an equivalent noise limit for SI.

$$\text{Pre} \, N_{SI} = 6 \sqrt{\frac{N}{n}} \tag{7}$$

This function is plotted in Figure 12 with the RMS amplitude level of SI (Pre N_{SI}) as vertical axis and the ratio between contaminated and uncontaminated samples as horizontal axis.

By plotting the amplitude level of SI and the calculated sample ratio, one can determine if the SI noise is within the ambient noise limit. Recalculating the SI in this manner implies higher tolerance level of SI noise than if only the RMS level of SI was considered. This higher tolerance level does not compromise the traditional random noise limits. Typically interference noise will contaminate between 10 and 30% of the trace samples. If 6 microbar is the ambient noise limit chosen, the RMS level of SI can measure between 19 and 11 microbars and still be within the noise limit. Although the proposed limit is based upon ideal conditions, it appears valid when tested by constructed data examples.

Noise limit based on FK-filtering

Statoil and others have suggested revised noise limits for SI noise, based on FK-filtering on shot. These limits are tolerant towards noise that can be

minimum distance possible to ensure that cable noise is uncorrelated between hydrophones (Karlson, 1985). The same surrounding ambient noise will therefore give higher readings from short group lengths than from longer. This effect should be compensated for when the noise limit is enforced. With the 3-microbar limit at 25 metre group length as reference then comparable noise limits for shorter groups could be:

25 m	12.5 m	6.25 m
3 microbar	4.3 microbar	6 microbar

(A)

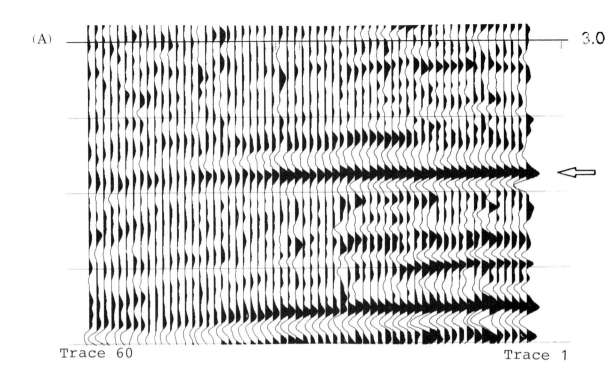

Trace 60   Trace 1

(B)

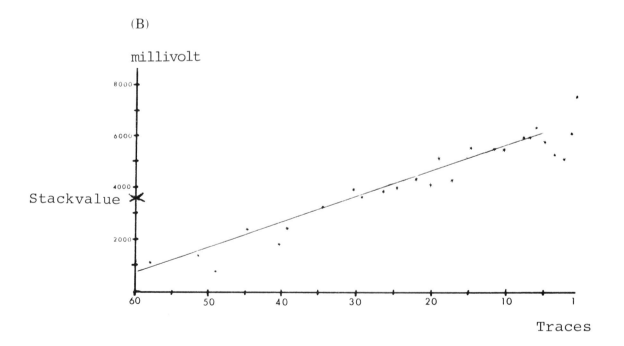

Fig. 11 (*a*) A reflector on a NMO-corrected CDP;
(*b*) the amplitude values in millivolts for every second
trace across the reflector.

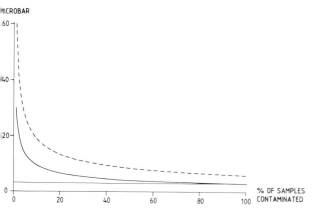

Fig. 12 The curves define suggested RMS noise limits for SI in microbar as a function of the percentage of samples contaminated on the CDP. The dotted line is based upon accepted random noise limit of 6 microbars. The continuous line on a 3 microbar limit.

CONCLUSION

SI noise appears with different characteristics on shot records and NMO-corrected CDPs. These characteristics can efficiently be exploited when filtering the noise. The number of trace samples contaminated by SI on the CDP can be estimated by its recurrent frequency and duration. By this number and the RMS level of SI an estimate can be made of the actual noise level given to the CDP as if there were random noise on all traces. By decomposing SI it can be compared with the random noise limit. SI noise with RMS levels many times higher than the random noise limit may give an actual combination of the stack that is within this limit.

A combination of the dip-related noise limit and recalculation of SI will be the most flexible. If the interference noise has a stable moveout along the profile and the noise can be separated from the primaries in the FK domain, the dip dependent limit will usually be the most tolerant. If the conditions are such that the use of dip-filtering is not desirable, the limit based on recalculation of SI may be used. This limit is independent of the choice of filtering method. Surveillance of the suggested noise limits is simple and only requires measurements that are presently carried out during quality control of the acquisition parameters plus a simple calculation. By using these noise limits, much crew downtime can be saved without compromising data quality.

separated from the primary events in the FK-domain.

The suggested dip related noise limit (Fig. 13) will tolerate interference noise with amplitude levels up to 24 microbar RMS. This seems to be conservative when tested by synthetic data samples.

NOISE FROM AHEAD

APPEARENT VELOCITY	NOISE LIMIT (RMS) MICROBAR	
GROUP LENGTH	25 m	6.25 m
∞ – 2200 m/s	3	6
2200 m/s – 2000 m/s	12	12
2000 m/s –	25	25

NOISE FROM BEHIND

APPEARENT VELOCITY	NOISE LIMIT (RMS) MICROBAR	
GROUP LENGTH	25 m	6.25 m
∞ – 24000 m/s	3	6
24000 m/s – 12000 m/s	12	12
12000 m/s –	25	25

Fig. 13 Noise limits for SI conditioning dip-filtering on shot-records.

REFERENCES

Bickel, P. J. (1965). On some robust estimates of location. *Ann. Math. Statist.* **43:** 847–858.

Bickel, P. J. and Doksum, K. A. (1977) *Mathematical Statistics: Basic Ideas and Selected Topics.* s. 369–404. Holden-Day: San Francisco.

Calvert, R. W., Schoelcher, G. J. H. and Blankespoor, H. J. (1984). Evaluation of a Controlled Seismic Boat Interference Experiment: a Final Report. Shell International Petroleum Maatschappij B.V. The Hague. Report EP-60410 Unpublished Handout.

Hatton, L. (1986). Weather and the 3-microbar limit on the North-West European Continental Shelf. *First Break* **4:** August 1986.

Karlsson, T. (1985). Seismic Examples Related to Streamer Parameters. Talk presented at the EAEG conference in Budapest, 1985.

Sherif, R. E. and Geldart, L. P. (1983) *Exploration Seismology. Volume 1, History, Theory and Acquisition.* Cambridge: Cambridge University Press.

Tukey, J. W. (1960). A Survey of Sampling from Contaminated Distributions. In: *Contributions to Probability and Statistics.* Standford University Press (Olkin, I. *et al.,* ed.) pp. 448–485.

Watt, T. and Bednar, J. B. (1983). Role of Alpha-trimmed Mean in Combining and Analysing Seismic Common-depth-point Gathers. Technical Program Abstracts and Biographies, SEG, Las Vegas 1983.

21

A Sonar Workstation

J. A. van Woerden, L. F. van der Wal, D. Ph. Schmidt and
C. van den Berg, TNO Institute of Applied Physics, Stieltjesweg 1,
PO Box 155, 2600 AD Delft, The Netherlands

In the Netherlands civil and hydraulic engineering have traditionally been important fields of research and development. The TNO-Institute of Applied Physics has been involved in (instrumentational) research and development for more than 25 years. A large number of special measuring systems have been developed ranging from optical, echoacoustical and conductivity sensors to the design of large data-gathering networks for the measurement of hydrological and meteorological parameters in the Dutch coastal waters.

Over the last few years the demand for underwater acoustical instrumentation has evolved from simple transit time gauges to complex imaging and feature extraction systems. The key phrase in this development has been 'full wave analysis'. Instead of simply measuring the elapsed time between the transmit pulse and the first bottom return, the echos from the water column and the sub-bottom are also processed and analysed.

Figure 1 shows a typical received echoacoustical signal. As can be seen, the same signal is used for a broad range of applications.

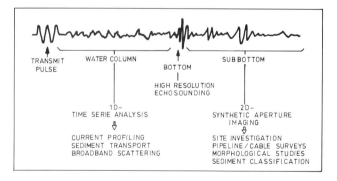

Fig. 1 Echoacoustic 'reference' signal.

In the computer industry another revolution has taken place in the same period of time. Computer systems with integrated signal processing and graphics capabilities, the so-called workstations, have become less expensive, gained in performance, and entered the personal computer market. Array processors are now available as plug-in boards that

Advances in Underwater Technology, Ocean Science and Offshore Engineering, Volume 16: Oceanology '88

can be added to standard Multibus, VME-bus, Q-bus and even PC-bus systems.

In this chapter a system architecture will be described in which state of the art developments of the computer industry are used as components to create an interactive, research environment with integrated data acquisition, processing and presentation facilities. The system to be described is called the sonar workstation. First the individual components within the sonar workstation will be described; next a few examples will be given of current applications.

GENERAL DESCRIPTION

The Sonar Workstation consists of three main parts: a transducer unit, a data acquisition and processing unit, and a host computer system. The data acquisition and processing unit is fully general purpose. It is configured on an open computer bus structure to offer full flexibility with commercially available hardware. Nevertheless special hardware modules had to be developed to create a useful instrument for different research purposes. The entire unit is housed in a reinforced cabinet which is easily handled.

The host computer is an IBM-PC/AT computer with a high resolution graphics display that acts as a user-interface for the workstation and a laser-printer output for quality presentation. A detailed block diagram of the sonar workstation is shown in Figure 2.

THE TRANSDUCER UNIT

The transducer unit consists of a cylindrical container with a removable transducer flange. The cylindrical container houses the preamplifiers, the matching networks, and switching electronics for up to four individual transducers. The preamplifiers typically yield 35 dB fixed linear gain over a frequency bandwidth which ranges from 100 kHz up to 10 MHz.

Two different transducer flanges have been constructed: one with only one 200 kHz transducer looking vertically downwards and one with four 1 MHz transducers in a Janus configuration. The four-transducer flange is used for current profiling. The transducer unit is connected to the sonar workstation with a two-wire cable.

DATA ACQUISITION AND PROCESSING UNIT

The main functions of this part of the Sonar Workstation are: signal generation, signal conditioning, and signal processing. All functions are fully software controllable. The user is free to set the system parameters as long as he obeys the overall bandwidth (100 kHz to 5 MHz) and dynamic range (80 dB). Figure 2 shows the components needed for the above mentioned functions. These are described in more detail below.

Programmable waveform generator

The Sonar Workstation contains two programmable waveform generators (PWGs): one is used for the

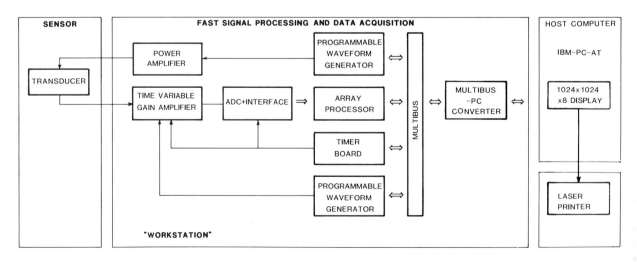

Fig. 2 Block diagram of the sonar workstation.

generation of transmission signals, and the other generates a user-defined control curve for the time gain amplifier. All transmission signals are calculated in a digital format with the desired amplitude and phase characteristics.

The transmission signal is then stored in the PWG data memory (4 kByte RAM; 10 bits), and using the clock frequency as a parameter the stored digital data can now be D/A-converted within any desired frequency range. The PWG allows for clock frequencies up to 20 MHz and is configured on a single multibus board.

Time gain amplifier

The sonar workstation contains a four-channel time gain amplifier (TGA). The TGA offers 80 dB of relative linear gain, ranging from -10 dB to $+70$ dB, over a frequency range of 100 kHz–5 MHz. Within this frequency band the signal to noise ratio is better than 80 dB while the common mode rejection equals ~ 70 dB. The operator can either apply a fixed or a time variant amplification to the received echoes. Both control voltages are supplied by the second programmable waveform generator.

Each TGA channel can be connected to an individual transducer–preamplifier combination. This allows for data acquisition over four channels in parallel. In addition two TGA-channels may be connected to the same transducer–preamplifier combination, for example, to study the effect of fixed v. time variant amplification. Finally the TGA allows for signal demodulation as well. Hence each channel contains two anti-aliasing filters, one for the original HF data and one for the demodulated HF data. Each TGA-channel is connected to an 8 bit A/D-convertor with a maximum sampling rate of 10 MHz.

Array processor

The implementation of a board-level array processor (Mercury ZIP 3232) forms a special feature of the sonar workstation. With the help of a special I/O-port the array processor facilitates a high data input rate of digitized samples (10 MHz for single channel acquisition). The digitized data are stored in 2 MByte of main data memory, which forms an integral part of the array processor. During data acquisition a limited number of preprocessing functions (e.g. digital filtering, envelope detection) can be performed by the array processor on line.

The I/O-port is driven by a specially designed digital interface, which connects the output of the four 8 bit A/D-convertors to the 32 bit wide I/O-port of the array processor.

Trigger and clock signals are supplied to two PWGs which act as signal sources. Timing signals, with respect to blanking and sample windows, are supplied to the A/D-convertors and the digital interface to the array processor. The timer board also generates the digital sample frequency and (if desired) a signal demodulation clock.

All timing functions are fully software controlled, and data acquisition parameters can be selected over a wide range of values. These design concepts of the timer board largely yield the desired system flexibility.

THE HOST COMPUTER SYSTEM

The host computer is a standard IBM-PC-AT with extensions consisting of a large winchester disk (120 MByte) and a high resolution graphics device ($1000 \times 1000 \times 8$) for gray level pixel displays, combined with a laser hardcopy unit for gray level pixel presentation. The system is run by the standard MS-DOS 3.3 operating system.

Application software makes the system highly interactive. First a user friendly menu-driven software package allows the user to configure the system with data acquisition, signal processing and display/presentation parts. Parameters for these tasks are easily inputted via the same user interface.

The software consists of a structure based on technical command language (TCL) a command decoder developed by TNO. TCL can be regarded as a 'software bus'.

Different components (software packages) are connected as system libraries to this bus and can be used by application programmes. A signal processing library (TCL/FPS), a file structure for generalized data acquisition (as time series data) and display features as TGS or GKS are available. Interactive processing on an instruction-for-instruction basis is possible to evaluate algorithms. Real time performance is reached by the introduction of macros and chaining of command sequences of evaluated algorithms.

The structure is illustrated in Figure 3.

PRACTICAL APPLICATIONS

Development of acoustic profiling instruments

Because acoustic doppler current depends on backscattering of sound from plankton, small particles and small scale inhomogeneties in the water, facilities to evaluate the backscattering strength as

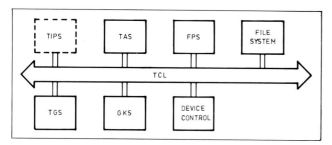

TCL	Technical command language
TCL/FILE SYSTEM	Data sequence organization
TCL/FPS	Signal processing library
TCL/TGS	Graphic functions
TIPS	Image processing library
TCL/Device control	Special functions to configure a data-acquisition system
TCL/TAS	Time sequence analysis library
TCL/GKS	Graphic primitives

Fig. 3 The software bus.

a function of frequency are very important. These are therefore available in the workstation.

Within the field of research for acoustic current profiling systems a conventional noncoherent doppler system with the following desired specifications is evaluated. The momentary doppler frequency is estimated via the first-order moment of the power density spectrum.

The desired specifications were as follows:

Current velocity (m s^{-1})
Range	0.0–5
Accuracy	0.04

Current direction
Range	0–360°
Accuracy	5°

Depth (m)
Range 1	2–10
Range 2	2–25
Range 3	2–50
Bins	25

Transducers
Number	4
Frequency	1 MHz
Configuration	Janus

The aim of this research is to evaluate system specifications for a so-called target system: a dedicated miniaturized low-cost current-profiling instrument for special applications in special areas.

Sub-bottom profiling evaluation based on seismic exploration schemes (synthetic aperture method)

The sonar workstation has recently been used for sub-bottom profiling based on seismic exploration methods. Due to the increased accuracy in position measurement during surveys (e.g. differential GPS), synthetic aperture processing techniques may become highly feasible in the near future. Extended features of the sonar workstation already include seismic migration in the wavenumber–frequency domain.

Trials were carried out to test the performance in cable and pipe searching applications as well as in morphological studies. Both commercially available subsurface radar and acoustic sub-bottom profiling equipment have been used for data-acquisition.

CONCLUSION

The system architecture described in this paper consists of an application specific sensor part, an 'intelligent front end' and a standard host computer system. It is optimized for interactive operation in programmes for high frequency, high resolution sonar research in the 100 kHz to 5 MHz area.

ACKNOWLEDGEMENT

The development of the sonar workstation is initiated and sponsored by the Hydro Instrumentation Department of the Dutch Rijkswaterstaat.

REFERENCES

1. Schmidt, D. Ph., Vogel, J. A. and van Woerden, J. A. (1983). A High Frequency Multichannel Subbottom Profiler. Proc. Ultrasonics Int. '83.
2. Van der Wal, L. E., Schmidt, D. Ph. and Berkhout, A. J. (1982). Acoustic Determination of Subbottom Density Profiles Using a Parametric Sound Source. Proc. Acoustical Imaging Vol 12, pp. 721–732, July 1982.
3. Schmidt, D. Ph., Vogel, J. A. and Verhulst, K. (1984). Development of ROV Operated Sonars for very high Resolution Imaging during Underwater Inspections. Proc. ROV '84, pp. 287–293, May 1984.
4. van Munster, R. J. and van Antwerpen, G. (ND). TCL (Technical Command Language) User Reference Guide. TNO Institute of Applied Physics.
5. Van der Wal, L. F. and van den Berg, C. (1986). Design and Construction of a Sonar Workstation. Proc. Oceans '86, pp. 472–475, September 1986.
6. Van der Wal, L. F. (1986). A Wave Theoretical Approach to Acoustic Current Profiling. Proc. of the IEEE Third Working Conf. on Current Measurement, pp. 203–212, January 1986.

22

Improving the Seismic Picture: the Latest Technique

R. L. French and D. C. N. Pratt, Racal Survey Limited, South Denes, Great Yarmouth, Norfolk, NR30 3NU, UK

It has become common practice to undertake detailed engineering geophysical investigations of areas around proposed offshore drilling locations with the intention of assessing potential construction, anchoring or drilling hazards. In this type of survey, sub-bottom profiling is regularly carried out, utilizing both single channel analogue and digitally recorded multichannel, reflection seismic acquisition techniques. The resolution and penetration of the typical systems used can be tabulated as in Table I.

It has become apparent that the quality of data acquired with surface towed sparkers is inadequate to provide information between the good quality data obtained from deep-tow sparkers and from digital seismic recordings. Typically, surface towed sparker data is totally unusable below the first multiple, where it is degraded by noise from other multiple events which follow. The data are further degraded by the variable nature of the sparker signature and by the poor attenuation of swell noise with the single fold technique. Racal Survey Limited has, over the past few years, attempted to improve the data quality by using alternative sources, for example small air-guns and waterguns. This has gone some way to

TABLE I
Resolution and penetration of typical systems

System	Frequency range	Typical penetration
Pinger	10–100 kHz	A few metres
Deep towed sparker	1000–7000 Hz	A few tens of metres
Deep-towed boomer	1000–10000 Hz	A few tens of metres
Surface towed sparker	200–1200 Hz	About 100 metres
Digital seismic system	10–250 Hz	1000 metres plus

improving the data quality but has still not overcome the noise and multiple problems.

Digitally recorded seismic data is usually band limited by the sampling rate used in recording and by the source spectrum. Data in the first hundred milliseconds sub sea bed are often degraded by source-generated noise and surface reverberation.

Racal Survey Limited have set out to develop an acquisition technique which will provide the highest possible resolution and signal to noise ratio in the first few hundred metres below the sea bed. The first attempts at solving the problem involved the use of

Advances in Underwater Technology, Ocean Science and Offshore Engineering, Volume 16: Oceanology '88

a single 15 cubic inch watergun as the source with a 16 trace, 6.25 m group streamer. The data was recorded at the very high sampling rate of 0.25 ms using TT Survey's recently developed TTS 96 system.

This chapter presents the results and compares them with conventional multi-channel site survey data and with conventionally recorded, single fold watergun data.

GEOLOGICAL HAZARDS

The geological features that may cause problems or hazards during structure emplacement or drilling operations may be subdivided as follows:

Gas hazards Pockets of shallow gas which are usually detected by the presence of amplitude, phase and velocity anomalies in the seismic records.

Structural hazards Faults, buried channels, slumps etc. which may cause differential settlement under a structure or deviate a well from its planned route.

Lithological hazards Gravel sheets, boulder beds etc. which may prevent penetration of legs or piles and which may also cause well deviation or loss of circulation of the drilling fluid.

EQUIPMENT REQUIREMENTS

In choosing sensors to resolve these features, the site survey contractor should take into account the work limitations.

(1) Features which are resolvable at frequencies lower than around 80 Hz will have already been noted on conventional seismic data, and so the system need not be too rich in low frequencies.

(2) To identify gas hazards by discovery of amplitude anomalies, the system should be capable of recording data in a form such that relative amplitudes can be recovered.

(3) To identify phase anomalies then the source needs a stable, clean pulse.

(4) To map subtle geological features in the first few hundred metres in commonly encountered conditions (in the North Sea), the source must be powerful (several bar metres) and with a broad bandwidth, rich in high frequencies.

(5) The ability to detect velocity anomalies will be improved by a multi-offset recording technique.

Technology to map the first 1500 metres or so below sea bed is now commonly available. Most contractors are able to offer a digital seismic spread which offers 'high resolution data' but perhaps a closer examina-

tion of some of the selected parameters should be made.

Seismic data is bandlimited by sampling at 1 ms, for historical reasons. This is the fastest sampling rate that could be supported on the 'hand me down' recording systems which were inherited from the exploration seismic world. This limits the highest frequencies to 250 Hz, or on the better systems to 360 Hz.

Care is often taken to use point sources but at the risk of spatial aliasing of the shallow, high dipping features that are located by the use of 25 m long receiver groups.

There is a common conception that the sparker sources regularly used contain high frequencies, but the multi-legged signature which may mask events underlying a strong reflector or create artificial anomalies by giving constructive interference between different legs from different reflectors in a thinly bedded sequence is very often ignored. Rarely is the shot-by-shot signature deconvolution which is necessary to collapse the sparker signature to an acceptable pulse undertaken.

These problems have not gone unaddressed.

The mini-sleeve exploder provides an acceptable pulse when working correctly but it requires continuous maintenance (recovery every 12 hours), and a certain amount of enthusiasm is needed to share a relatively small boat with large volumes of liquid petroleum gas and liquid oxygen. 'Clean' gas supply may also be a problem in remote locations.

It is possible to make a DFS V (and other field systems of that generation) sample at sampling rates of less than 2 ms, but this is done by electronic trickery which takes the system beyond its design specification and takes many hours of conversion time and careful testing.

DATA PROCESSING

The problems with fast sampled seismic data do not end in the field. Most seismic data processing centres have some problems with sample rates of less than 1 ms and usually overcome these by tricking their software: playing with sample rates and record lengths to allow the software to function with what it thinks is a longer record length at slower sampling rate. This can for example cause preset anti-alias filters to have their frequencies doubled.

Problems are also encountered on all but the best laser plotters when attempting to plot densely spatially sampled data at normal horizontal display scales.

THE SOLUTION

In an attempt to provide better geophysical data in the area where it appears to be most needed (the first few hundred metres below sea bed) Racal Survey Limited offer the following solution.

Source

Early attempts to find a better source led to the use of small chamber (1–10 cubic inch) airguns. Although these produced a more powerful and more repeatable pulse than sparkers, the energy spectrum was weighted towards low frequencies and was sadly lacking in energy above 300 Hz. Because of the resonant nature of the signature (Fig. 1) either arrays had to be formed, which led to field complications, or wave shape kits had to be used, which dramatically lowered the power of the gun.

Finally airguns were abandoned in favour of the 15 cubic inch watergun, which generates a relatively flat frequency response between 0 and 1000 Hz (Fig. 2). Operationally the gun has proved reliable and easy to use and handle and can be fired as fast as once per second.

Streamer

Earlier work carried out by the authors had shown that the high frequencies within the seismic signal are rapidly attenuated with offset (Figs 3(a,b)). High frequency energy is further degraded on stacking by residual moveout caused by inexact velocity correction; these errors increasing in size with offset.

As the main aim is improved resolution, a maximum offset limit of 100 m has been set and the basic group length chosen at 6.25 m with the intention of retaining a good degree of in-line noise cancellation. A further development will be to acquire comparable data with a 32 channel, 3.125 inch group system to evaluate whether resolution is compromised by the 6.25 m group length.

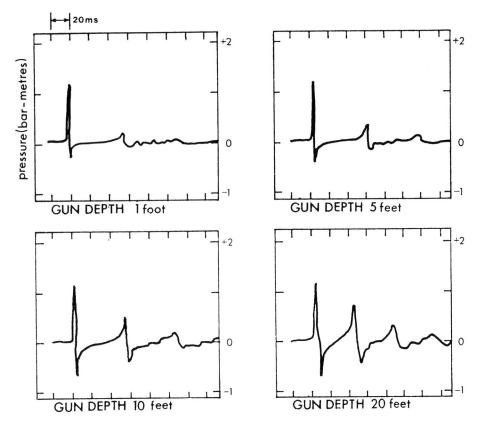

Fig. 1 Signature of a single 20 cubic inch airgun at various depths. Note the resonant nature of the signature and its change with depth. All shots fired at 2000 psi with hydrophone 45 feet below the gun.

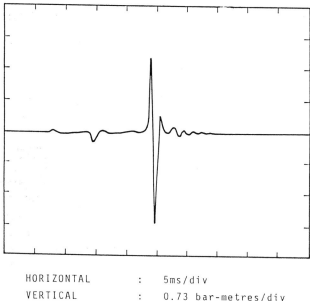

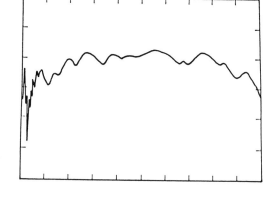

HORIZONTAL	:	5ms/div
VERTICAL	:	0.73 bar-metres/div
FILTERED	:	0-1000Hz

| HORIZONTAL | : | 100Hz/div |
| VERTICAL | : | 10dB/div |

Fig. 2 Signature of the 15 cubic inch watergun.

Recording system

A number of new systems are available which are specifically designed for high resolution seismic work and which can provide fast digital sampling and recording of a number of seismic receiver channels.

The system chosen is the TTS 96. This advanced data acquisition system is microcomputer controlled, allowing recording parameters such as filters, preamp gains and sample rate configuration to be quickly changed merely by loading new software into the processor. The processor is also able to execute extensive quality control functions and, with further development, to undertake various degrees of data processing in the field.

The system is robust and compact, enabling it to be rapidly deployed to hulls of convenience, which is useful when undertaking geophysical surveys in the more remote parts of the world.

COMPARISON DATA

The comparisons presented are all from the North Sea and recorded in the 1987 operating season. Examples from the new technique are compared with data gathered using a 48 channel, 1 ms sampled recording system and high resolution airgun array and data gathered by the NSRF deep-tow sparker

system. This latter combination of tools is considered representative of the systems normally used in undertaking site surveys.

Comparison 1: *Does the short offset multichannel technique provide any improvement over single channel, analogue recording of the same source?*

Figure 4 shows a small channel feature in the North Sea. The dataset on Figure 4(a) is recorded from a single hydrophone group and has undergone band pass filtering, swell filtering, and time varying gain. These parameters were chosen by an experienced operator to optimize the data quality in 'normal' North Sea conditions—a two metres swell. Figure 4(b) shows data recorded simultaneously, sharing the same shots from the same watergun source. The data have then been stacked using a single velocity function and undergone time variant filtering; finally they have been displayed with automatic gain control.

The difference is remarkable. The multi-channel process has surpassed the high noise levels present at time of recording and more refined gain control for final presentation has allowed the seismic characteristics of individual stratigraphic units to be

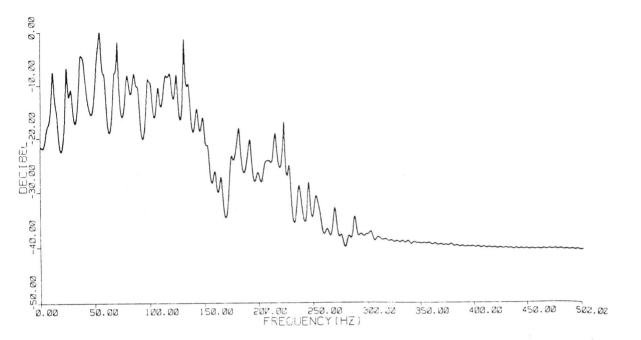

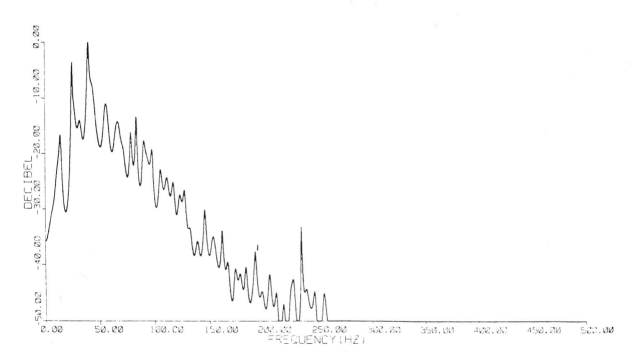

Fig. 3 (*a*) Amplitude spectrum of data from a single 12.5 m
streamer channel at 75 m offset from the source; (*b*)
amplitude spectrum from the same shot, from an identical
streamer channel at 600 m offset from the source.

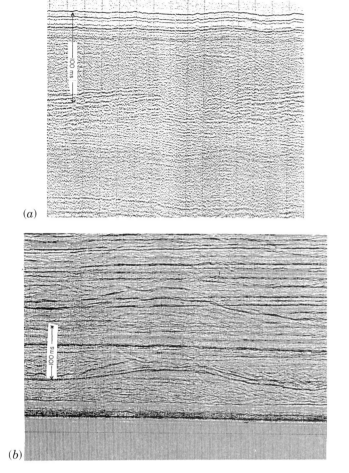

(a)

(b)

Fig. 4 (a) Single channel analogue data over a channel feature in rough sea conditions; (b) simultaneously recorded data using the new technique.

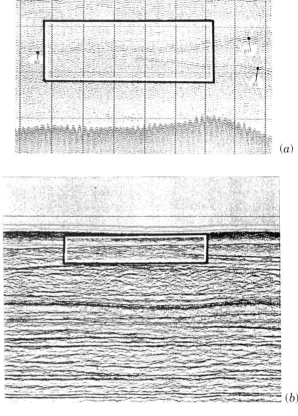

(a)

(b)

Fig. 5 (a) Deep tow sparker profile of a 'punchout' feature; (b) the same feature using the new technique.

retained. An added bonus is that it is possible to choose both the horizontal and vertical scales of display to suit the interpreter rather than being constrained by a limited number of settings on the field recorders.

Comparison 2: *Does the system give better very shallow data than correctly used analogue systems?*

Figure 5(a) shows a 'pinch-out' feature examined with the deep-tow sparker. Figure 5(b) shows the same feature using the new technique. Both datasets have their merits. The deep-tow sparker certainly provides superior resolution of the feature and shows laminations in the overlying sediments which are not well resolved with the new technique. However, the deep-

tow sparker data do not maintain the seismic character of the unit as well, owing to energy loss and less sophisticated processing techniques.

Figure 6(a) shows a deep-tow sparker profile through a channelled sequence. Although the deep-tow sparker maps the structure within the channel very well, it has insufficient energy to resolve the base of the feature and the underlying material. Conventional 1 ms digital data (Figure 6(b)) do little to help, as the features are not well resolved, although they suggest that the channelling may go some way deeper than the 30 ms below the sea bed observed on the sparker. The data gathered with the new technique (Figure 6(c)) provide much greater detail and allow the base of the channel feature to be accurately picked. Both digital data sets also highlight a

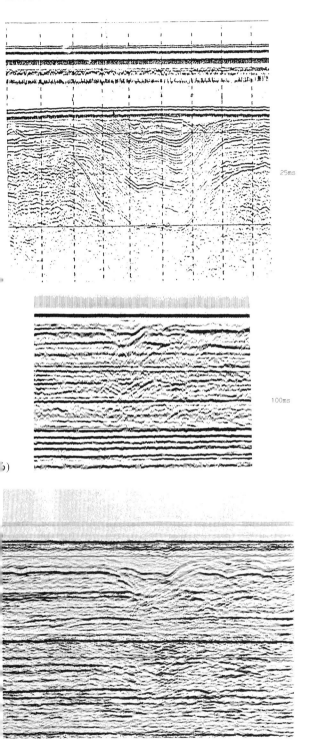

higher, inter-channel reflector, which is the feature mapped from the deep-towed sparker data.

Comparisons 3: *Does the new technique offer any advantages deeper in the seismic section?*

Figure 7(a) shows data from the North Sea shot with a conventional 1 ms sampled spread. The highlighted reflector is a sand horizon known to be gas charged. Looking in detail at the 1 ms data (Figure 7(b)), it is hard to define any structure in the reflector which appears as a strong doublet. The 0.25 ms sampled data show much more detail of the reflector and suggest that the reflector may be a set of discrete sandy hummocks rather than the continuous unit suggested by the 1 ms data.

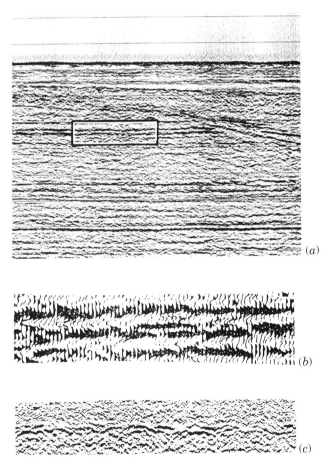

Fig. 6 (a) Deep tow sparker data over a North Sea channel feature; (b) conventional 1 ms sampled data over the same feature; (c) data gathered over the channel feature using the new technique.

Fig. 7 (a) Conventional 1 ms sampled data over a known gas-bearing horizon; (b) detail of gas-bearing horizon using 1 ms sampling technique; (c) detail of same horizon with ¼ sampling technique.

CONCLUSIONS

Field trials with the new system have shown that a gap existed in the spread of geophysical tools available to the market to provide high quality engineering data. The system has shown that improved data can be provided in the 25–500 m sub-sea bed range.

Data gathered using the system exhibit improved penetration over currently used deep towed analogue techniques, improved signal to noise enhancement and weather tolerance compared to higher powered single channel analogue sensors, and improved resolution over existing 1 ms sampled multi-channel techniques.

Field experience has shown that the source and streamer are reliable and easily deployed. The system components are light and robust and could easily be packaged to be transported to support survey work in remote areas.

ACKNOWLEDGEMENTS

The authors wish to thank Racal Survey Limited for allowing this material to be presented; the crew of the *Bon Espoir* for acquiring it, T T Surveys Limited and Ensign Geophysics Limited for their technical assistance; and BP Petroleum Development Limited, McClellands Engineering Limited, Norsk Hydro and Statoil for their cooperation.

23

Offshore Drilling and Safety: the Role of Digital High Resolution Site Surveys

John C. R. Arthur, J. Arthur and Associates, Richmond, Surrey, UK

Offshore drilling for oil and gas was first developed in the Gulf of Mexico as an extension of the highly successful exploration which has taken place onshore in the southern states of the USA. The first wells in the late 1950s were in shallow water from simple platforms in an environment where weather and sea conditions, if not always clement, were predictable and manageable from the operational point of view.

Thirty years later, exploration offshore has extended into provinces where, despite advances in meterological forcasting techniques, weather conditions are no longer so well defined. Correspondingly, with the move into deeper waters, drilling methods have become more sophisticated, hole depths have increased, and platforms for field developments have shown themselves to be marvels of engineering technology—mini-cities contained within a remarkably small sea bed 'foot print'.

With the increasing exposure of men and equipment to hostile environmental conditions have come advances to protect those men and structures from 'hazards' existing beneath the seafloor. The encounter of the unexpected geological condition—either by the initial exploration drill bit probing a completely unknown 'geomechanical' environment or by the placing on the seafloor of an immense additional load and the subsequent development of the field by development wells—is now acknowledged to deserve a degree of a study hitherto unknown.

Encouragement for this, for those companies less enlightened with regard to 'prevention rather than cure', has been the shallow gas blow-out. Shallow gas accumulations have caused conditions hazardous to oil and gas operations right from the start of offshore drilling operations. The violent nature of these encounters has caused several fatalities, many injuries, pollution events, and drilling production equipment losses amounting to hundreds of millions of pounds. As the search for oil and gas moves further into undeveloped areas, encounters with shallow gas are expected to become increasingly numerous, either because of the greater accumulation of shallow gas in the more recent sediments in basin centres or because of

Advances in Underwater Technology, Ocean Science and Offshore Engineering, Volume 16: Oceanology '88

©Society for Underwater Technology (Graham & Trotman, 1988)

lower levels of knowledge of regional shallow geology.

This is undoubtedly highlighted by a comparison of drilling statistics from the Gulf of Mexico and the Norwegian Continental Shelf. Between 1979 and 1982, 4449 wells were drilled in the Outer Continental Shelf USA lease area of the Gulf of Mexico. Sixteen blowouts occurred, of which eight could be attributed to shallow gas occurrences (ref. 1). During the earlier period of 1971–1978, 7553 new wells were started and one blowout occurred for every 250 wells drilled. Of these, 30 blowouts were during drilling of either exploratory or development wells (ref. 2), and 19 of these were directly attributed to shallow gas.

Only 558 wildcat and appraisal wells have been drilled offshore in Norwegian waters since 1966. Of these, 150 have encountered shallow gas, and seven have blown out (ref. 3). Statistics for other parts of the North European Continental Shelf were not available to the author at the time of compilation of this chapter.

Concern in the Gulf of Mexico was sufficient in the early 1970s for a study to be initiated of the shallow sub-sea bed geology at ten such blowout locations (ref. 4). Drilled prior to the inauguration of multichannel digitally recorded and processed high resolution seismic survey methods, these wells had relied on 'seat of the pants' drilling and come unstuck! The research showed, albeit post occurrence, that the majority of these blowouts could have been avoided if current technology had been available.

From these early beginnings by isolated contractors (ref. 5) and researchers (ref. 6) has sprung a whole industry of site survey contractors with a variety of seismic sources, streamers, recording equipment, and experience. At its height, before the last recession, there were ten contractors based in the United Kingdom alone, offering up to 18 vessels. A healthy industry, it would have seemed, and yet in many areas it was still wide of the ultimate goal—a totally trouble-free spud and tophole drilling programme.

Why was this? Equipment required for the acquisition of geological data is varied, and proper selection is highly dependent on the geological character of a site, stratigraphic resolution and acoustic energy penetration requirements, and the detail and quality of the interpretation required. However, under the usual commercial pressures, contractors cannot offer an infinite variety of tools. In fact, there are considerable operational and financial reasons why it is preferable for a con-

tractor to offer a 'standard' operating package as a dedicated spread of equipment on a proven, stable reliable vessel. Not all packages are the same. Most have the same *basic* capability. How much care therefore is taken in selecting a spread appropriate to the task in hand? What is the task? What are the problems? These are the questions that this chapter sets out to answer.

THE TASK

A safe and efficient well-drilling operation from the time of arrival on site until the spud cans are lifted or the anchors recovered is the objective. Most of this operation is outside the scope of this chapter; for the majority of time in drilling a well is in drilling from beneath the 20 inch casing target depth (TD). That initial period of setting up and initial drilling can cause some of the worst problems, as evidenced above in the emotive area of blowouts. However, this is not the full picture. Site surveys are conducted by enlightened operators to ensure that all detectable geological features that could cause hold up to the spudding or drilling operation are detected not only before the drilling work on site but before formulation of the drill plan and finalization of the surface drilling location. Where potential problems are identified, modifications to the plan or position can be considered and acted upon if considered necessary.

Here lies the rub.

Geophysicists are engaged to carry out the survey (with the help of electronics engineers and surveyors), geologists are (sometimes) utilized to evaluate the data and present the results, but does anybody bother to ask the driller what his problems are? Will he tell you anyway? Does he know which ones can be solved? With regard to tophole drilling, prediction ahead of the bit remains a grey area which needs to be clearly addressed. The responsibility for this lies in several areas. For instance in a recent well documented blow out, several aspects of the acquisition (a contractor responsibility?) and programme timing (an operator responsibility?) fell short of the ideal.

Clearly this does not have to be the outcome of a site survey and thankfully is generally not. How many near misses does the industry have? Owing to a somewhat natural reticence to speak about these things, we are unlikely to learn. This is at best a pity, at worst unforgiveable in terms of learning from past experience.

Steps are now being taken to rectify this situation. In Norway, the Norwegian Petroleum Directorate has proposed that because the shallow part of the well is without any commercial interest, information on the tophole and occurrence of shallow gas should be released as soon as possible. Linked to an understanding of the regional shallow geology this may enable better predictive methods to be developed. However, whilst regional studies have been continuing in certain shelf regions around the world for some years, notably the UK (ref. 7), Canada (ref. 8) and USA (ref. 9), Norway has recently abandoned (ref. 10) its regional shallow seismic programme (ref. 11) into which this new source of knowledge could have been fitted. However, the reverse problem exists in the UK, where shallow hole data are sparse in detail and only available on open file after the statutory period for well data release, although the regional shallow geology programme is virtually complete albeit not yet fully published.

THE PROBLEMS (Geological)

It is important to appreciate that not all geologically interesting sea bed and sub-sea bed features form what may be called 'hazards'. In fact the very use of the word is misleading. A geological hazard is normally understood by the public, as well as many agencies, to be an 'active' geological phenomenon that can be or is dangerous to offshore activities. The site investigation industry usually apply the term to any geological phenomenon which may pose an engineering constraint (ref. 12). Most geological hazards exist as a potential rather than an actual and continuous threat and consequently do not always lend themselves to direct measurement or short-term observation. As geologists typically describe geological features in a qualitative way, this is frequently the initial approach to interpretation of high resolution site surveys (ref. 13). The perception of circumstances which may be regarded as pertinent to the safety of offshore drilling is an even more subjective matter and requires an understanding of drilling processes and the requirements for a stable hole.

Briefly these may be reviewed as follows, bearing in mind the subject of this chapter—the special role that may be played by digital studies. All seismic studies, whether presented in realtime on hard copy as single channel analogue paper profiles or recorded on magnetic tape from a multi-channel streamer as 'digital' samples for subsequent processing in a computer before display, rely on the generation of an acoustic wave front and the return of energy from discontinuities in the sub-surface.

The difference lies in the ability of the latter to be analysed mathematically and re-formed to provide a less ambiguous geological display. In the course of this manipulation and subsequently with the 'final' product, certain additional pieces of information become available which, if carefully utilized, can provide clues to the nature of the lithologies (and their properties) through which the seismic wave front has passed. Thus there needs to be an intimate relationship between the geological interpreter and the processing analyst if the full potential of the required data is to be realized. This fact is frequently overlooked by acquisition contractors undertaking interpretation at some geographical distance from the processing house employed to 'manage' the data. More seriously perhaps, the lack of input on the likely regional geological picture from a geologist can lead to incorrect parameters being selected by the analyst, thus giving a poor display into which geophysical artefacts may even have been introduced.

The initial encounter of drilling hardware, be it spud cans at the feet of a jack up or a temporary guide base, is with the sea bed itself. Usually described in some detail from high frequency analogue profiling tools such as echo sounder, side-scan sonar and sub-bottom profilers, there is not a great deal more that digital coverage can provide directly. However, whilst surface materials and shallow structure can usually be adequately described with the assistance of 'ground truth' from sea bed samples, the context in which these sediments occur may well be obscured by highly reflective bottom materials which return most of the incident energy to the surface. All seismic profiling relies on a certain proportion of energy being transmitted through upper layers and a proportion being successively returned at deeper interfaces. Whilst the return of most energy at sea floor can in itself be diagnostic of 'hard' bottom conditions, the possibility of subsequent 'punch-through' by jack-up rig spud cans is a very real danger. Hence the necessity of identifying the local sedimentological environment, first by reference to local setting and secondly by adequate geological evolution from unambiguous high resolution sub-surface data.

These considerations are also clearly of importance in the evaluation of platform locations (ref. 14). With the advent of digital sampling techniques some of the ambiguities of analogue-only techniques may be removed. For enhancement of data acquisition in the first 100 m of sediment below sea bed high digital sampling rates are required if any detail is to be achieved. Quarter millisecond sampling in prime weather conditions provides a practical signal recovery capability of at least 800 Hz which, with an appropriate source, gives a potential bed resolution of 0.7 m. Early single channel data recording in this mode was shown to be very effective (refs 15,16) in improving thin bed resolution but lacked the quantitative seismic velocity information now available at these higher frequencies from the use of multi-channel ¼ ms sampling (refs 17,18). Only with this sophistication can shallow high quality geophysical profiles begin to anticipate the 'punch-through' situation by identifying low density, low velocity layers immediately beneath a highly reflective sea bed.

Further work is needed in this area, and there will be no chance of supplanting the geotechnical borehole until multi-channel streamers and high frequency stable sources can be taken down close to sea bed. Nevertheless an improvement in identification of shallow-sea bed conditions over analogue methods is now available, giving early warning of soft or unstable conditions which can be quantified by, and correlated with, well-placed soils boreholes.

This level of resolution, given adequate penetration is also providing better identification of likely problems in setting 30 inch casing. Generally placed within 100 m of the seafloor, the drilling or driving and setting of this initial string of casing can be held up by many factors such as boulders at or near sea bed, unconsolidated and/or granular deposits within glacially cut and infilled fossil sub-sea bed channels and unsuitable conditions for cementing. Given good data and an understanding of the depositional/erosion processes, an experienced interpreter can highlight these and other conditions. If drilling is taking place in areas of rapid deposition or high sea bed gradient, sea bed slope stability is high on the list of enquiry (ref. 19). In areas of even low seismicity and low slope (1–2°), this topic deserves closer attention and emphasizes the setting of the shallow site geology in its regional context (ref. 11, p. 31) through tie lines to adjacent wells and full use of all background data.

The need for an intelligent evaluation by an experienced geophysical interpreter with geological training is even more critical when it comes to the intermediate part of the section where the drilling and setting of the 20 inch casing (generally less than 100 m sub-sea bed) is being considered. At the field acquisition level it will have been critical for the programme initiator to have appreciated the differing capabilities of seismic sources available. Unless plenty of time is available between end of data acquisition and finalization of well plan, the use of sparker (although cheap and cheerful) may result in without adequate time to source signature deconvolve on a line by line basis, data which cannot adequately address the main purpose of the survey is anything other than very crude terms. 'Bright spot' highlighting by a trainee interpreter in a two day rush is not an adequate answer to evaluation of shallow, local mini-reservoirs, possibly under pressure. Correct source selection needs also to be matched by adequate moveout control on the events identified, to provide, as a minimum, correct RMS seismic velocities for reduction of the time data to a depth scale of relevance to drilling. Various attempts have been made in the past, and undoubtedly deserve further study (ref. 20) with the higher sampling rate and high fold of data now becoming available, to determine the changes in seismic interval velocity in this 'mid' section and relate this to normal or over-pressured gas occurrences (ref. 21). Coupled with this is the need for further advances in mathematical data evaluation to identify gas occurrences correctly, measure their thickness, and map their extent. The driller can evolve techniques and make plans to handle shallow gas (ref. 22), but the best piece of information he can be armed with is a categoric statement that shallow gas is or is not present from a source that can be trusted.

At depth, the value of high resolution data acquisition merges into the conventional seismic section, mainly owing to the natural attenuation of high frequency energy with depth. However, in 1972, many believed that 70 Hz was unattainable at 1 s. Better sources with more energy in the higher part of the spectrum and improved streamers with high sampling rates have now proved this an artificial limit. Better resolution at depth, however, will need to be paralleled by an improved approach to data interpretation, an appreciation of the full value of offshore drill site surveys and a realistic view of the time scale needed to process and interpret such data. Education is needed on both sides of the fence.

REFERENCES

1. Outer Continental Shelf Oil and Gas Blowouts 1979–1982. US Geological Survey Open File Report 83-562. August 1983.
2. Outer Continental Shelf Oil and Gas Blowouts. US Geological Survey Open File Report 80-101. 1980.
3. Mathieson, R. (1987). Shallow Gas: Safety Aspects and the Technological Challenge. Shallow Gas Seminar, Norwegian Petroleum Directorate. August 27–28, 1987.
4. FA10 Shallow Seismic Safety Study. Proprietary report by Fairfield Industries.
5. Arthur, J. C. R. (1979). The Application of High Resolution Multi-channel Seismic Technique to Offshore Site Investigation Studies. Proceedings of Society of Underwater Technology Conference. March, 1979.
6. Lucas, A. L. (1974). A high resolution marine seismic survey. *Geophysical Prospecting* **22:** 667–682.
7. Fannin, N. G. T. (1979). The Use of Regional Geological Surveys in the North Sea and Adjacent Areas in the Recognition of Offshore Hazards. Proc. Society of Underwater Technology Conf. March 1979.
8. King, L. H. (1979). Aspects of Regional Surficial Geology related to Site Investigation Requirements—Eastern Canadian Shelf. Proc. Society of Underwater Technology Conf. March 1979.
9. Tirey, G. B., Zinzer, D. and Fulton, P. (1985). Geological and Geophysical Data Acquisition, Outer Continental Shelf Through Fiscal Year 1985. Resource Evaluation Programme Report MMS 87-0003.
10. Aamodt, F. R. (1987). Shallow Gas: The Extent of Its Spread. Shallow Gas Seminar, Norwegian Petroleum Directorate. August, 1987.
11. Gunleiksrud, T. and Rokoengen, K. (1979). Regional Geological Mapping of the Norwegian Continental Shelf with Examples of Engineering Applications. Proc. Society of Underwater Technology Conf. March, 1979.
12. Offshore Geologic Hazards. Education Course Note Series No. 18. AAPG. Rice University. May, 1981.
13. Trabant, P. K. (1984). Applied High Resolution Geophysical Methods (Offshore Geoengineering Hazards). D. Reidel Publishing Company.
14. Huysinga, J. K. and Biddle, A. R. (1978). Shallow Seismic as an Aid to Locating Offshore Installations. European Offshore Petroleum Conf. and Exhibition EUR 113, 1978.
15. Stokes, A. W., Tilston, C. G. and Stirling, R. M. (1982). A Geophysical Survey Aid Foundation Design and Cuts Costs. Offshore Technology Conf. OTC 4342, 1982.
16. Ray, C. H. (1978). Thin Bed Resolution Using the Fairflex Marine Energy Source *EAEG* June, 1978.
17. Games, J. P. (1985). High Resolution Geophysical Surveys for Engineering Purposes. *Offshore Site Investigation,* Vol. 3, Chapter 7.
18. Games, K. P. (1987). A breakthrough in the detection of subtle structures and stratigraphic features. *The Hydrographic Journal* No 46, October.
19. Plyssel, M. R. and Campbell, K. J. (1979). North Western Gulf of Mexico – Engineering Implications of Regional Geology. Proc. Society of Underwater Technology Conf. March, 1979.
20. Sutton, G. R. and Moore, B. D. (1987). Inversion on an unmitigated stacked section to determine an interval velocity model. *Geophysical Prospecting* **35:** 895–907.
21. Waters, S. and Moore, N. (1978). Pore Pressure Predictions from High Resolution Seismic Data. OTC Paper, 1978.
22. Sandlin, C. W. (1986). Drilling Safely Offshore in Shallow Gas Areas, Society of Petroleum Engineers, European Petroleum Conf. October, 1986.

24

The Ocean Drilling Program

Philip D. Rabinowitz, Louis Garrison, Audrey Meyer and Jack Baldauf,
The Ocean Drilling Program, Texas A. & M. University, College Station,
Texas 77843, USA

The Ocean Drilling Program (ODP), an international program of scientific ocean drilling, is the successor program to the Deep Sea Drilling Project (DSDP). ODP began its field operations with a shakedown and sea trials cruise in January 1985 in the Gulf of Mexico. Approximately every two months since that time, an internationally staffed expedition of our drilling research vessel the SEDCO/BP 471 (Fig. 1), better known to the scientific community as the *Joides Resolution,* has taken place in very remote but geologically important areas of the world's oceans (refs 1–4).

The *Joides Resolution* has drilled at over 100 sites (as of November, 1987) in the North Atlantic, including the Norwegian Sea, Labrador Sea and Baffin Bay, and Mediterranean Sea; in the eastern Pacific off the Galapagos and coast of Peru; in the Weddell Sea and Subantarctic South Atlantic; and in regions of the Indian Ocean (Fig. 2).

FACILITIES

The *Joides Resolution* has a seven-storey laboratory structure located forward of the derrick on the starboard side. Within this unique structure are separate laboratory spaces containing state-of-the-art equipment for shipboard analyses of cores (ref. 5). Instrumentation is available for study of the physical properties of sediments and rock, palaeomagnetics, palaeontology, and petrology. There are well-equipped laboratories for inorganic and organic chemical analyses and for analyses by X-ray diffraction and X-ray fluorescence. The scientific tasks are also supported by a photographic laboratory, an electronics repair shop, the capability for gathering analog and digital underway geophysical data, a global positioning navigation system, a central computer processing unit with 50 microcomputers distributed throughout the laboratory spaces, and an excellent reference library. The ship has living quarters for up to 50 scientists and technical support staff and a crew of 65. It can drill in water depths exceeding 8 km.

MAJOR THRUSTS OF THE ODP

The major thrusts thus far within the Ocean Drilling Program can be summarized as follows.

Operational
    Testing operational limits of *Joides Resolution*
    High latitude operation

Advances in Underwater Technology, Ocean Science and Offshore Engineering, Volume 16: Oceanology '88

Technical
  Retrieval of sedimentary rocks
  Retrieval of basement (basaltic) rocks

Scientific
  Palaeoenvironments
  Continental margins
  Ocean crust

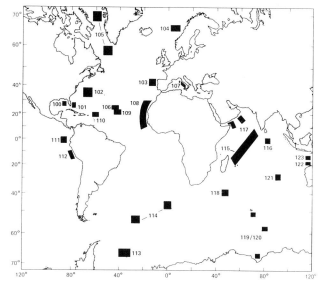

Fig. 2 Site locations for Legs 100–123.

Fig. 1 *Joides Resolution*, drillship for the Ocean Drilling Program. Note laboratory module on starboard side forward of the derrick.

OPERATIONS

The *Joides Resolution*'s dynamic positioning system includes capability for station keeping using long-baseline, short-baseline, and ultra-short-baseline systems. It is supported by 12 powerful 800 hp retractable thrusters, as well as by two main propellers, each driven by six 750 hp motors. In the operational mode, the station keeping system should hold the ship to ±2% of water depths, with wind limits of 45 knots, gusts to 60 knots, significant wave height of 15 feet, maximum wave height of 27 feet, and surface currents of 2.5 knots, provided the prevailing environment is within 30° of the bow or stern (ref. 6).

Because of its size and displacement, *Joides Resolution* has proven itself a remarkably stable drilling platform for work in hostile sea conditions. Previous predictions based on stability calculations and computer modelling had indicated that the vessel would be little affected by wind and swell in such areas. The operational limits would be dictated by the need to restrict the bending moment of the drillstring immediately beneath the hull. During our first two years of field operations, our most hostile environment was in Baffin Bay and the Labrador Sea (Leg 105). Here, even the worst wind and sea conditions experienced (wind speeds in excess of 35 knots and wave heights of 9–10 feet) caused little pitch or roll.

The drilling capabilities of the vessel were put to their most severe test at the start of our third year of field operations in the sub-Antarctic South Atlantic Ocean (Leg 114; ref. 7). Here, strong gale forces (greater than 41 knots) were encountered during 29 days of the cruise (maximum winds 86 knots; maximum seas 40–50 feet; maximum roll 26°). More than one-third of the cruise had combined seas exceeding 18 feet. Despite these extreme conditions, the vessel remained remarkably stable, and drilling operations were suspended for a total of only 1.5 days because of adverse weather.

The ship also has what we believe to be the world's largest heave compensator, a 400 ton 'hydraulic shock absorber' that minimizes the relative motion of the drillstring. This has enabled us to retrieve relatively undisturbed cores suitable for high resolution geological studies in these hostile environments.

On Leg 105 (ref. 8), *Joides Resolution* also tested her high-latitude capabilities by drilling in Baffin Bay in a region surrounded by icebergs. The services of a Canadian ice patrol vessel was used to track the movement of individual icebergs in order to warn us when it was necessary to relocate the vessel to a position of safety (ref. 9). Ice management also became an integral part of drilling operations during Leg 113 in the Weddell Sea. The use of the state-of-the-art support vessel *Maersk Master* (Fig. 3) enabled drilling operations to continue in ice laden areas due to the *Master*'s capability in towing and pushing ice.

Fig. 3 *Maersk Master* towing large iceberg off Antarctica. This is probably the largest object ever moved by Man.

Judging from our experience in both northern and southern polar regions, the Ocean Drilling Program can look forward to a future filled with important scientific achievements in high latitudes.

TECHNICAL THRUSTS OF THE ODP

The drilling systems used in scientific ocean drilling and those used for industrial drilling are similar. However, the objectives of these two drilling practices differ. In industrial drilling the prime objective is to reach the target depth as efficiently as possible with little or no coring. A major requirement for scientific ocean drilling is to drill deeper into both sediments and hard rock while improving the percentage and quality of the recovered samples.

Within ODP, systems have been developed and are still developing that allow retrieval of relatively undisturbed samples of rocks from beneath the sea floor (ref. 10). Specialized coring tools have been developed for coring the harder formations (Wireline Coring System; ref. 11), for overcoming problems related to core disturbance in soft sea floor sediment (Advanced Piston Corer; refs 12–14) and for coring areas where lithologies alternate between hard and soft beds (Extended Core Barrel; ref. 15). We now feel that, in general, we can routinely retrieve very high quality undisturbed and continuous core, for the upper 200–300 m of sediment in the deep ocean basins. This allows for palaeoenvironmental studies at scales not previously available.

One of our most important technical challenges has been obtaining good core recovery in the basaltic rocks that constitute the upper part (Layer 2) of the hard rock oceanic crust. In particular, an important scientific objective has been to recover cores in unsedimented newly formed oceanic basalt (zero-age crust) beneath the sea. The sea floor in this environment does not have the thick (~100 m) layer of sediment necessary to provide stability for the drill string before it enters hard rock. For this purpose a 40 000-pound guide and re-entry system base was designed and constructed; an additional 100 000 pounds of cement pumped into bags attached to the underside of the base provides further stability. A television camera attached to the outside of the drillstring and lowered to the sea allows scientists to view the normally rugged sea floor in mid-ocean ridge environment and enable them to lower the guidebase accurately to its desired location (Fig. 4; ref. 16).

SCIENTIFIC THRUSTS OF THE ODP

The Ocean Drilling Program has addressed many of the scientific objectives resulting from the Conference on Scientific Ocean Drilling (ref. 17).

We have made important inroads to our study of passive type continental margins—those margins believed to have been formed by processes associated with the rifting of continental crust and embryonic emplacement oceanic crust during the early formation of ocean basins—by coring in Legs 103, 104, 107 and 117. For example, Leg 104 drilled more than 900 m into a sequence of seaward dipping seismic reflectors on the continental margin off Norway and recovered predominantly volcanic rocks (ref. 18). On Leg 103 scientists were able to collect a near continuous section of pre-syn and post-rift sediments

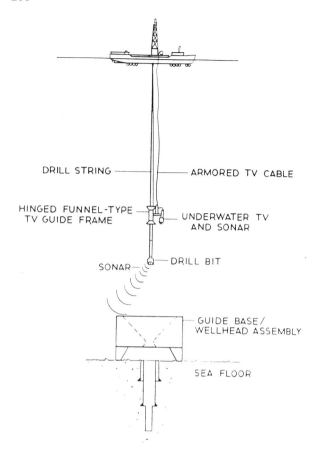

DRILL STRING ——— ARMORED TV CABLE

HINGED FUNNEL-TYPE
TV GUIDE FRAME

UNDERWATER TV
AND SONAR

SONAR — — DRILL BIT

GUIDE BASE /
WELLHEAD ASSEMBLY

SEA FLOOR

Fig. 4 Deployment of underwater TV and sonar equipment on drillstring for Leg 106 bare-rock drilling operation.

holes and conducting geophysical experiments. For example, during Leg 102 (ref. 24), we re-entered a hole (DSDP Hole 418) that previously drilled over 500 m into the ≈110 MY-old, basaltic oceanic basement in 5500 m of water. Hydrophones clamped at discrete intervals within the hole recorded seismic waves from a surface shooting geophysical vessel in order to study anisotropy of the ocean crust.

Palaeoenvironmental objectives were addressed on Legs 100, 101, 104, 105, 107, 108, and 112–117. For example, Leg 108 (ref. 25) collected over 3800 m of core in order to investigate the late Neogene history of oceanic circulation and climate system (ice sheets, sea ice and polar oceans) and the low-latitude ocean atmosphere components (surface ocean, upwelling regions, wind circulation and land climate). Leg 117 collected over 4300 m of core in order to decipher the record of the Indian Ocean SW monsoon and the evolution of the Himalayas and their impact on global climate. The cores recovered to study these objectives must be continuous and of high quality. In many cases we double or triple piston cored the sites. In order to understand key palaeoclimatic responses that may occur during earth orbital changes of 20 000–100 000 years, the core obtained must be investigated in pains-taking detail (order of 10 cm intervals).

CONCLUDING REMARKS

The two years following our Indian Ocean campaign (post Leg 123; November 1988 to November 1990) will see the *Joides Resolution* retrieve cores in the Northwest and Central West Pacific Ocean in order to address problems relating to island-arc/fore-arc processes, mountain-building processes and formation of marginal basins.

Many of the scientific objectives there will present challenges in the field of deep-ocean technology far beyond those available from conventional offshore drilling practice. The deep-water challenges will most likely be in areas of deeper penetration of thick sedimentary layers as well as deeper penetration and better core recovery within the basaltic basement complex.

The state-of-the-art onboard scientific, drilling and operational equipment have been fully tested in the challenging environments of some of Earth's most remote geographical areas.

ACKNOWLEDGEMENTS

The Ocean Drilling Program is funded by the Joint

(ref. 19). Active type continental margins—those margins associated with deep sea trenches, active volcanos, belts of high seismicity and subduction of oceanic plates—have been cored on Legs 107, 110 and 112. In particular, Leg 110 (ref. 20) resulted in the first successful penetration of a décollement zone. This zone separates the underthrusting Atlantic Ocean crust of the American Plate from the over-riding Barbado Ridge at the eastern edge of the Caribbean Plate in an area of prominent plate convergence.

The origin and evolution of ocean crust have been addressed in Legs 102, 106, 109, 111, 115 and 118. During Leg 106 (ref. 21) we cored, for the first time, young or 'zero age' volcanic rock on the mid-Atlantic ridge. A hot spot trace was sampled at discrete locations along its length on Leg 115 (ref. 22), and the deepest hole ever drilled into the oceanic crust was deepened to 1288 m sub-basement on Leg 111 (ref. 23). We have also established 'natural laboratories' by cleaning and/or deepening previously drilled DSDP

Oceanographic Institutions, Inc. (JOI), a non-profit corporation comprising the ten major US oceanographic institutions which is in turn funded by the National Science Foundation (NSF), an independent US federal agency. NSF receives contributions from six non-US members (France, West Germany, Canada, Japan, United Kingdom and the European Science Foundation Ocean Drilling Consortium, comprised of 12 countries). Texas A. & M. University (TAMU) is Science Operator of the program and as such ensures that adequate scientific analyses are performed on the retrieved cores by providing and maintaining shipboard science laboratories, providing logistical and technical support for shipboard science teams, managing post cruise activities, curating and distributing core samples, and coordinating the editing and publishing of the scientific results. Lamont-Doherty Geological Observatory of Columbia University (L-DGO) is the prime logging contractor for ODP and supplies a complete suite of geophysical and geochemical services including the acquisition and processing of *in situ* logging measurements. The scientific advice and direction comes from the (JOIDES) Joint Oceanographic Institutions for Deep Earth Sampling, an international consortium of the ten JOI institutions and members from the six non-US members.

The contribution of the large numbers of scientists, engineers, technicians and administrators (at TAMU, L-DGO, JOI, JOIDES, NSF, and SEDCO) who are responsible for ensuring the success of ODP is greatly appreciated.

REFERENCES

1. Rabinowitz, P. D., Carlson, R., Gartner, S. *et al.* (1984). The Ocean Drilling Program: The Next Phase in Scientific Ocean Drilling. Proc. Offshore Tech. Conf., OTC 4698: 443–449.
2. Rabinowitz, P. D., Garrison, L., Baldauf, J. *et al.* (1987). The Ocean Drilling Program: Results from Second Year of Field Operations. Proc. Offshore Tech. Conf., OTC 5460: 343–355.
3. Rabinowitz, P. D., Garrison, L. E., Harding, B. W. *et al.* (1986). The Ocean Drilling Program After One Year of Drilling Operations. Proc. Offshore Tech. Conf., OTC 5184: 283–296.
4. Rabinowitz, P. D., Garrison, L., Herrig, S. *et al.* (1985). Scientific Ocean Drilling: An Overview of the Ocean Drilling Program. Proc. Offshore Tech. Conf., OTC 4989: 279–286.
5. Kidd, R., Rabinowitz, P. D., Garrison, L. *et al.* (1985). The Ocean Drilling Program III: The Shipboard Laboratories on JOIDES RESOLUTION. Proc. Marine Tech. Soc. Conf., *Ocean Engineering and the Environment* 1: 133–145.
6. Foss, G. N. (1985). The Ocean Drilling Program II: JOIDES RESOLUTION, scientific drillship of the '80's. Proc. Marine Tech. Soc. *Ocean Engineering and the Environment* 1: 124–132.
7. Ciesielski, P., Kristoffersen, Y., Clement, B. *et al.* (1987). Paleoceanography of the Subantarctic South Atlantic. *Nature* 328: 671–672.
8. Arthur, M., Srivastava, S., Clement, B. *et al.* (1986). High latitude paleoceanography. *Nature* 320: 17–18.
9. Harding, B. and Rabinowitz, P. D. (1986). High Latitude Drilling in the Ocean Drilling Program. In: POLARTECH '86 International Offshore and Navigation Conference and Exhibition, Helsinki, Finland, 27–30 October 1986. Vol. 1, p. 83–106.
10. Huey, D. P. and Storms, M. A. (1985). The Ocean Drilling Program IV: Deep water coring technology, past, present, and future. Proc. Marine Tech. Soc. Conf. *Ocean Engineering and the Environment*. p. 146–159.
11. Larson, V. F. and Serocki, S. T. (1974). Deep Water Coring for Scientific Purposes. Soc. of Petr. Eng. of AIME, SPE 5171.
12. Storms, M. A., Nugent, W., and Cameron, D. H. (1983a). Design and operation of the hydraulic piston corer. Deep Sea Drilling Project, Tech. Report No. 12, Scripps Institution of Oceanography, University of California at San Diego.
13. Storms, M. A., Nugent, W., and Cameron, D. H. (1983b). Hydraulic piston coring—a new era in ocean research. Proc. Offshore Tech. Conf., OTC 4622: 369.
14. Huey, D. P. (1984). Design and operation of an advanced hydraulic piston corer. Deep Sea Drilling Project Tech. Report No. 21, Scripps Institution of Oceanography, University of California at San Diego.
15. Cameron, D. H. (1984). Design and operation of an extended core barrel. Deep Sea Drilling Project, Tech. Report No. 20, Scripps Institution of Oceanography, University of California at San Diego.
16. Howard, S. P., Serocki, S. T., and Brittenhauss, T. (1986). Development of a scientific drilling and coring system for the Mid-Atlantic Ridge, Ann. Mtg of ASME—Feb. 1986, handout, 11 pp.
17. COSOD (1981). Report of the Conference on Scientific Ocean Drilling. Joint Oceanographic Institutions, Inc., Washington, D.C., 112 pp.
18. Eldholm, O., Thiede, J., Taylor, E. *et al.* (1986). Formation of the Norwegian Sea. *Nature* 319: 360–361.
19. Boillot, G., Winterer, E., Meyer, A. *et al.* (1985). Evolution of a passive margin. *Nature* 317: 115–116.
20. Moore, J. C., Mascle, A., Taylor, E. *et al.* (1987). Expulsion of fluids from depths along subduction-zone decollement horizon. *Nature* 326: 785–787.

21. Detrick, R. S., Honnorez, J., Adamson, A. C. *et al.* (1986). Mid-Atlantic bare-rock drilling and hydrothermal vents (Leg 106), *Nature* **321:** 14–15.
22. Backman, J., Duncan, R., MacDonald, A. *et al.* (in press). Hotspot volcanism and the neogene carbonate budget in the Indian Ocean. *Nature*.
23. Becker, K., Sakai, H., Merrill, R. *et al.* (1987). News from a deepening hole. *Nature* **325:** 484–485.
24. Salisbury, M., Scott, J., Auroux, C. *et al.* (1985). Looking down an old hole: Leg 102 of the Ocean Drilling Program. *Nature* **316:** 682.
25. Ruddiman, W., Sarnthein, M., Baldauf, J. *et al.* (1986). Paleoclimatic linkage between high and low latitudes (Leg 108). *Nature* **322:** 211–212.

25

Measurement of Chain Geometry in the Ground: a Comparison of the Effects on Long and Short Piled Anchors

E. Burley, Queen Mary College
and *R. Martin,* Anchortech Limited UK

Piles are being increasingly used as permanent anchors for single point mooring systems, and to achieve the high resistances required in weak soils, it is invariably necessary to position the chain connection to the pile at some depth below the seabed The chain, therefore, has to cut down through the seabed soils to achieve an equilibrium configuration under the action of the applied tension.

Published information on this aspect of chain geometry is restricted to an analytical treatment of the problem and no comparison with measured values has so far been reported.

The experimental work presented in this chapter coupled to a comprehensive investigation of the effects of variation in soil resistance normal to the chain, will enable a more complete analysis of the anchor chain to be undertaken.

EXPERIMENTAL STUDY

Laboratory tests were carried out to measure the profile of both a chain and cable, anchored at a fixed point below the soil surface, subjected to a horizontal tension at the surface. These tests were performed in a small rectangular water channel, equipped with horizontally graduated top rails supporting a mobile depth probe. This enabled direct measurement of the horizontal and vertical coordinates of the chain. To determine contact between the probe and chain, an electrical circuit was arranged so that the circuit was closed when contact was made, thus causing a buzzer to sound. Measurements were carried out in soft clay and loose dry sand of depth 310 mm, with a uniform cohesion of 2.4 kPa and angle of friction 30°, respectively.

GEOMETRY AND EQUILIBRIUM OF CHAIN IN SOIL

Reese (ref. 1) discussed the equilibrium of a chain which is tangential to the seabed and considered resistance to movement of the chain normal to the chain path only. Gault and Cox (ref. 2) have extended this work to include the effects of chain weight and tangential resistance to movement of the chain. They develop their equations by considering equilibrium

Advances in Underwater Technology, Ocean Science and Offshore Engineering, Volume 16: Oceanology '88

of forces about two rectangular axes and assume the cable geometry as a circular arc. By so doing, however, they ignore moment equilibrium of the system and this equilibrium can only be achieved for the assumed circular geometry when both the tangential resistance and cable weight are zero, thereby reducing the system to that considered by Reese.

The moment equilibrium equation for Gault and Cox's assumptions is derived in the appendix, and use of this enables moments to be calculated for any point along the cable profile. Figure 1 shows the effects of layer thickness on the value of this moment on an example with uniform soil strength and shows the need to keep to suitably thin soil layers to achieve the true momentless geometry that a chain must assume.

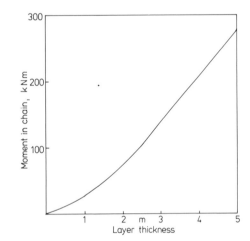

Fig. 1 Theoretical moment in chain v. soil layer thickness.

VARIATION IN SOIL RESISTANCE NORMAL TO THE CHAIN

The well recognized formula for shallow strip foundations due to Casman *et al.* (ref. 3) is:

$$q = cN_c + 0.5\gamma BN_\gamma + \bar{q}N_q$$

where

q is the ultimate bearing pressure

B is the width of foundation

γ the effective unit weight of soil with friction angle ϕ and cohesion c

$\bar{q}$ unit surcharge at level of foundation

The exact formulas for N_q and N_c were given by Prandtl (ref. 4), and Brinch Hansen (ref. 5) recommends an empirical relationship for N_γ which, he

points out, corresponds closely with calculations made by Lundgren-Mortensen and Odgaard and Christensen.

The soil resistance normal to the chain differs from the above in two respects. The depth of the chain below the surface varies over the length of foundation, and the chain is likely to be completely surrounded by soil. Nevertheless, soil resistance normal to the chain should be approximated by bearing capacity theories.

COHESIVE SOILS

For purely cohesive soils, i.e. $\phi = 0$ the shallow foundation equation reduces to:

$$q = cN_c + \bar{q}$$

and

$N_c = 5.14$ from Prandtl's equation.

However, Meyerhoff (ref. 6) while agreeing with this figure for smooth sided foundations, has shown the factor to be higher by 0.57 for foundations with rough sides. The normally accepted deep bearing capacity factor is 9 to 11 and Meyerhoff showed that theoretically the maximum resistance is reached at a depth of twice the width, whereas his experimental results indicated that the maximum was reached at about twice this figure, with a parabolic increase from the surface value to the deep value. Meyerhoff has also shown that for a completely embedded anchor beam, the equivalent theoretical bearing capacity factors

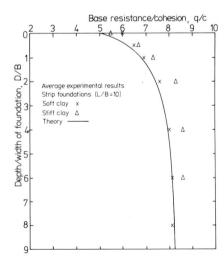

Fig. 2 Bearing capacity of buried model foundations in clay due to Meyerhoff (ref. 6) compared with Brinch Hansen theory.

re 11.42 and 11.99 for smooth and rough sides respectively, an increase of 38% over the value for a deeply embedded strip footing, which he calculates as 8.28 and 8.85 respectively.

Brinch Hansen gives the relationship:

$$q = (\pi + 2)(1 + d_c^a)c$$

here

$$d_c^a = 0.4 \arctan D/B$$

This is independent of the roughness of the foundation sides and agrees well with Meyerhoff's experimental results, plotted against depth as can be seen in Figure 2. The limiting value is also in good agreement with Meyerhoff's theoretical values.

Taking into account the completely buried nature of the chain, a realistic relationship for N_c, in terms of depth and width, would therefore be:

$$N_c = 5.14 + 2.93D/B - 0.36625(D/B)^2 \text{ for } D/B < 4$$

and

$$N_c = 11 \text{ for } D/B \geqslant 4 \qquad \text{(Eqn A)}$$

where D is the depth and B is the effective width of a chain or actual width of a cable.

In soft clays, the compressibility of the material leads to local shear failure. For such conditions a reduced bearing capacity factor is required. In this case Terzaghi suggested the use of a reduced cohesion, $c^1 = \frac{2}{3}c$. Meyerhoff, however, concluded that the reduction factor should be 1 for shallow foundations and 0.9 for deep foundations.

A reasonable estimate for the chain in soft clay would, therefore, be:

$$N_c = 3.2 + 2.9D/B - 0.3625(D/B)^2 \text{ for } D/B < 4$$

or

$$N_c = 9 \text{ for } D/B \geqslant 4 \qquad \text{(Eqn B)}$$

Since anchor chains and particularly cables are relatively small, the critical value of $D/B = 4$, occurs at shallow depth. The analysis is, therefore, also carried out for the constant values of N_c, given by equations (A) and (B) respectively, taken over the whole depth of profile considered.

These four theoretical results are superimposed on the measured chain and cable profiles in Figures 3 and 4. It can be seen that both versions of equation (A) give too steep a profile, particularly near the surface, whereas equation (B) gives good agreement over the whole depth. The results using a constant $N_c = 9$ are too steep in the surface layer above the specified critical depth.

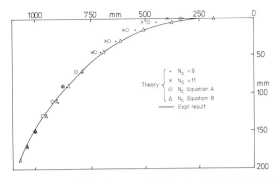

Fig. 3 Chain profile in clay: applied load = 196.2 N (20 kg).

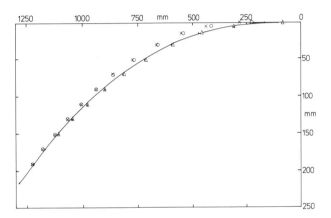

Fig. 4 Cable profile in clay: applied load = 196.2 N (20 kg).

NON-COHESIVE SOILS

For non-cohesive soils the general shallow strip foundation equation now reduces to:

$$q = 0.5\gamma B N_\gamma + \bar{q} N_q$$

Various authors (refs 3–9) have derived bearing capacity factors for shallow strip footings, and several (refs 5,6,9 and 10) have produced similar results for deep pile foundations. Few, however, have shown the relationship between bearing capacity and depth for strip foundations. Meyerhoff (ref. 6) derived curves for strip footings in terms of the resultant bearing capacity factor $N_{\gamma q}$, plotted against the non-dimensional parameter (depth/width), whilst Brinch Hansen (ref. 5) derived a depth factor, by which to multiply the surface value of N_q, the value of N_γ remaining constant with depth. A comparison of the ultimate pressures given by these two methods (Table I) shows that for depths greater than nine times the breadth, Meyerhoff's values are considerably greater than those predicted by Brinch Hansen.

Incorporating the above two theories into the

TABLE I
Meyerhoff and Brinch Hansen theories: ultimate
pressure v. depth (units: kPa)

D/B	Meyerhoff	Brinch Hansen
0	22	22
2	113	162
5	309	392
10	793	777
20	2102	1547
40	4752	3086

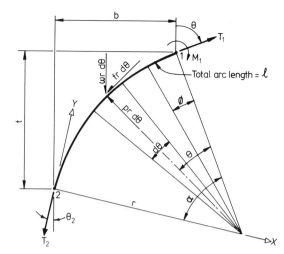

Fig. 7 Equilibrium geometry of arc of chain in soil.

analysis of chain profiles gives the results shown in Figures 5 and 6 superimposed on the measured values for the chain and cable in loose sand. These results show that Brinch Hansen's variation of ultimate foundation pressure with depth gives better agreement than Meyerhoff's values, but that both underpredict the resisting pressure for shallow depth. Meyerhoff discovered a similar discrepancy when comparing his theory with experimental strip foundation results. Much better agreement is reached if this shallow resisting pressure is increased, without significantly increasing pressures at greater depth. This can be achieved by using a value of N_γ four times the value given by Prandtl.

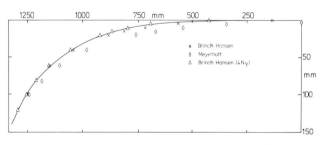

Fig. 5 Chain profile in sand: applied load
$H = 196.2$ N (20 kg).

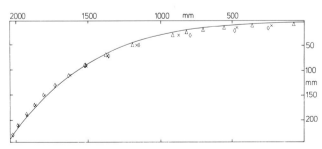

Fig. 6 Cable profile in sand: applied load
$H = 588.6$ N (60 kg).

The results are in agreement with Brinch Hansen's suggestion that the friction angle for plane strain is the appropriate value and that a reasonable estimate of this is $\phi_{pl} = 1.1\phi$.

Practical application

The variation of angle to the horizontal against depth for a chain and an equivalent strength wire rope in a typical North Sea soft soil location is shown in Table II together with the resultant vertical component of tension. This shows that as the attachment point of the cable to the pile is taken deeper into the soil the vertical component of force on the pile becomes of increasing importance and may reach the point at which vertical failure of the pile could occur.

Six alternative solutions to a typical anchoring problem are given in the appendix to illustrate the problem of the risk of vertical failure occurring under horizontal load. In all cases, vertical failure will occur before horizontal failure, but solution (1A) is completely inadequate. Substituting a wire rope in place of the buried chain, improves the situation markedly by virtue of the reduced frontal area, resulting in solutions (2B) and (3B) being acceptable for both horizontal and vertical components with a load factor close to 2.

Installation of piles

One of the principal problems of driving pile anchors, as opposed to jacket bearing piles which are driven through conductors, is that of support before the pile has been driven sufficiently far to be self-supporting in the seabed. Because pile anchors are *de facto*

TABLE II
*Angle of chain and vertical component of chain
tension v. depth*

Depth	Chain		Rope	
	Angle to horizontal (degrees)	Vertical component(kN)	Angle (degrees)	Vertical component(kN)
0	90	0	90	0
2	84.3	238	86.2	163
4	81.2	361	84.1	248
8	75.3	577	80.3	398
12	69.5	770	76.6	537
16	63.6	944	72.9	666

Horizontal tension at seabed = 2500 kN
Chain size 70 mm, Rope size 76 mm
Soil: Clay, cohesion increasing linearly from 4 kPa at seabed to
20 kPa at 20 m depth

driven in isolated positions, there is no convenient structure to provide support.

The hammer required to drive the pile will generally weigh as much or more than the pile itself. When located on top of the pile, the pile-hammer assembly is highly unstable, and clearly instability is greater the longer the pile.

It is also apparent from the chain profiles in the soil that the longer pile (solution (1) in the previous section) with a deeper pad eye requires a greater vertical capacity than the shorter solution. Since this vertical capacity is achieved by adhesion which is also present during driving, so a larger hammer is required to drive the longer pile, thereby further increasing instability. Indices of installation performance are given in Table III for the three piled solutions previously discussed. These show that overall pile weight is similar for the two piles, with driving time 40% greater overall. However, the individual pile weight of the very short piles is smaller, so less craneage is required. Furthermore, driving time only represents a small proportion of overall installation time, and if the small piles are quicker to handle, overall installation time can be less than for the single piles. It will also be seen that the long piles are ten times as unstable as the very short piles.

Installation vessels

The only practical means of installing long piles in isolated locations is to use a follower from the surface. The vessel is generally moored in position, because vessels that are large enough to perform such operations in deep water are not generally dynamically positioned. The process is cumbersome, time-consuming, and requires a large vessel which is slow to move around the field between anchor locations. One pile per day is good progress in such conditions. Short piles can be installed from smaller vessels with the advantage of greater speed around the field and lower day-rates. It is a practical proposition to support the pile and submersible hammer assembly in a free-standing frame on the seabed, the hammer and frame being recovered on completion of driving. Speed of pile handling is greatly increased, and the installation can then be performed from a dive support vessel (DSV) operating on dynamic positioning (DP), and three piles per day should be achievable without difficulty.

A further point in favour of the DSV is that it is a versatile vessel capable of performing a number of subsea tasks during the hook-up of new fields, particularly single point moorings and floating production systems. The cost of mobilizing the vessel to the field can therefore be shared between piling and hook-up tasks, thereby reducing overall project costs and the number of vessels working in the field.

Large derrick barges and DSVs are not generally set up for handling mooring chains, so the pile would be driven with a short length of wire or chain attached, the remainder of the chain being deployed by an anchorhandling tug. Connection to the pile would be by diver, so the DSV on location is beneficial.

An alternative method of installation is to deploy the piling system from the anchorhandler itself, thereby avoiding subsea chain connections. This

TABLE III
Relative performance indicators

	Solution 2B Short wide pile	Solution 1B[a] Long pile	Solution 2B 2 very short wide piles
Length	13 m	26 m[b]	9 m
Hammer size	1.0	1.6	0.75
Pile weight	1.0	3.5	0.5×2 Nos
Instability	1.0	5[b]	0.5
Driving time	1.0	2.5[b]	0.7×2 Nos

Note: [a]The pile in solution 1B requires lengthening by 6 m to give
      the same load factor against vertical failure as the other two
      solutions
      [b]Pile driven 3 metres below mud line by follower

poses an upper limit on the size of pile that can be driven, and it is in these circumstances that the very short piles coupled together may be attractive. As long as the length of chain connecting the piles is somewhat greater than the water depth, this can be accomplished without divers. The deck space for storing large numbers of piles may be limited, and chain locker size limits the number of piles that can be interconnected in one session; however, piles per day should be achievable.

Costs

Ultimately the best combination of cost and technical performance will be the most attractive, so cost indices are given in Table IV for the three piled solutions. These costs may vary substantially depending on field location, distance to harbour, other work in the field, etc. This table assumes a field location about a day's steaming from harbour, doubled for a derrick barge, and no other work other than piling on location. An 8-leg mooring system is assumed. When the overall cost index of 1.0 is equivalent to £300 000 the variations represent substantial sums.

CONCLUSIONS

The chain profile below the seabed may be predicted to a reasonable accuracy using Gault and Cox's equations, provided the correct soil resistance values are used. For soft clay the relationship for N_c given in equation (B) is recommended, whilst for sands Brinch Hansen's depth correction factor applied to N_q together with a value of N_γ four times Prandtl's figure is most appropriate.

The angle of inclination of the chain at the pad eye is of paramount importance on the design of the piled anchor. Failure to allow for this effect may result in vertical failure of the pile. A pile anchor designed to restrain horizontal load achieves its maximum restraint when the load is applied at the centre of soil pressure. A shallow pad eye minimizes

TABLE IV
Relative costs of piled mooring

	8 long piles	8 short piles	16 very short piles
Vessel day-rate	2.0	1	0.3
On location	3.0	1	1.5
Mobilization/ demobilization	2.0	1	1.0
Piles	1.8	1	1.0
Overall cost	3.35	1	0.72

the length of buried chain and uplift loads on the pile, but can only be combined with maximum restraint if the pile itself is short.

Using a long pile deepens the pad eye and increases uplift loading, and the combination of these effects and installation requirements results in greatly increased cost compared with the short pile. It can also be economic to couple very short piles together, even if it means doubling the overall number of piles driven.

REFERENCES

1. Reese, L. C. (1970). A Design Method for an Anchor Pile in a Mooring System. Reprints Vol I. Fifth Annual Offshore Technology Conf., Houston, Texas, pp. 209–214.
2. Gault, T. A. and Cox, W. R. (1974). Method for Predicting Geometry and Load Distribution in an Anchor Chain from a Single Point Mooring Buoy to a Buried Anchorage. (Paper OTC 2062). Sixth Annual Offshore Technology Conf., Houston, Texas.
3. Terzaghi, K. (1943). *Theoretical Soil Mechanics*. Wiley, New York.
4. Prandtl, L. (1920). Uber die Harte plastischer Korper. (Math. Phys. Kl) Nachr. Kgl. Ges. Wiss. Gottingen.
5. Brinch Hansen, J. (Oct. 1968). A Revised and Extended Formula for Bearing Capacity. (Bull. No. 28). Danish Geotechnical Institute.
6. Meyerhoff, G. G. (1951). The ultimate bearing capacity of foundations. *Géotechnique* **2**.
7. Reissner, H. (1924). Zum Erddruckproblem. Proc. First Int. Congr. Appl. Mech.
8. Meyerhoff, G. G. (March 1965). Shallow Foundations. (Vol 91 No. SM2.) Journal of Soil Mechanics and Foundation Division, ASCE.
9. Terzaghi, K. and Peck, R. B. (1967). *Soil Mechanics in Engineering Practice* 2nd Ed. Wiley, New York.
10. Berezantzev, V. G., Khristoforov, V. S. and Golubkov, V. N. (1961). Load bearing capacity and deformation of piled foundations. Vol. 2 Proc. 5th International Conf. on Soil Mechanics.

APPENDIX A

From Figure 7 moment at X is given by:

$$M_x = wr^2 \int_0^{\theta} (\cos(\theta_1-\theta)-\cos(\theta_1-\phi))d\phi + T_1 r(1-\cos\theta)$$

$$-pr^2 \int_0^{\theta} \sin(\theta-\phi)d\phi - fr^2 \int_0^{\theta} (1-\cos(\theta-\phi))d\phi$$

from which:

$$M_x = T_1 r(1-\cos\theta)+wr^2(\theta\cos(\theta_1-\theta)+\sin(\theta_1-\theta)$$
$$-\sin\theta_1)-pr^2(1-\cos\theta)-fr^2(\theta-\sin\theta)+M_1$$

For 1st layer from seabed $M_1 = 0$.
For subsequent layers $M_1 = M_2$ calculated for $\theta = \alpha$ from previous layer.

APPENDIX B

Example A Mooring chain horizontal at seabed with tension of 2500 kN.

Solution 1A Conventional long slim pile 20×1.52 m diameter driven 3 m below mud line with pad eye connection at 15 m below seabed.

Solution 2A A short wide anchor pile 13×6 m with pad eye connection at 8 m below seabed.

Solution 3A Two shorter wide anchor piles 9×3 m bridled together, with pad eye connections at 5.7 m below seabed.

All solutions are designed to have a load factor of 2 against horizontal failure.

Example B (i.e. solution (1B)–(3B))
As above but with an equivalent strength wire rope instead of chain into the ground.

	Solutions					
	1a	2a	3a	1b	2b	3b
Vertical component of tension at pad eye (kN)	902	587	561	635	402	383
Available adhesion on pile (kN)	650	767	727	650	767	727
Load factor against vertical failure	0.72 PILE FAILED	1.3	1.3	1.02	1.9	1.9

26

The Tyro 86 Penetrator Experiments at Great Meteor East

T. J. Freeman, Building Research Establishment *C. N. Murray,* Joint Research Centre, *and R. T. E. Schuttenhelm,* Rijks Geologische Dienst

The penetrator is a torpedo-shaped object which is allowed to free fall through the water column and penetrate the sea bed by virtue of its own momentum. During the past five years, the Building Research Establishment (BRE), in collaboration with the EEC Joint Research Centre (JRC) and various oceanographic institutions, has been developing the use of penetrators for the measurement of sediment properties in deep water (ref. 1). Sediment strength can be deduced from the measured deceleration in the sediment, and other properties can be measured by onboard sensors and relayed to the ship acoustically. The instrumented penetrator has been particularly relevant to the site investigation of the areas being studied to assess the feasibility of deep ocean disposal of radioactive waste, because large scale penetrators have been suggested as a possible technique for burying the waste (ref. 2).

INSTRUMENTATION SYSTEMS

The acoustic doppler shift system (ADSS) uses the observed doppler shift of a constant 12 kHz transmitter to measure the velocity of the penetrator. ADSS was used for the first deep ocean tests of large, instrumented penetrators (ref. 2), which demonstrated that a 1800 kg penetrator can reach burial depths of over 30 m and can be tracked in 5400 m of water using acoustic instrumentation. Further deep ocean trials, performed in 1984 and 1985 (refs 3 and 4), used ADSS to measure the behaviour of penetrators of different sizes and shapes.

The signal from the ADSS transmitter is attenuated as the penetrator enters the sediment; this reduces the quality of the data and in some cases results in total loss of signal. Measurements made during the 1983 tests, which used a commercially available transmitter, indicated that the signal strength after the penetrator had come to rest in the sediment was marginal for successful operation of the system. To improve the performance of ADSS, a more powerful transmitter was designed and built by the University of East Anglia (ref. 5).

The pulsed acoustic telemetry system (PATSY) is a more versatile, 3.5 kHz system, which encodes data as the time intervals between a series of pulses so that up to eight channels of data can be transmitted. PATSY, which is being developed by the

Advances in Underwater Technology, Ocean Science and Offshore Engineering, Volume 16: Oceanology '88

Institute of Oceanographic Sciences (ref. 6), was tested during the 1984 and 1985 penetrator trials (refs 3 and 4).

PATSY consists of three packages: a communication package at the tail, a sensor package at the centre of gravity, and an optical switch (used to detect the instant of impact) at the nose. For the *Tyro 86* experiments, the sensor package comprised two accelerometers, two inclinometers, and two pressure transducers to measure the pore pressures in the surrounding sediment. Data collection, storage, and transmission are controlled by a microprocessor; there are four modes of operation, which are entered sequentially: (*i*) quiescent mode, where no data are collected or transmitted; (*ii*) slow data collection (SDC) mode, where no data are transmitted, but all sensors are read 10 times every second and the data stored in memory; (*iii*) fast data collection (FDC) mode, where no data are transmitted, but accelerometer data only are collected at 500 Hz for 2.5 s; and (*iv*) transmit mode, where a period of live data transmission (typically all channels are scanned every 2 s) is followed by transmission of data stored during the SDC and FDC modes.

The FDC mode is used to collect data during the penetration event and can be activated either by the optical switch being blocked or by one of the accelerometers recording a deceleration greater than 3 *g*; to minimize the danger of premature triggering, the quiescent mode is used during the launch preparations and the SDC mode is only activated at the last possible moment by removing a shutter from the optical switch. The transmit mode is entered automatically after the FDC mode has timed out and is normally repeated until the battery power expires. However, PATSY can be programmed to interrupt the transmit cycle to allow a period of transponding, i.e. transmission of signals in response to acoustic pulses from the ship; this allows the position of the penetrator relative to the ship to be fixed and the depth below the seafloor to be confirmed. Because of its ability to transpond, PATSY was known originally as LFTS (low frequency transponder system).

Potential applications of PATSY include the long term monitoring of sea bed experiments. Murray *et al.* (ref. 7) describe a system where data transmission is controlled by an instrumented surface platform linked to a land-based control centre via the European Space Agency's Meteostat satellite communications network; battery power needed for acoustic transmission is conserved by onboard data processing which allows the system to transmit data in real time over long periods.

Super-doppler (ref. 8) was developed by the University of East Anglia (UEA) as a method of sending data using the ADSS transmitter. During free fall, Super-doppler transmits at 12 kHz allowing the doppler shift to be measured as it is for ADSS, but once the penetrator has come to rest, Super-doppler switches mode and starts to transmit data as a binary string by frequency shift keying (FSK) between 11.3 and 12.7 kHz. Unlike PATSY, Super-doppler has no memory and can only transmit data in real time; the prototype system as tested on the Tyro is capable of transmitting data from two channels at a rate of approximately 1 reading per second with 7 bit accuracy.

PENETRATOR DESIGN

The type X penetrator was developed to obtain sediment property data to greater depths; it consists of a steel shell 356 mm in diameter and 3.56 m long, with a cylindrical recess in the tail to house the ADSS transmitter (Fig. 1(*a*)). The density can be varied by using different ballast materials and the drag factor varied by using different surface finishes. Using solid lead ballast, the type X weighs approximately 3.2 tonnes but costs no more to produce than a 1800 kg solid steel penetrator.

The type 1B penetrator is a modified version of the type 1 penetrators used in the March 1983 experiments (ref. 2); both these penetrator designs are manufactured from solid steel and weigh approximately 1800 kg in air. However, the type 1B is shorter and fatter to accommodate the more bulky PATSY instrumentation and has a central 160 mm diameter hole to house the three component packages of PATSY (Fig. 1(*b*)).

The SV penetrator is being developed by BRE to measure the properties of the sediments which fill the pathway created as the penetrator enters the sea bed; this information is needed to evaluate the degree of sealing that would take place behind a disposal penetrator. The SV penetrator is based on the same steel shell used for the type X penetrators, but has a modified tail section to house a Super-doppler transmitter. Ballast is provided by lead shot in the cavity between the steel shell and the central mechanism, which deploys a shear vane from the tail of the penetrator (Fig. 1(*c*)). Once the penetrator has come to rest in the sediment, a cycle of six tests is performed, each consisting of a 300 mm push followed by a 90° rotation.

The vane mechanism is powered hydraulically and requires seawater to be drawn into the penetrator as the vane is deployed. A pressure transducer monitors the ambient pressure inside the tail of the

Fig. 1(a) The type X penetrator.

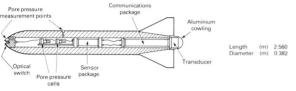

Fig. 1(b) The type 1B penetrator.

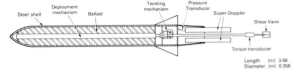

Fig. 1(c) The type SV penetrator.

penetrator giving an indication of the permeability of the surrounding sediment, no change from hydrostatic indicating an open pathway back to the water column. The ambient pressure, together with the torque measured by a load cell mounted directly behind the vane, are transmitted to the ship by Super-doppler.

THE GREAT METEOR EAST STUDY AREA

Great Meteor East (GME) is an area at the western extremity of the Madeira Abyssal Plain, which is being studied to assess the feasibility of ocean disposal of radioactive waste (ref. 9). Although its geographical boundaries have never been precisely defined, most of the studies have been performed between latitudes 30° to 33°N and longitudes 23° to 26°W. As the studies progressed, a detailed study area, the so-called 10 km box, was selected as the area which appeared to have the most suitable geology and sediment properties for containing radioactive waste. The 10 km box is centred at 31° 17′ N and 25° 24′ W (Fig. 2).

Great Meteor East is generally very flat; the depth of the abyssal plain is between 5430 and 5450 m, with a slope of less than 1 in 3000 (ref. 10). However, about 30% of the study area is covered by abyssal hills, formed from the basement rock, which rise to a few hundred metres above the plain. Seismic reflection profiles suggest that the sedimentary section below the abyssal plain, which is typically 500 to 1000 m thick, can be split into two units (ref. 10). The uppermost unit, which varies in thickness from 150 to 560 m and contains many continuous, parallel, and relatively flat-lying reflectors, is thought to be made up predominantly of turbidites interlayered

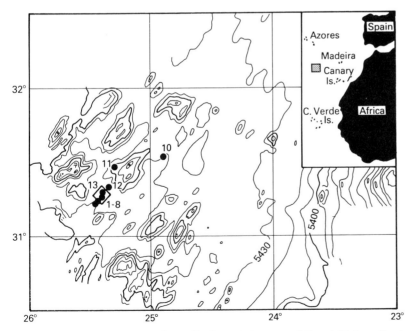

Fig. 2 Bathymetric map of Great Meteor East study area showing penetrator test sites.

with thin pelagic beds (refs 4, 10 and 11). Individual reflectors within this unit show a high degree of continuity across the abyssal plain, except where they are offset by faulting. Where abyssal hills stand out above the plain, they are usually covered by draped pelagic sediment about 100 to 200 m thick, although basaltic basement occasionally outcrops along steep sides or at the crests.

Detailed knowledge of the properties of the sediments is restricted to the uppermost 20 to 35 metres, which have been sampled using various devices (refs 4, 12–14). This interval consists of thick (typically 1.5 m) turbidites interlayered with thin (typically 0.1 m) pelagic beds. While most of the turbidites have come from the continental rise to the east, a few originated at the seamounts to the west. The thicker turbidites have sandy or silty bases ranging in thickness from a few centimetres to a few tens of centimetres. The fine-grained parts of the turbidites are marly oozes, generally containing 30–50% silt, 35–90% carbonate, 10–20% clay minerals and 0.1–1.0% organic carbon, with a mean grain size of 3.5 μm. The silty basal layers contain 30–90% carbonate intermixed with basaltic fragments and have a mean grain size in the range 2.5 to 200 μm.

The pelagic sediments have variable carbonate contents and range from pelagic clays to calcareous clays, marls, and oozes. The average carbonate content in the 10 km box is 44% and the mean grain size 3.5 μm. The superficial pelagic sediments on the abyssal hills are more indurated than those interbedded with the turbidites and contain several pumice layers. Manganese nodule fields are sometimes found on top of the pelagic sediments.

The seismic profiling records have identified numerous discontinuities in the reflectors, and these have been interpreted as faults (ref. 11). The faults are most prevalent in the central part of GME (ref. 14), where the upper sedimentary unit is thickest. No faults have been detected in the uppermost 200 m of sediment at the 10 km box, and few deeper ones have been identified with any degree of confidence.

TABLE I
Details of penetrator experiments

Test no.	Type	Telemetry system	Weight (kg)	Ballast	Surface finish	Density ($t\,m^{-3}$)	Test result
8601	X	ADSS[a]	1870	Steel shot	Unpainted	6.03	Good test
8602	X	ADSS[a]	2345	Lead shot	Rusty	7.55	Failed in water at 2050 m
8603	X	ADSS[a]	2345	Lead shot	Unpainted	7.55	Good test
8604	X	ADSS[a]	2300	Lead shot	Sand textured[c]	7.41	Signal level drop at 1200 m
8605	X	ADSS[a]	3165	Solid lead	Matt paint	10.19	Signal level drop at 4500 m
8606	X	ADSS[a]	2345	Lead shot	Sand textured[c]	7.55	Good test
8607	X	ADSS[b]	2325	Lead shot	Sand textured shaft/matt nose[c]	7.48	Failed in water at 955 m
8608	X	ADSS[b]	2330	Lead shot	Rusty	7.50	Poor signal quality
8609	1B	PATSY	1800	Solid steel	Gloss paint	7.86	No data transmission
8610	X	ADSS[a]	3130	Solid lead	Gloss paint	10.07	Good test
8611	X	ADSS[a]	3210	Solid lead	Matt shaft/ sand textured nose	10.33	Signal lost at 28.8 m penetration
8612	X	ADSS[a]	3145	Solid lead	Gloss paint	10.13	Good test
8613	X	ADSS[a]	1755	Steel shot	Matt paint	5.65	Good test
8614	1B	PATSY	1800	Solid steel	Gloss paint	7.86	Data not processed
8615	SV	Super-doppler	1370	Lead shot	Unpainted	4.77	Failed in water at 1375 m

[a]Uprated transmitters built by University of East Anglia
[b]Commercially available transmitters
[c]The tail sections of these penetrators were finished in matt paint

THE TYRO 86 EXPERIMENTS

The type X penetrators and uprated ADSS transmitters were tested for the first time from the MV *Tyro* at the Great Meteor East test site in October 1986; these trials were performed by BRE/JRC in collaboration with RGD (Rijks Geologische Dienst—Geological Survey of the Netherlands), who were using the *Tyro* to carry out geological, geochemical and geophysical testing at GME. Fifteen penetrators were tested: 12 type X penetrators with a variety of ballast materials and finishes, and instrumented with ADSS (ten of the uprated UEA transmitters were used); two type 1B penetrators instrumented with PATSY and one SV penetrator instrumented with Super-doppler. Details of the penetrator tests are given in Table I.

The tests were carried out at three sites (Fig. 2): penetrators 8609 and 8610 were deployed in an area known to contain faults, penetrator 8611 at a site flanking an abyssal hill, where the sediment cover was about 50 m, and the remainder at or near the 10 km box. All the penetrators were launched using the method shown in Figure 3. Signals from the three acoustic systems were recorded in two containerized laboratories using hydrophones deployed on davits or suspended from spar buoys. Details of the data collection and processing techniques are given in refs 3 and 4.

Fig. 3 Launching a penetrator from the stern of the MV *Tyro*.

Prior to their use on the type 1B penetrators, the PATSY units were tested on a torpedo-shaped, recoverable instrumentation sledge (ref. 15), which slides down a 19 mm diameter plastic-coated steel wire deployed from the stern of the ship (Fig. 4). The 1200 m cable is sufficiently long to ensure the sledge attains its terminal velocity before impacting the braking system which consists of four deceleration discs followed by a 4.5 m long spring-loaded damper filled with seawater. The tension between the discs and the cable is adjusted so that the deceleration of the sledge is similar to that experienced by a penetrator when it impacts the sea floor.

RESULTS

The results of the penetrator tests are summarized in Table II. Data were obtained from all but two of the penetrators instrumented with ADSS. The results of the ten successful tests are shown in Figure 5 as velocity versus the depth of penetration. The reason for the failure of the uprated transmitter used in test 8602 and the decrease in signal strength observed in tests 8604 and 8605 is thought to be leakage around the transducer; this was prevented on subsequent tests by the application of additional potting compound. The signal from test 8611 was lost at a burial depth of 25 m. This test was performed at the abyssal hill site, and the measured attenuation of the signal by the sediment was nearly four times greater than the largest attenuation measured in the other tests. The cause of the failure of the commercially available ADSS unit used in test 8607 remains a mystery.

No data were received from the first PATSY unit; the second unit functioned, but entered the fast-data-collection mode during the descent. It is thought that this was triggered by a fault in one of the electronic components resulting in the generation of a false reading on one of the accelerometer channels; details of the penetration event were therefore lost. However, the PATSY unit functioned successfully in the sediment, and inclinometer and pore pressure data were received for three hours before it was necessary to leave the site. These data have not yet been processed.

The SV penetrator was lowered to the sea bed twice on the RGD containerized winch prior to launch. On both tests the Super-doppler instrumentation functioned correctly and data were transmitted to the ship with a very low error rate. The shear vane mechanism, however, only performed part of its cycle, and it is suspected that one of the hydraulic components was leaking under pressure. Nevertheless, it was decided to launch the SV penetrator to test the ability of the Super-doppler unit to transmit through sediment. Unfortunately, the unit failed in the water column and it is thought that the cause of the failure

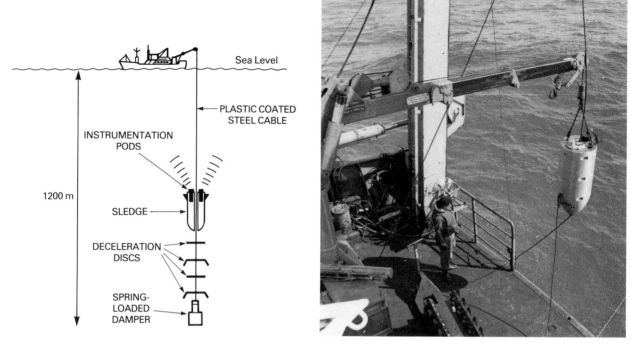

Fig. 4 The recoverable instrumentation sledge.

TABLE II
Summary of penetrator test results

Test no.	Impact velocity ($m\,s^{-1}$)	Drag factor	Nose penetration (m)	Signal attenuation ($dB\,m^{-1}$)	Latitude (°N)	Longitude (°W)	Site	Shear strength profile[a] (z = depth in m) (kPa)
8601	46.1	0.137	30.8	0.18	31° 16.1′	25° 24.7′	10 km box	$1.7+1.0\,z$
8602	44.5[b]	0.196	—	—				
8603	51.7	0.142	39.4	0.36	31° 14.9′	25° 23.4′	10 km box	$0.4+1.0\,z$
8604	43.8	0.193	34.0	0.10	31° 14.6′	25° 23.9′	10 km box	$-0.5+1.0\,z$
8605	64.5	0.128	53.4	0.31	31° 13.8′	25° 24.7′	10 km box	$-2.5+1.2\,z$
8606	43.1	0.204	33.4	0.08	31° 13.4′	25° 25.5′	10 km box	$-1.0+1.1\,z$
8607	43.0[b]	0.207	—	—				
8608	46.3	0.175	35.4	0.27	31° 14.0′	25° 24.8′	10 km box	$0.5+1.0\,z$
8610	68.1	0.113	56.2	0.32	31° 31.9′	24° 53.7′	Fault zone	$0.5+1.1\,z$
8611	63.0	0.136	—	1.34	31° 27.1′	25° 17.7′	Abyssal hill	$6.6+2.4\,z$
8612	67.6	0.116	58.4	0.34	31° 19.3′	25° 22.2′	10 km box	$0.4+1.0\,z$
8613	43.7	0.140	29.0	0.18	31° 18.9′	25° 23.7′	10 km box	$3.2+0.8\,z$
8615	30.2[b]	0.245	—	—				

[a]Calculated from equations (1) and (2) using: $N_1 = (12+v^{0.2})$; $N_2 = 2.0$, and $N_3 = N_1/45$
[b]Peak velocity

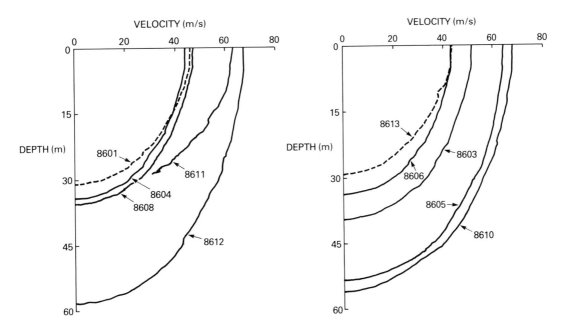

Fig. 5 Velocity versus depth of penetration for 10 successful
ADSS tests.

was the same as that experienced by the ADSS transmitter used in test 8602 (i.e. leakage around the transducer) as it had not been possible to add extra potting compound to the Super-doppler transmitter.

SIGNAL ATTENUATION

For a homogeneous deposit, the drop in signal power per metre (i.e. the attenuation of the signal measured in dB m^{-1}) is constant and proportional to the transmission frequency. Published data for abyssal plain sediments (ref. 16) suggest that the attenuation of the ADSS signal should be in the range 0.3 to 1.8 dB m^{-1}. The signal attenuations measured during the ten successful ADSS tests are listed in Table II; for the tests carried out in abyssal plain sediments, the attenuation varied between 0.08 and 0.36 dB m^{-1}, suggesting that the entry pathways may not have been completely full of sediment in some of the tests.

Coates *et al.* (ref. 17) discuss the relative merits of operating at 3.5 and 12 kHz; because the attenuation of the signal is proportional to the frequency of transmission, the attenuation of the PATSY signal will be only 30% of that for the signal from ADSS or Super-doppler. However, for the abyssal plain sites, signal attenuation is unlikely to be a critical consideration in the choice of instrumentation system. At a depth of 5400 m, the signal power from the UEA ADSS

transmitter is about 35 dB above the background noise level. Assuming an average attenuation of 0.35 dB m^{-1}, it should be possible, therefore, to track penetrators to burial depths of at least 100 m with this system. However, for other sediment types such as those encountered at the abyssal hill, PATSY may offer the only viable technique of transmitting data acoustically.

COMPARISON OF INSTRUMENTATION SYSTEMS

The *Tyro* 86 experiments have confirmed that ADSS is an excellent method of measuring the motion of the penetrator as it free falls through the water column and impacts the sea bed. Since it measures velocity, only one integration is required to produce measurements of penetration depth, and identification of the point of contact with the sea floor is facilitated by the step change in frequency caused by the difference in sound velocity between sediment and seawater (ref. 2). However, no other measurements are possible with ADSS, and the determination of the deceleration in the sediment requires differentiation of the raw data which accentuates any random noise.

PATSY is a far more versatile system, allowing the transmission of data captured at rates of up to

500 Hz and long term monitoring of sea bed experiments for periods of months or even years. However, PATSY is a complex and sophisticated electronic package and, as a result, has suffered reliability problems in the deep ocean environment. A disadvantage of PATSY is that, should a fault occur before the transmit mode is entered, no data are recovered and even the point at which the fault occurred cannot be established.

Super-doppler offers the benefits of velocity measurement during the free fall together with limited communication capabilities after the penetrator has come to rest. The ability of Super-doppler to transmit through sediment has not yet been demonstrated, although the attenuation of the signal should be simlar to that experienced by ADSS. Hence the attenuation should not prevent reception of the signal, although it may limit the data transmission rate.

SHEAR STRENGTH PROFILES

If the soil forces are assumed to be proportional to the undrained shear strength of the sediment (c_u), the measured rate of change of velocity (i.e. acceleration) of the penetrator can be used to derive a profile of sediment strength with depth. The motion of the penetrator in the sediment is governed by the force balance equation:-

$$\text{Mass} \times \text{acceleration} = \text{Buoyant weight in soil} - \text{Soil force} \qquad (1)$$

The BREPEN method of analysis (ref. 18) assumes the soil force to be made up of four components: end-bearing, tail losses, skin friction, and drag. The drag component is assumed to have the same form as hydrodynamic drag, and to be governed by the same drag factor, so that:

$$\text{Soil force} = N_1 A(c_u)_n + N_2 A(c_u)_t + N_3 A_s(c_u)_s + Vg(\rho_p - \rho_w)\rho_s v^2/(\rho_w v_i^2) \qquad (2)$$

where $(c_u)_n, (c_u)_t$ and $(c_u)_s$ = sediment strength at the nose, tail and averaged over the surface area,

A and A_s = cross-sectional and surface area,

N_1, N_2 and N_3 = coefficients (functions of v),

V = volume of penetrator,

ρ_p, ρ_w and ρ_s = density of penetrator, water and sediment,

v and v_i = current velocity and impact velocity, and

g = acceleration due to gravity.

Starting at the moment of impact, a time stepping approach is used to calculate the profile of c_u with depth. For each time step, the soil force acting on the penetrator is calculated from the measured deceleration using equation (1); this value of soil force is then substituted into equation (2) to determine $(c_u)_n$, using the values of $(c_u)_n$ calculated in previous steps to estimate $(c_u)_t$ and $(c_u)_s$.

For a normally consolidated soil, the variation of strength with depth is generally approximately linear; hence, a convenient method of comparing the results of different penetrator tests is to use a linear regression analysis to reduce the strength profile to a straight line. The results from the ten successful ADSS tests are summarized in Table II and it is evident that the agreement between tests is excellent, with the exception of test 8611. This was the test at the abyssal hill site and indicates that the sediments at this site are significantly stronger than the abyssal plain deposits.

INFLUENCE OF SURFACE FINISH AND DENSITY ON PENETRATION

In water depths greater than a few hundred metres, the penetrator will achieve its terminal velocity and the impact velocity can therefore be calculated from the equation:-

$$v_i = \{2Vg(\rho_p - \rho_w)/(C_d A \rho_w)\}^{1/2} \qquad (3)$$

where C_d is the drag factor and the other terms have the same meaning as before.

Figure 6 shows the measured variation of terminal velocity and penetration depth with penetrator density. The predicted variation of terminal velocity with density based on equation (3) is shown for three values of drag factor. The results indicate that the penetrators fall into three groups: those with a glossy finish have a drag factor of 0.113 to 0.116; those with a matt or unpainted finish have a drag factor of 0.128 to 0.142; and those with a rusty or sand-textured finish have a drag factor of 0.175 to 0.204. The one exception to this pattern is the test with the SV penetrator, which produced a drag factor of 0.245. This illustrates the sensitivity of the overall drag factor to poor streamlining of the tail section.

The predicted variation of penetration depth with density based on the BREPEN soil model and assuming a sediment strength gradient of $c_u = -0.6 + 1.1 z$ (the mean of the profiles given in Table II, excluding test 8611) is shown for penetrators with drag factors corresponding to the three different types of finish; the BREPEN program uses the drag factor to calculate the impact velocity, but assumes the soil forces

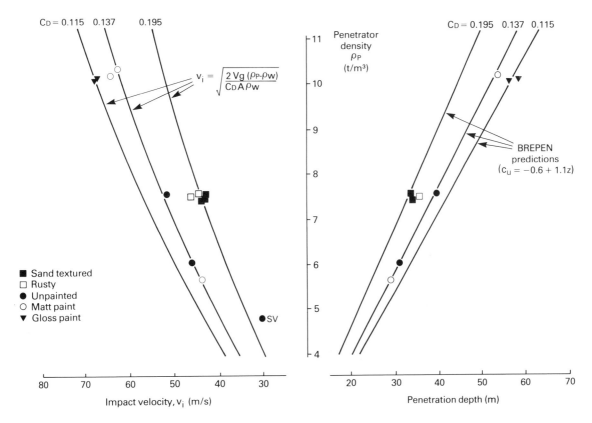

Fig. 6 Variation of terminal velocity and penetration depth
with penetrator density.

to be independent of surface finish. The good agreement between the predicted and measured penetration depths suggests that this assumption is justified.

CONCLUSIONS

The *Tyro* 86 experiments have demonstrated that penetrations of over 50 m in soft sediments are feasible and that, in abyssal plain sediments, an acoustic link with the penetrator can be maintained.

The reliability of the doppler instrumentation makes it the most practical and cost effective method of monitoring penetrator motion; signal attenuation measurements suggest that penetrations of up to 100 m could be monitored using the UEA transmitters.

Pulse interval telemetry and frequency shift keying are both viable techniques of data telemetry from deep ocean experiments, although transmission of FSK signals through sediment has not yet been demonstrated.

Consistent strength profiles were derived from the penetrator tests performed at the abyssal plain sites and were apparently unaffected by variations in penetrator density on surface finish. There were no measurable differences between the abyssal plain sites visited, but one test indicated the sediments at the abyssal hill site to be some 2.5 times stronger.

The drag factor of the penetrators varied from 0.113 for a smooth surface to 0.207 for a sand-textured finish. The motion in the sediment appears to be insensitive to surface finish and can be accurately predicted using the BREPEN soil model.

ACKNOWLEDGEMENTS

The authors wish to thank the captain, officers and crew of the MV *Tyro* for their help in performing the experiments. Particular thanks go to Rokus Hoogendoorn and his staff from the Rijks Geologishe Dienst who performed the difficult and labour-intensive operations involved in the launching of the penetrators safely and efficiently. The acoustic signals were recorded and processed by Ken Bromley of the

Building Research Establishment, Nick Cooper of Fugro Ltd, Chris Flewellen of the Institute of Oceanographic Sciences, and Richard Harvey of the University of East Anglia; their help in performing these experiments and writing this report is gratefully acknowledged.

The experiments described in this paper were performed as part of the radioactive waste management programmes of the UK Department of the Environment, the Commission of the European Communities and the Dutch Ministry of Economic Affairs. The results will be used by the UK, CEC and Netherlands in the formulation of waste management policy, but do not necessarily represent their respective policies. This paper appears with the permission of the Director of the Building Research Establishment.

REFERENCES

1. Freeman, T. J., Carlyle S. G., Francis, T. J. G. and Murray, C. N. (1984). The Use of Large Scale Penetrators for the Measurement of Deep-Ocean Sediment Properties. Proc. Oceanology Int. 1984, Brighton, March, OI 1.8.

2. Freeman, T. J., Murray, C. N., Francis, T. J. G., McPhail, S. D. and Schultheiss, P. J. (1984). Modelling radioactive waste disposal by penetrator experiments in the abyssal Atlantic Ocean. *Nature* **310**: 130–133.

3. Freeman, T. J. and Burdett, J. R. F. (1986). Deep Ocean Model Penetrator Experiments. Commission of the European Communities Nuclear Science and Technology series, report no. EUR 10502.

4. Shuttenhelm, R. T. E., Auffret, G. A., Buckley, D. E. *et al.* (ed). (in press). Geoscience Investigations of Two North Atlantic Abyssal Plains: the ESOPE International Expedition. Vols I and II, CEC–Joint Research Centre, JRC report.

5. Coates, R., Matthams, R. F. and Owens, A. R. (1985). The Design of an Acoustic Data Link for a Deep Sea Probe. Report to the Department of the Environment no. DOE/RW/85.130.

6. Flewellen, C. G. (1987). Development of a Pulsed Acoustic Telemetry System for Penetrators. Report to the Development of the Environment no. DOE/RW/87.097.

7. Murray, C. N., Weydert, M. and Freeman, T. J. (in press). An ocean satellite link for long term deep deployed instrumentation: concepts, experimental development and applications. *Deep Sea Research*.

8. Coates, R. (1987). The Super-doppler Penetrator Telemetry System. Report to the Department of the Environment, no. DOE/RW/87.031.

9. Nuclear Energy Agency (1984). Seabed Disposal of High-Level Radioactive Waste. Status report on the NEA coordinated research programme, OECD, Paris, ISBN 92-64-12576-0.

10. Searle, R. C. (1987). Regional Setting and Geophysical Characterisation of the 'Great Meteor East' Area in the Madeira Abyssal Plain. *In* Weaver, P. P. E. and Thompson, J. (eds). *Geology and Geochemistry of Abyssal Plains*, Spec. Publ. 31 Geol. Soc. London, pp. 49–70.

11. Duin, E. J. and Kok, P. (1984). A geophysical study of the Western Madeira Abyssal Plain. Mededelingen Rijks Geologische Dienst, vol 38–2, pp. 67–89.

12. Kuijpers, A., Schuttenhelm, R. T. E. and Verbeck (1984). Geological Studies in the Eastern North Atlantic. Mededelingen Rijks Geologische Dienst, vol. 38–2, 233 p.

13. Searle, R. C., Schultheiss, P. J., Weaver, P. P. E. *et al.* (1985). Great Meteor East (Distal Madeira Abyssal Plain): Geological Studies of Its Suitability for Disposal of Heat Emitting Radioactive Wastes. Institute of Oceanographic Sciences, Wormley, UK, report no. 193, 161 pp.

14. Shephard, L. E., Auffret, G. A., Buckley, D. E., Schuttenhelm, R. T. E. and Searle, R. C. (1987). Geoscience Characterisation Studies. *In* Anderson, D. R. and Hinga, K. (eds) *Feasibility of Disposal of High Level Radioactive Wastes into the Seabed*. OECD/Nuclear Energy Agency, Paris, vol. 3, 256 pp.

15. Murray, C. N. and Jamet, M. (in press) Development of a deep water high speed retrievable instrument sledge. *Ocean Engineering*.

16. Clay, C. S. and Medwin, H. (1977). *Acoustical Oceanography*, Figure 8.23, p. 260, Wiley, London.

17. Coates, R., Bromley, A. K. R. and Flewellen, C. G. (1987) Penetrator Telemetry. Proc. 6th Int. Conf. of Electronics for Ocean Technology, Heriot-Watt University, Edinburgh, March 1987.

18. Cooper, N. and Freeman, T. J. (in press). The BREPEN Software Package for the Prediction and Analysis of Penetrator Behaviour, Report to the Department of the Environment on contract PECD7/9/096.

27

The Disposal of Heat-Generating Nuclear Waste in Deep Ocean Geological Formation: A Feasible Option

C. N. Murray, Commission of the European Communities, Joint Research Centre, Ispra Establishment, 21020 Ispra (Va), Italy.

Since 1977 the International Seabed Working Group, under the auspices of the OECD Nuclear Energy Agency, has been carrying out a coordinated research programme to investigate the feasibility and safety of the disposal of heat-generating wastes into deep ocean abyssal plain formations. The main objectives of the research are to assess the long-term safety of the option, to identify characteristic study zones in the North Atlantic and Pacific Oceans, and to demonstrate the necessary engineering emplacement capability within oceanic geological formations.

The production of energy by nuclear power plants covers about 20% of the electricity generation of the European Community member states. One of the larger Community countries, France, has reached the point of producing over half of its electricity consumption by nuclear sources. As happens with all types of industrial production, waste is produced at all stages of the process, from uranium mining and refining to the final dismantling of the power plant after having reached the end of its service life. However, radioactive waste is not only generated by power production: industrial and medical uses of radioisotopes generate in the order of 15% of the amount of low level waste, and military radioactive waste production is probably in the same order of magnitude as the civil nuclear one in the countries having a nuclear defence weapon system.

Methods for the safe disposal of solid low, medium and high level radioactive wastes are therefore the subject of investigations in many countries.

A number of options for the final disposal of high level wastes in different kinds of deep geological continental formations using the concept of a mined repository are being actively studied. To assess the technical feasibility and long-term safety of these options a major effort is being placed on studying the natural and artificial barriers that isolate the wastes from man. A demonstration of engineering capability is also required at sites which have the required characteristics of isolation and long-term stability. The types of formations being considered for high level radioactive wastes within the European Communities are salt formations, clay and hard rock (granite), and sub-sea bed sediments.

Advances in Underwater Technology, Ocean Science and Offshore Engineering, Volume 16: Oceanology '88

Since 1977 an international coordinated programme has been developed under the auspices of the OECD Nuclear Energy Agency to study the feasibility and safety of emplacing vitrified nuclear wastes or unreprocessed fuel elements into deep ocean sedimentary formations (Fig. 1). The objective of the research programme of the Seabed Working Group is to provide scientific and technical information to enable international and national authorities to assess the long-term safety and engineering feasibility of emplacing high level radioactive wastes packaged in suitable containers at some depth into the sedimentary geological formations of the deep ocean floor.

The characteristics that potentially make this option attractive are that the formations under investigation are stable and large in extent; they contain little or no strategic materials and consist of sedimentary material which has been shown to strongly retain most radionuclides. The nature of this material is such that it tends to seal itself after natural or artificial disturbances. The fact than many of the areas under investigation are at water depths of greater than 5 km ensures that accidental human intervention is extremely unlikely.

A common factor in the studies of the different geological forms is the identification of natural and artificial barriers which will retard the dispersion of

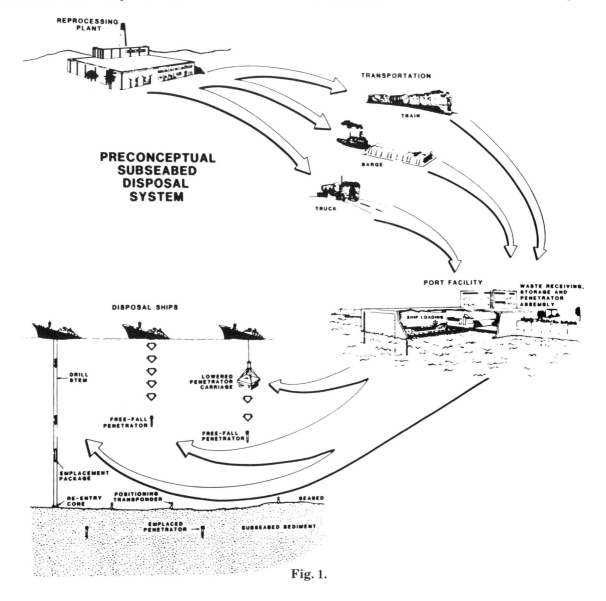

Fig. 1.

the waste radionuclides and assure the degree of isolation needed for their safe long-term storage.

The barriers that have been identified are:

- *The waste form* Waste can be produced in a highly stable and insoluble form.
- *The container* This can be designed to assure containment of the waste during its most active phase where heat production is greatest.
- *Backfilling material* This can be made with materials which have similar or complementary properties of the geological environment.
- *Geological isolation* Geological formations can be chosen so as to minimize the possible contract between the waste and the biosphere.
- *Geochemical retention* Many of the radionuclides of importance which have long half-lives are strongly bound to the geological media of the formation.
- *The environment* Owing to the large capacity of the environment to dilute and isolate radionuclides, levels of any nuclides that may eventually find this way to the biosphere will be very significantly lower than natural radioactivity levels.

Investigations are thus being carried out to identify the major mechanisms which operate in these barrier and control the rate of dispersion of radionuclides. The development of a mathematical description of these processes to model realistically the dispersion of materials in each compartment of the overall system can then be used to investigate the long-term safety of the option.

Although the extrapolation of data into the far future cannot be substantiated by direct experimental verification, the use of historical data allows an understanding to be gained about the likely natural processes that could interact with the repository and the rates that have occurred in the past. By using a probability distribution for the controlling parameters it is possible to estimate the effect and magnitude of these processes in the future on the ability of the geological formation to safely segregate high level wastes.

RADIOLOGICAL SAFETY

To assess the feasibility of the disposal option a clear understanding of the major mechanisms that will interact with the emplaced nuclear waste must be obtained. The safety of the option consists in estimating the radiological detriment to man and the environment which could result from the disposal of high level waste within the sea bed sediments of the deep ocean (ref. 1). To make this

assessment models have been developed and data obtained on the major components of the system, including corrosion of the waste package, leaching of vitrified glass or unprocessed fuel rods, transport of radionuclides in the sediments, dispersion and mixing in ocean waters of nuclides escaping from the sediments, processes of scavenging of radionuclides from ocean waters by sorption of particulates and sedimentation, and the pathways to man via consumption of seafood, sea salt, desalinated water, by inhalation of airborne sediments or marine aerosols as well as by external irradiation.

The assessments have been made using both deterministic and stochastic methodologies for three scenarios: (1) a base case or 'normal' case where the waste is buried at its prescribed depth and all the barriers behave as anticipated, (2) an abnormal scenario where one or several of the components behave in a non-normal fashion, and (3) scenarios of transportation accidents occurring in coastal and in deep sea areas.

The use of two independent methodologies makes it possible not only to determine the most likely dose resulting from each scenario but also the range of uncertainty of this estimation, given the uncertainty in the available data. The Radiological Assessment Task Group calculate for the base case that the peak dose to the maximally exposed group is on the order of 10^{-9} Sv a^{-1}. This is about 10^5 times smaller than the ICRP recommended limit of 10^{-3} Sv a^{-1} and 10^5 times smaller than natural background doses. The uncertainty in the value ranges between 3×10^{-15} and 3×10^{-8} Sv a^{-1}.

Only three abnormal scenarios were found to be able to increase doses significantly: the emplacement of penetrators of depths of less than 10 m in the sediment, the existence of an upward water velocity on the emplacement site, and a change in the retention properties of the sediments for radionuclides. Individual doses were found to be increased by factors of 10^2 to 10^4; however, their probability of occurrence remained extremely small. The calculation did nevertheless underline the need for a very clear understanding of site sediment characteristics.

For transportation accidents it was found that doses could be very severe, especially for accidents in coastal waters. However, the design of the vessel and organization of recovery actions could make probability of the occurrence of such doses extremely small.

SITE ASSESSMENT

Another important aspect of the international

investigation on the sub-sea bed option are the geo-science characterization studies which have been carried out on a selected number of zones in both the Atlantic and Pacific Oceans. The objectives of these investigations were to develop site assessment guidelines and study procedures which could form the basis of evaluating geoscience feasibility.

Some fifteen study areas have been evaluated to varying levels of detail. Because of the need to focus available resources on a limited number of areas, three areas were retained for more critical inves-tigation: two Atlantic areas (Great Meteor East in the North-east Atlantic and the Southern Nares Abyssal Plain in the north-west Atlantic) and one north Pacific study Location (E2) approximately 1100 km east of Hawaii. Detailed geoscience studies were conducted at these locations over approxi-mately the last ten years.

Three main factors have been identified (ref. 2) for establishing feasibility: areas must have predict-able characteristics both in time and space to ensure long-term stability and to enhance the level of confidence in nuclide transport risk assessment models; the areas must be geologically stable with a minimal probability of tectonic activity disrupting the integrity of the sediment barrier; and chemical transport by pore water advection must be less than the rate of chemical diffusion.

The Site Assessment Task Group concludes on the basis of all available data that sites can be found that will satisfy at least the minimal geoscience requirements for a potential sub-sea bed high-level waste repository.

ENGINEERING FEASIBILITY

In parallel to the studies of the long-term safety of the option provided by barrier properties which will retard the dispersion of waste and site specific investigations which are aimed at characterizing study areas in the Atlantic and Pacific Oceans having properties such as to ensure safe long-term containment, major efforts have been made to demonstrate the feasibility of critical portions of engineering emplacement system.

To undertake this task, the Seabed Working Group formed the Engineering Studies Task Group in 1981 which was given the objective of identifying potential engineering methods for sub-sea bed dis-posal and to establish order of magnitude costs (refs 3,4). This work has to be coordinated with other task groups already studying the option.

In investigating these tasks, the ESTG recognized that a variety of methods might be capable of being

used to emplace heat-generating wastes. The group did not consider it their duty to identify the optimum concept but rather to identify the major engineering obstacles and to attempt to provide plausible methods for dealing with them. It was recognized that an extensive programme of analysis and experimentation would be necessary and that this would very probably require the development of special equipment and instrumentation.

Early investigators of sea bed disposal developed over twenty concepts for emplacing high level waste. These have been reviewed by a number of researchers (refs 5–8). In general all the options identified can be split into two classes: those methods in which the waste package is driven directly and those in which a hole or trench is first formed and into which the waste package is then inserted.

On the basis of the above reviews of the different options, two emplacement methods were chosen for further study (ref. 8) which it was believed well represented the two classes of methods identified. These were insertion of waste into drilled holes and by free-falling sea bed penetrators (Fig. 2). Because it was felt that much less was known about pene-trators, an experimental programme was defined by the ESTG to investigate this option in detail. The two options will be briefly described in the next sections.

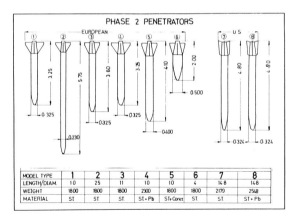

Fig. 2.

EMPLACEMENT BY DRILLING

Drilling at a depth of 6000 m is established tech-nology and has been performed in a number of cases for relatively small diameter holes. A study of the drilling operation to obtaining holes of nearly 1 m in diameter has shown that no major technical dif-ficulty exists for the drilling of a hole, its reaming ut to the required diameter, the installation of a casing, or the removal of the casing from the upper part of the hole.

A study by Taylor Woodrow in the United Kingdom (ref. 9) has allowed the development of a referred solution for emplacement, which will be escribed briefly

Holes will be drilled into the selected ocean sediment layers, which are up to 1000 m thick, stopping approximately 200 m short of the bedrock; the holes which are spaced horizontally at 400 m, are to be drilled by a conventional drill-ship which is hired only for these operations and works quite separately from the disposal operations.

Waste canisters will be brought from land by specialized supply vessels in shielded flasks weighing about 150 t to a disposal platform which is a semi-submerged concrete structure with a displacement of 480 000 t; unloading of the flasks will probably be done using a transfer bridge between supply vessel and the platform.

On the platform, the canisters will be handled in a submerged cell. A number of canisters are placed by remote control in steel pipes and the annulus between canisters and pipes is filled with a suitable filler material; the pipe loaded with canisters is then lowered and a second pipe is attached to it; the operations are repeated until the desired number of canisters for emplacement are assembled, forming a string of up to 600 m length.

The string of canisters is then lowered to the well-head by dynamic positioning of the platform and using sonar guidance and TV cameras. The remote entry operation continues until the load reaches its destination in the well.

After filling the holes, the annulus between the pipes which hold the cylinders is filled with a suitable grout. The casing in the top 200 m will then be removed by reaming and the top section of the hole backfilled, creating the required barrier to the sea bed.

A study carried out (ref. 10) on the reliability of sposal operations for both drilled and penetrator nplacement identifies factors that play a major le in reliability. These are the severity of the nsequences of a fracture, which can perhaps be presented by maximum effective dose equivalent; e probability values assigned to a specific enario; and the values assumed to represent izard to operators. Other factors, such as cost and st emplacement differences, must also be taken

into account when comparing the two technical operations.

FREE FALL PENETRATOR OPTION

The ESTG considered that because little was known about the behaviour of these vehicles, a sustained experimental effort would have to be made to develop a realistic assessment of this technology (ref. 11). The main components of the programme consisted of mathematical modelling of sea bed penetrator behaviour, development and testing of penetrators, *in situ* testing of site geotechnical sediment properties, and *in situ* investigations of penetrator hole closure.

Modelling studies

To better understand the behaviour of free-fall penetrators in the water column and to predict the depth of penetration for a given load of solidified waste, modelling studies have been developed by UK, USA, France and the CEC to optimize penetrator design (refs 12–14).

The depth of burial of a penetrator is directly related to its velocity at impact with the sea bed. The terminal velocity of the vehicle is reached when the drag forces are balanced by the gravitational force accelerating the body. Thus the velocity is related to the vehicle buoyant weight and the frontal area drag coefficient. In other words the depth of penetration is a variable of weight of vehicle and its density, other factors being kept constant. As an example of attainable penetration depths Figure 3 shows the variation in emplacement depth for penetrators against two variables: overall weight and density.

Another important factor is the accuracy with which a penetrator can be emplaced at a given site.

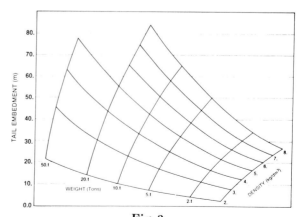

Fig. 3.

Analysis has shown that the main variables controlling the flight path of a penetrator are ocean currents, launch angle, fin alignment, and stability (a variable of secondary construction misalignments and drag effects). Fin alignment appears to be the most critical design variable, but it appears that this can be considerably reduced by designing the fins deliberately to produce spiralling.

In situ penetrator testing

From 1983 to 1986 over fifty large model penetrators have been used to investigate their behaviour in the water column and during impact and penetration in deep ocean sediments. The main objectives of the tests, at both Great Meteor East and Nares Abyssal Plain sites in the North Atlantic, were to compare model predictions on penetration depth with different forms of penetrators and to test new instrumentation and telemetry systems developed to gain in situ data on the geotechnical properties of the sediments. Results of tests carried out in 1986 on the MV Tyro will be discussed elsewhere (see Freeman et al., Murray et al., this volume).

In general terms it has been successfully demonstrated that free-fall penetrators can be emplaced at great depth within the sediment formations studied (≥ 60 m) and that in conjunction with instrumentation specially developed for these investigations, the penetrator is a potential tool for obtaining in situ information on selected geotechnical and eventually geochemical properties of marine sediments.

Hole closure field trials

The question of hole closure and entry-path sealing has been a dominant part of ESTG activities since it began its work. On the basis of modelling studies of the interaction of marine sediments during and after the passage of free fall penetration it became clear that shallow water experiments, at depths of greater than 200 m, would produce the conditions necessary to observe most of the phenomena. This was fortunate, as it had been also recognized that deep ocean experiments were beyond the current technical and financial abilities of the group in the time frame envisaged by the SWG for the study.

The results of the modelling suggested that the large negative Bernoulli pressure created behind the penetrator, due to its high forwarded velocity, would induce the collapse of the sediment wall into the impact zone.

In situ experiments were therefore planned to investigate the disturbed area caused by the passage of a penetrator. This experiment was undertaken near Antibes in November 1986 on the southern coast of France. The site possessed the appropriate water depths and a homogeneous but stiff clay. On the basis of preliminary experiments it was decided that the impact point would be acoustically marked by a sonar reflector slid down a trailing wire. A hydrostatic corer platform equipped with a sonar system, low light level video camera, piezo cone probe etc. was positioned over four impact sites determined by the sonar reflector. The geotechnical probe made measurements of cone penetration resistance, friction and pore pressure down to 9 m behind two penetrators, and down to 30 m at a location 5 m away. Two 6.5 m cores were recovered over impact sites, and a third core of undisturbed soil was obtained. Video inspection of the impact sites appeared as shallow depressions 10–20 cm deep and of a slightly smaller diameter than the penetrator. The bottom was visible and appeared closed.

The geotechnical probes detected no differences in the strength and pore pressure behaviour between disturbed and undisturbed sites. The results of the experiments suggest that the properties of the sediment filling the entry path are similar to intact non-disturbed soil.

ECONOMIC CONSIDERATIONS

Costs for the whole disposal operation have been estimated for a number of terrestrial repositories. The initial investment and cumulated operation costs are high in absolute amounts, ranging from one to several billions of US dollars, depending on the size of the repository. In terms of the current electricity generation, these costs represent only 1% to a maximum of 3% of the electricity-generating costs.

First estimates of the costs for emplacement in the deep sea bed sediments (refs 9,11) give figures of 0.2% to a maximum of 0.3% of the generating cost for drilled emplacement. The option of emplacement by free-fall penetrators will lead to even lower values. However, any changes in the assumed methods of emplacement or in the requirements imposed by regulatory agencies could grossly affect the results. For example, a reduction in the number and type of vessels required for drilled emplacement, now assumed to be a drill ship, a semi-submersible and a transport ship, could greatly reduce overall costs. On the other hand, more stringent requirements for accident-resistant structures in penetrators could significantly increase the overall cost of penetrator emplacement due to the large numbers required.

The ESTG report concludes that, regardless of adjustments that may be made in design or cost estimates, either emplacement method would probably be affordable and would compare favourably with other geological waste disposal methods.

SUMMARY

The Seabed Working Group has carried out a coordinated research programme into the feasibility and safety of disposing of heat-generating nuclear waste into the deep ocean sedimentary formations. The results of these investigations are to be published by the Nuclear Energy Agency of the OECD.

On the basis of the data available at present from this programme it appears that the option is technically feasible and safe, and that economic engineering emplacement systems can be developed. Political considerations are probably the most outstanding factor for its further investigation.

ACKNOWLEDGEMENT

This present brief overview draws strongly on the investigations carried out by the Seabed Working Group and especially on the results and conclusions presented to the NEA by SWG Task Groups in their Final Reports on the Feasibility and Safety of Sub Seabed Option.

REFERENCES

1. Radiological Assessment Task Group Final Report. Radiological Assessment NEA Seabed Working Group (in preparation).
2. Site Assessment Task Group Final Report. Geoscience Characterization Studies. NEA Seabed Working Group (in preparation).
3. Anderson, D. R. (Ed.) (1980). Proceedings of the Fifth Annual NEA-Seabed Working Group Meeting, Bristol, UK, March 3–5, 1980. SAND 80-0754, Sandia National Laboratories, Albuquerque, USA.
4. Anderson, D. R. (Ed.) (1981). Proceedings of the Sixth Annual NEA-Seabed Working Group Meeting, Paris, France, February 2–5, 1981. SAND 81-0427, Sandia National Laboratories, Albuquerque, USA.
5. Talbert, D. M. (1980). Subseabed Radioactive Waste Disposal Feasibility Programme: Ocean Engineering Challenges for the '80's. SAND 80-0304, Sandia National Laboratories, Albuquerque, USA.
6. Atkins Planning (1981). Concepts for the Disposal of High Level Radioactive Waste: Task 7, the Deep Ocean Bed. DOE/RW/82/015. Department of the Environment, UK.
7. Manschot, D. and Van Stein-Callenfels, J. E. (1981). A Survey of Emplacement Methods for Radioactive Canisters in Deep Sea Ocean Sediments. Marine Structure Consultants Report to Deep Ocean Research Apparatus (DORA) Partners, under contract to the Commission of European Communities.
8. Engineering Studies Task Group Final Report: The Emplacement of High Level Radioactive Waste in Deep Ocean Sediments. NEA Seabed Working Group (in preparation).
9. Bury, M. R. C. (1985). The Offshore Disposal of Radioactive Waste by Drilled Emplacement: a Feasibility Study. Published for the Commission of the European Communities by Graham and Trotman, EUR 9754 EN.
10. Sarshar, M. M. (1986). Reliability of Sub-seabed Disposal Operations for High Level Waste. Published for the Commission of the European Communities by Graham and Trotman, EUR 10542 EN.
11. Ove Arup and Partners (1985). Ocean Disposal of Radioactive Waste by penetrator emplacement. Published for the Commission of the European Communities by Graham and Trotman, EUR 10170 EN.
12. Freeman, T. J. and Burdett, J. R. F. (1986). Deep Ocean Model Penetrator Experiments. Nuclear Science and Technology, EUR 10502.
13. Boisson, J. Y. (1985) Etudes des conditions d'enfouissement de pénétreurs dans les sédiments marines. Nuclear Science and Technology, ERU 9667.
14. Visintini, L. and Marazzi, R. (1985). Hydrodynamic Analysis and Design of High-Level Radioactive Waste Disposal Model Penetrators. In: JRC Report Series Study of the Feasibility and Safety of the Disposal of Heat Generating Wastes into Deep Oceanic Geological Formations, SP I.07.C2.85.47, 60 pp.
15. Geoscience Investigations of Two North Atlantic Abyssal Plains: The ESOPE International Expedition. To be published by the Commission of the European Communities.

Part V

Marine Environmental Considerations

28

Mathematical Modelling of Sewage Discharges into Tidal Waters

T. Keating, S. Steines and *H. G. Headworth*
The Southern Water Authority, UK

Southern Water is one of ten regional water authorities in England and Wales, and as part of its river basin management role it is responsible for sewage disposal in the counties of Hampshire, Sussex, Kent and the Isle of Wight. The volume of sewage effluent which SWA has to deal with amounts to some 1100 megalitres per day (ml d^{-1}) from a population of 3.7 million, and 70% is discharged directly into estuaries and coastal waters following treatment or partial treatment.

The standards to which each individual discharge must conform are set having regard to the nature of the receiving waters, whether they be rivers, estuaries or the sea. Effluent discharges entering rivers and estuaries are generally controlled in terms of their biochemical oxygen demand (BOD), suspended solids, and ammonia loads. Appropriate discharge consents for these are set having regard to river and estuary quality. Rivers and estuaries are accorded a quality standard based on the 1979 National Water Council Classification (ref. 1). Some 2200 km of rivers in Southern Water's area, represented by some 540 river stretches, come within this classification system, of which 1460 km meet the standards of Classes 1A and 1B (good quality) and 700 km meet Class 2 (fair quality). Likewise, 380 km of its estuaries lie within the NWC Estuary Classification (ref. 2), of which 320 km are assigned to Class A (good quality) and 60 km to Class B (fair quality).

Unlike discharges to rivers and estuaries, those made to the sea seldom receive any treatment other than screening and maceration of solids. The design and siting of an outfall must ensure that the EC directives on the quality of bathing waters (ref. 3), the discharge of dangerous substances (ref. 4) and the quality of shell fish waters (ref. 5) are complied with.

The effects of proposed discharges on non-tidal rivers are usually assessed on the basis of river flow, existing quality, and the proposed effluent loadings. Stochastic techniques can be used to assess more precisely the effluent quality required to ensure 95% overall compliance. However, assessing the effects of discharges into tidal waters is more complicated because of the tidal regimes and the complex biochemical processes that occur. Increasingly, math-

Advances in Underwater Technology, Ocean Science and Offshore Engineering, Volume 16: Oceanology '88

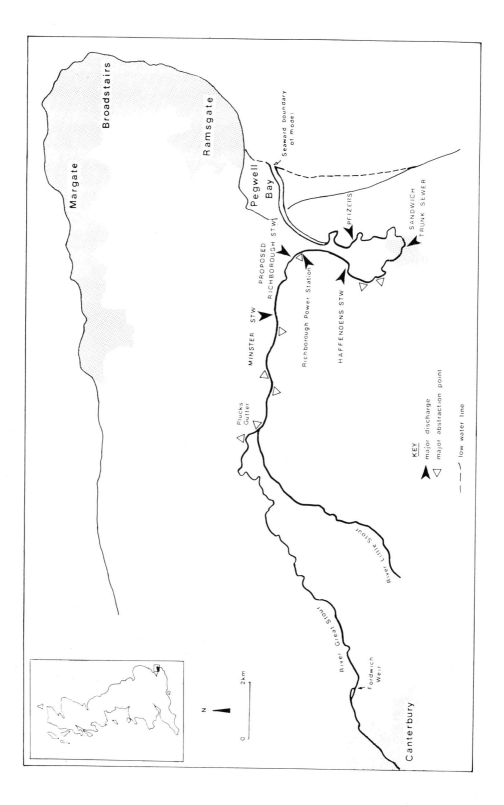

Fig. 1 The Stour Estuary, Kent.

ematical models are employed to help assess these effects, and the nature of the receiving waters determines the type of model which is needed.

This chapter describes the environmental effects of sewage discharges into estuarine and coastal waters. Two case studies illustrate the mathematical modelling techniques currently being used within Southern Water to assist the design process. The first case study describes a model of the River Stour Estuary, and its use in determining appropriate BOD and ammonia standards for a new sewage works serving Ramsgate. The second case study describes a tidal model of the Solent, which has been used to simulate the dispersion of sewage from a new sea outfall at Cowes on the Isle of Wight. This is being constructed as part of a major scheme to improve the water quality in and around the River Medina Estuary.

STOUR ESTUARY MODEL

Several large holiday resorts lie on the promontory at the eastern extremity of southern England, known as the Isle of Thanet (Fig. 1). In the summer months the population of the coastal towns swells from some 150 000 to around 230 000. It has long been recognized by Southern Water that unsatisfactory arrangements exist there for sewage treatment disposal, and a major programme of construction has been started to remedy this situation. For the towns of Margate and Broadstairs schemes are under way which will dispose of screened and macerated sewage into the sea via outfalls up to 2 km long. Of the schemes which have been studied for dealing with Ramsgate's sewage, the discharge of a fully treated effluent into the tidal reaches of the River Stour at Richborough is considered the most suitable. Such a works would cater for a 19.5 Ml d^{-1} discharge, or 21.8 Ml d^{-1} if sewage from the town of Sandwich is also treated at the works.

The need to assess the impact of an effluent discharge from an inland works into the River Stour led to the use of a numerical model of the river and estuary. This model had been constructed some years earlier by Hydraulics Research (ref. 6) in connection with the promotion of a storage reservoir near Canterbury.

Definition of effluent standards

To ensure that a discharge from the proposed Richborough Sewage Treatment Works does not lead to a deterioration in water quality in the River Stour and its estuary the key aspects of water quality must be evaluated. In the freshwater part of the River

Stour the existing water quality criteria is the NWC Class 2 designation which has 95 percentile limits for BOD of 9 mg l^{-1}, and for DO of 40%. The tidal Stour has Class A status in the NWC Estuary classification scheme which has regard to biological, aesthetic and chemical quality, although the model only has regard to chemical parameters.

The minimum dissolved oxygen in the estuary and the length of time over which it persists determine the health of the resident fauna and the movement of migratory fish. It is considered that the minimum DO in the Stour should be approximately 40%. This is considered more critical than the NWC criterion of a mean DO of 60%, since this ignores the condition of low DO most harmful to estuary fauna. The ammonia standard for the river needs to comply with the European Inland Fisheries Advisory Commission (EIFAC) limit for un-ionized ammonia which is aimed at protecting coarse fish and permitting movement of migratory fish through the estuary. The EIFAC limit for a Class 2 freshwater river is 0.025 mg l^{-1}, and with the relatively high pH of the River Stour (around pH 8), this corresponds with a total ammoniacal nitrogen concentration of approximately 0.7 mg l^{-1} (NH$_3$-N) at 20°C.

Model basis

The Stour Estuary Model is a one-dimensional (depth and width averaged) fixed element model originally designed to simulate tidal propagation. The model accommodates the steep salinity gradient which characterizes the estuary as well as the complex tidal patterns which occur. It also models the supply of river water by gravity to adjacent low lying farm land for crop irrigation and wet fencing. Water quality subroutines to determine oxygen and nitrogen balances have now been coupled to the hydraulic equations. Schematically, the estuary is divided into 77 elements each 500 m long, covering the 35 km length of the River Stour from its mouth to the upstream tidal limit, and 2.5 km of a tributary stream from its confluence with the main river.

Water and solute movements are governed by the laws of conservation of mass and momentum. The processes simulated by the model are described by a series of differential equations which are solved numerically using a finite-difference method. The hydraulic equations are solved for water levels and velocities in each section at discrete time intervals and the results used in the mass balance equations for organic carbon, organic nitrogen, ammonia, oxidized nitrogen, and dissolved oxygen. The equations specify the rate at which substances are changed by advection and diffusion and contain terms for the rate of addition of pollutants from outfalls, the rate of decay by biological or chemical degradation, and the rate of formation of other substances as a result of breakdown.

Field data

The information required to run the model are: tidal levels in Pegwell Bay; river flow and temperature; river solute concentrations; effluent discharges (flow and composition); and abstractions from the river.

Tidal levels in Pegwell Bay were generated using 24 tidal harmonic constituents derived from a 1-year analysis of tides at Ramsgate by the Institute of Oceanographic Sciences in 1963. Model runs were made for a 30-day period in July and August 1985, which coincides with the peak of the crop irrigation season. The first 15 days were treated as a run-in period to stabilize the model; only the results of the last 15 days were used.

A uniform 95 percentile low river flow of 177 Ml d^{-1} was assumed throughout the 30-day period (i.e. the flow which is exceeded for 95% of the time). This compares with the average flow in the Stour of 400 Ml d^{-1}, and the 1 in 50 year minimum 10-days flow of 112 Ml d^{-1}. The concentrations of BOD, ammonia, organic nitrogen, nitrate and chloride used were based on analyses for the period 1975–85. Dissolved oxygen concentrations in the upper reaches of the river average between 90% and 105%. 100% saturation was assumed at the upstream boundary. In reality, the river is often supersaturated during the daylight hours of the summer owing to algal photosynthesis, but the effect of algal respiration on the night-time DO is not known. The model, though time-dependent with regard to tidal movements, treats the river and discharge loadings as constant so the effects of diurnal fluctuations on water quality are not simulated. River quality in the summer is exacerbated by the reduced solubility of oxygen in warmer water. Therefore to avoid an over-optimistic result a high water temperature of 20°C was assumed in the model for the river and the effluent discharges.

Six effluent discharges to the river are included in the model simulations: three are from sewage treatment works, two are from industrial plants manufacturing paper and pharmaceuticals, and one is cooling water from a power station. The loadings of BOD, ammonia, organic nitrogen, nitrate, dissolved oxygen, and chloride were obtained from historic data, where available, or from their consented flows and concentrations. The biochemical oxygen demands exerted by river bed sediments were also estimated in the model.

The three principal abstractions from the river were also included in the model. These are by Richborough Power Station for cooling water, by Southern Water for public water supply, and by riparian landowners for irrigation of the low-lying farmland.

Use of the model

The main use of the model in this study has been to predict the effect on river quality of various standards of effluent quality from the proposed Ramsgate sewage works to enable the design of a suitable treatment plant. Other effluent discharges to the river are made downstream of the proposed new works. These have a polluting effect on the river and they interact with one another, as they will with the proposed new discharge, and the model has been used to assess the most acceptable arrangement for effluent treatment to minimize the overall effects on river quality.

As a follow-up to this work the model has been used to assess the possible water resources benefit of the proposed Richborough discharge and whether the Stour can support greater abstraction by Southern Water upstream at Plucks Gutter for treatment and public water supply.

Results of the study

Outputs from the model are solute concentrations for each of the 77 elements at 15 minute intervals over 15 days. Results are expressed graphically as (1) profiles along the river showing concentrations of ammonia, dissolved oxygen, etc. for selected time-steps and (2) concentrations of these determinands for a selected model element over the 15-day period. The form these results take are shown on Figures 2 and 3 for the two most important criteria, ammonia and dissolved oxygen.

The study shows that the lowest DO occurs at low water on spring tides, although the minimum DO at all tides during the month varies only by about 5%. Conversely, the highest ammonia concentration generally occurs at low water on neap tides or shortly afterwards. However, again, for any set of effluent loadings the difference in maximum concentration for all tides varies only by about 0.2 mg l^{-1} over the month.

Under present conditions, the effect of existing discharges from the two industrial plants and the raw sewage discharge from the town of Sandwich is to cause DO to fall to around 63% saturation with a maximum ammonia concentration of 0.7 mg l^{-1}. (As with all other results they relate to the conditions of flow, temperature, abstraction, etc. described

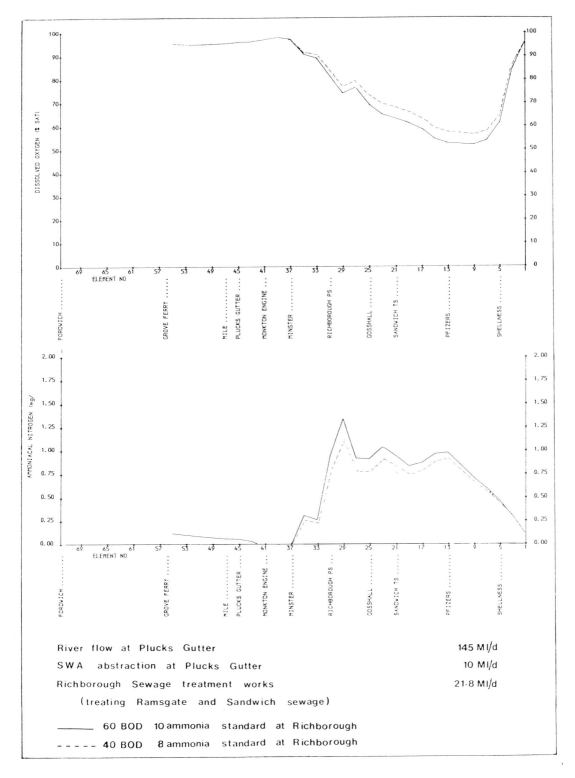

Fig. 2 Dissolved oxygen and ammonia profiles at low water
as predicted by the Stour model.

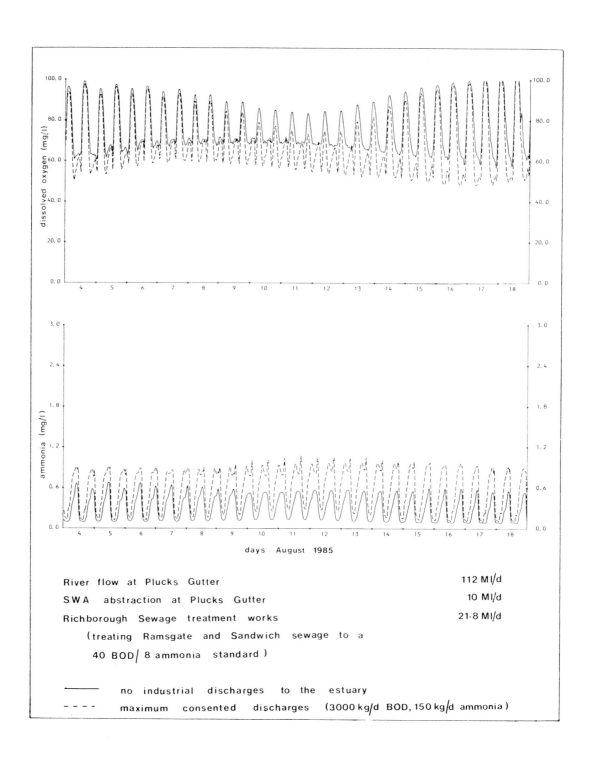

Fig. 3 Dissolved oxygen and ammonia profiles at element
11 of the Stour Estuary model.

earlier.) These figures are similar to what is known to occur in the river in practice.

When the anticipated increase in effluent from Sandwich by the turn of the century is taken into account, the resulting minimum DO would be 47% and the maximum ammonia concentration would be 1.32 mg l^{-1}. The latter figure would be unacceptably high, and it shows that any scheme for treating and discharging Ramsgate sewage effluent into the river should include the treatment of the crude sewage from the town of Sandwich. If effluent from Sandwich were to be treated and disposed of elsewhere than in the River Stour, the effluents from the industrial plants would still remain, and these themselves have sufficient effect on river quality as to make the removal of Sandwich effluent of little benefit.

Various degrees of treatment for the Richborough effluent were examined by the model. These varied from a relatively low level of treatment producing an effluent having a BOD of 60 mg l^{-1} and ammonia concentration of 50 mg l^{-1} to effluents resulting from a high level of treatment, having a BOD of 30 mg l^{-1} and ammonia concentration of 5 mg l^{-1}. In addition, the effects of lower river temperatures were also examined. The results showed that an effluent with an ammonia concentration much below 10 mg l^{-1} did not lead to a significant improvement in quality in the river, and that the difference in DO levels in the river from effluent BODs between 30 and 60 mg l^{-1} was only between 63 and 67%. However, the treatment of Sandwich effluent with that of Ramsgate at the proposed Richborough site does have a noticeable benefit to river quality with the minimum DO improved from 44% to 59% and the maximum ammonia concentration reduced from 1.2 mg l^{-1} to 1.0 mg l^{-1}.

Consequently it was concluded that the most suitable arrangement is to design and construct a works to treat Ramsgate and Sandwich sewage together to produce an effluent with a BOD of 40 mg l^{-1} and ammonia level of 8 mg l^{-1}. This arrangement would produce a minimum DO and maximum ammonia concentration of 59% and 0.96 mg l^{-1} respectively, and this would not be detrimental to the river quality.

Resource benefit of the Richborough discharge

The model shows that the proposed discharge of some 22 Ml d^{-1} of treated sewage into the river at Richborough acts like a small impoundment and raises river levels upstream. Southern Water is authorized to abstract 10 Ml d^{-1} at Plucks Gutter provided the river flow is in excess of 145 Ml d^{-1}: this condition ensures that the level of water in the river is sufficient for crop irrigation. Since flows less than this occur in many years, it means that the abstraction cannot be considered a 'reliable yield'. The addition of the Richborough effluent discharge to the river should raise river levels and so make it possible for the removal of the low flow restriction to permit abstractions in all but the very driest years. Southern Water is currently examining the practical implications of this in terms of environmental and conservation interests.

SOLENT TIDAL MODEL

Cowes, Isle of Wight (Fig. 4) is a world-famous centre for sailing and it attracts many tourists, particularly during the summer months when the population may double. At present Cowes sewage is discharged untreated through numerous outfalls into the Medina estuary and the Solent which results in visible slicks and solid deposition on the foreshore. To overcome these problems Southern Water are constructing a series of interconnecting gravity sewers and pumping stations which will transfer the sewage to a new sea outfall extending 700 m into the deep water channel of the Solent. This project is expected to cost some £6 000 000 and is scheduled for completion in 1990. The siting and length of the outfall were determined after numerous field studies and extensive simulations using tidal hydrodynamic and pollution models.

Field studies

In 1984, Hydraulics Research were commissioned by Southern Water to investigate the advection and tidal mixing characteristics at a number of sites within the Solent using radioactive tracers. Following this work, sites to the east of the Medina estuary were selected for further study and in 1986 Hydraulics Research were again commissioned to undertake the work. Tidal current measurements and float tracking from a number of sites extending northwards from the shore into the deep water channel were carried out under neap and spring tide conditions. Tide levels at sites throughout the Solent were obtained from staff gauges and permanently installed water level recorders, and meteorological data were obtained from the survey vessels and from shore-based stations. This work showed that flow paths were parallel to the coast and floats were not drawn into the Medina estuary. The work also produced an excellent database for calibrating the mathematical models.

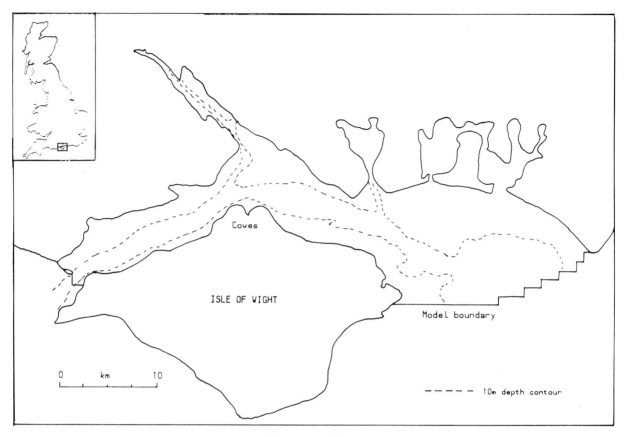

Fig. 4 The Solent.

Tidal model

The tidal hydrodynamics of the Solent were derived using the TIDEWAY-2D model which is a constituent part of the Hydraulics Research TIDEWAY system and is used under licence (ref. 7). The model solves the hydrodynamic equations over a rectangular grid using the finite-difference method. Inputs to the model are the sea bed topography, obtained from Admiralty charts, supplemented by local survey data, and tide levels along the sea boundaries of the area. The model predicts tide levels and currents at each grid point throughout the tide cycle. Figure 5 shows neap tide levels observed at Cowes on July 17, 1986, and the corresponding model predictions, while Figure 6 shows the predicted magnitude and direction of the tidal currents around the Medina estuary at one instant during the tide cycle.

To simulate the movement of floats, the computed depth-averaged velocities were corrected for the effects of wind, typically 0.5% of the wind speed in the sheltered waters of the Solent. Figure 7 shows a comparison between the predicted motion of floats released 6 hours apart, 650 m offshore with the observed paths of two floats released at 600 m and 700 m, from sites along the proposed line of the outfall, east of the Medina estuary.

Pollution model

The quality of bathing waters are usually assessed by the enumeration of coliform bacteria, which are an indicator for the presence of sewage in the water. Sewage typically contains 5×10^{7} faecal coliforms per 100 ml. The EC Directive on the quality of bathing water (ref. 3) requires that waters in which bathing is traditionally practised should have faecal coliform concentrations less than 2000 per 100 ml in 95% of samples taken over the bathing season (typically May to September in the UK). Thus dilutions greater than 25 000 must be achieved if long sea outfalls are not to cause beach failures in terms of the Directive.

Experimental evidence suggests that the numbers of coliform bacteria present in sea water decay exponentially with time from release although the rates of decay vary both diurnally and seasonally.

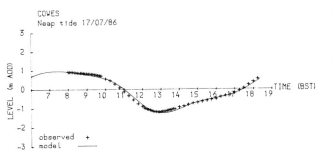

Fig. 5 Observed and predicted neap tide levels at Cowes on July 17, 1986.

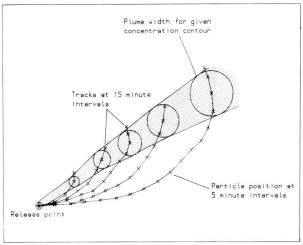

Fig. 8 Method of sewage plume construction.

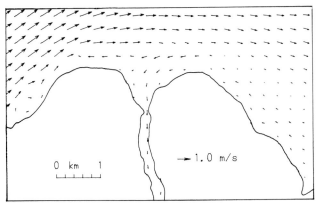

Fig. 6 Tidal currents around the Medina Estuary at HW− 2 h.

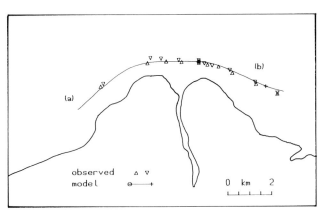

Fig. 7 Observed and predicted float tracks, released on spring tide at (*a*) HW+2 h and (*b*) HW−4 h.

The rate of decay is usually expressed in terms of the time for a 90% die-off, the T_{90} time, and values in the range 4 to 10 h are often quoted, although the figure to be used for design purposes is not clear and a sensitivity analysis should be carried out. Predicted particle tracks obtained from the tidal model, bacterial decay rates, and a model of the lateral spreading or dispersion about the particle track due to turbulence were used to compare the extent of the sewage plume arising from different sites under varying meteorological conditions.

Figure 8 shows one method of plume construction. A series of particle tracks at 15-minute intervals was generated by the tidal model with the positions of the particles recorded at 5-minute intervals. If the particle release time within the tide cycle and hence the initial water depth and tidal current are known, the initial dilution at the surface can be calculated. The concentration along the track can therefore be calculated for specific bacterial mortality and dispersion rates. Assuming that the lateral spread of sewage from its mean path follows a gaussian or normal distribution, the distance to a specific concentration contour can be calculated and this is shown plotted as a circle (ref. 8).

If all 50 tracks covering one tidal cycle are computed, the extent of the 2000 coliform/100 ml contour, for example, at specific times during the tide may be displayed. Alternatively the total area swept out by the discharge at concentrations greater than this limit can be shown on one diagram.

Different tidal ranges, outfall locations, or wind conditions can be tested by computing sets of particle tracks for each particular case, while the effects of different mortality and dispersion rates can be examined using any particular set of tracks.

The design parameters for simulating the Cowes outfall were:

discharge = 5.9 Ml d^{-1},
initial concentration of faecal coliforms = 5×10^7/100 ml, dispersion rate = 0.5 m^2 s^{-1} under a neap tide with a 5 m s^{-1} north wind, time for 90% bacterial mortality, $T_{90} = 10$ h.

A number of sites to the east of the Medina estuary were examined under the design conditions, and the sensitivity of the selected site to variations in the design parameters was also examined. The results showed that the dispersion characteristics from a site 700 m offshore were such as to ensure that the local beaches and designated shellfish waters comply with the relevant EC Directives. A discharge consent for this outfall has now been granted by the Department of the Environment, and the outfall is due to be pulled out in 1988.

CONCLUSIONS

Mathematical models describing the hydrodynamics and the effects of sewage discharges into the Stour Estuary and the Solent have been described. Following extensive validation, the models have been used to examine design proposals for new sewage works. The constraints imposed by the current anti-pollution legislation mean that the environmental consequences of proposed discharges should be thoroughly examined before construction begins. In this field, mathematical models have an increasing role to play.

REFERENCES

1. National Water Council (1978). River Water quality; the next stage. Review of discharge consent conditions. National Water Council, London.
2. National Water Council (1981). River quality: the 1980 Survey and future outlook. National Water Council, London.
3. Council of European Communities (1976). Directive 76/160/EEC concerning the quality of bathing water. *Official Journal of the European Communities* L31/1.
4. Council of European Communities (1976). Directive 76/464/EEC on pollution caused by certain dangerous substances discharged into the aquatic environment of the community. *Official Journal of the European Communities* L129/23.
5. Council of European Communities (1979). Directive 79/923/EEC on the quality required of shellfish waters. *Official Journal of the European Communities* L281/47.
6. Hydraulics Research (1979). Tidal Irrigation in the Stour Estuary, Kent. Report Ex 952.
7. Hydraulics Research (1982). HRS tidal modelling system. Report IT 238.
8. Keating, T. (1987). Mathematical modelling of sea outfall performance. The *Public Health Engineer,* **14**: 53–58.

29

The (NERC) North Sea Community Project, 1987–1992

M. J. Howarth, Proudman Oceanographic Laboratory, Bidston Observatory, Birkenhead, UK

The aim of the North Sea Community Project is to develop a water quality model of the North Sea. Within this broad aim three more specific objectives have been identified: the development of a three-dimensional transport model, the better understanding and quantification of non-conservative processes, and the definition of a seasonal cycle. The project is interdisciplinary, covering physical, chemical, biological and sedimentological processes and their interactions, and will act as a focus for UK shelf seas oceanography. This paper describes the scientific background to the project and outlines its implementation.

WHY A FURTHER STUDY OF THE NORTH SEA?

The continental shelf seas around the British Isles have a major impact on life in the UK and on the Continent. The seas are a resource, not just for hydrocarbons and fish, but also for aggregates and renewable energy; are important for commerce through shipping and for communications through tunnels and seabed telephone cables; are a potential threat through coastal erosion and flooding and to coastal and offshore structures; are a receptacle for natural and man-made wastes, including heat from power stations; affect our climate, both locally, mellowing the extremes of winter and summer, and globally, transporting heat and as a source for water; and finally provide extensive recreational amenities. For the well-being of all it is vital that these activities, separately and as a whole, do not stress the environmental health of the North Sea. Although potential threats are associated with all these activities two wide scale threats are posed by contaminants and eutrophication—the depletion of dissolved oxygen caused by the decay of unusual plankton blooms encouraged by high levels of nutrients, of which one source is the increasing application of fertilizers for agriculture.

Accordingly the goal of this study is to develop a water quality model in order to provide the accurate quantitative assessments and long term predictions necessary for a rational management strategy for the North Sea. Such a model will be complex since it will involve inter-related physical, chemical, biological and sedimentological processes. However, the

Advances in Underwater Technology, Ocean Science and Offshore Engineering, Volume 16: Oceanology '88

foundations already exist since the dominant physical processes of waves, tides and surges which underpin the others are mainly understood. The distributions of the chemical, biological and sedimentological parameters are largely known, the relevant processes have been identified, and quantification of their rates is beginning to be estimated.

A prerequisite to achieving this goal is the development of a water transport model able to predict the horizontal and vertical variations of currents in the North Sea as well as the evolving density field. This is a logical extension to the present tried and tested two-dimensional tide and surge models and to experimental three-dimensional models. Progress has been made in this direction (ref. 1). The objective is feasible given that powerful computers, such as the CRAY XMP at Rutherford, are available in the UK. Such a model can be applied immediately to predicting the movement and distribution of quasi-conservative tracers, such as salinity or caesium 137. However, most contaminants are non-conservative, their distribution and movement being controlled by complex inter-related chemical, biological and physical processes, the relative importance of the different processes varying from contaminant to contaminant. Hence, a second major objective is to identify and quantify these processes. The third aim in any scientific study, complementing theory and modelling, is observations, in this case over a seasonal cycle. High quality observations are necessary to improve theoretical understanding of the processes, against which theories and models can be tested and to provide initial and boundary conditions for the models. Here again we are well placed with new instrumentation and sampling and analytic techniques holding great promise, such as acoustic doppler current profilers, HF radar for spatial coverage of surface currents, purposeful tracers which can be detected at very low concentrations, *in situ* measurement of light and fluorescence, trace metal and gas analysis in ultra clean conditions and with proven techniques achieving greater automation, reliability and accuracy such as CTDs, nutrient analysis, vector averaging current meters, and drogued buoy tracking.

A water quality model developed for the North Sea will be applicable to many shelf seas since the North Sea (Fig. 1) encompasses a wide variety of conditions: of tidal range, from 6.5 m near the Wash and the Straits of Dover to zero at the amphidromes in the Southern Bight and off Denmark and Norway (Fig. 2); of tidal current strength, in excess of 1.5 m s^{-1} off Lowestoft and the Straits of Dover down to a maximum less than 0.3 m s^{-1} off Denmark (Fig. 3); of winter storms with maximum hourly wind speeds

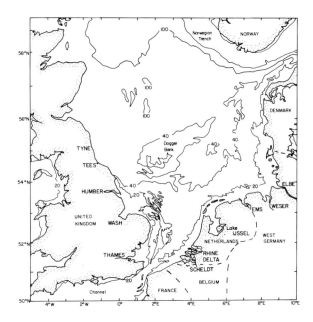

Fig. 1 Map of the North Sea; depths in metres.

at 10 m above still water up to 30 m s^{-1} in the south and 40 m s^{-1} in the north (Fig. 4) with corresponding maximum significant wave heights between 8 and 16 m (Fig. 5) and with maximum surge levels in excess of 2 m in the southern North Sea (Humber to Denmark) and in excess of 4 m in the German Bight (Fig. 6) and depth averaged surge currents up to 1 m s^{-1} (Fig. 7); a range of sediment type from sand and gravel to mud (Fig. 8) and of bottom roughness from smooth to regions of extensive sandbanks, off the Thames, the Straits of Dover and Norfolk; of density stratification with regions homogeneous throughout the year separated by frontal regions from regions which stratify either because of solar heating in summer (Fig. 9) or because of fresh water river discharges; of water masses (Fig. 10) ranging from water of oceanic origin via the northern North Sea and via the Channel to water of coastal origin, the North Sea is surrounded by industrialized coasts with several major estuaries providing a significant source of fresh water, contaminants, and nutrients.

In the rest of this chapter the present knowledge of shelf seas processes is outlined and some problems identified before describing the aims and objectives, administrative framework and implementation of the project. As is appropriate for this conference, the paper is dominated by the physics with reference only to the salient features of the chemistry, biology and sedimentology.

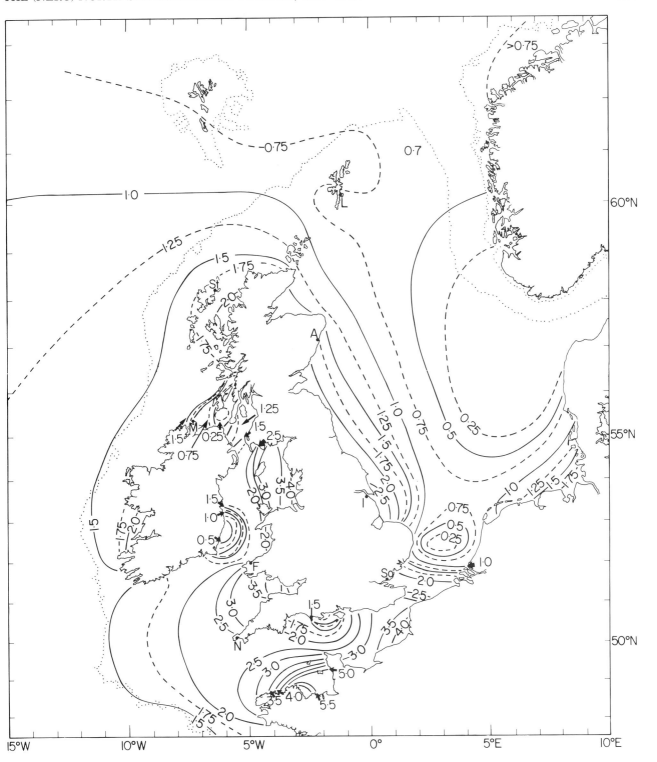

Fig. 2 Tidal amplitudes for an average spring tide (m). To obtain the maximum tidal range in the North Sea multiply by a factor which varies spatially between 2.8 in the north and 2.4 in the south (ref. 15).

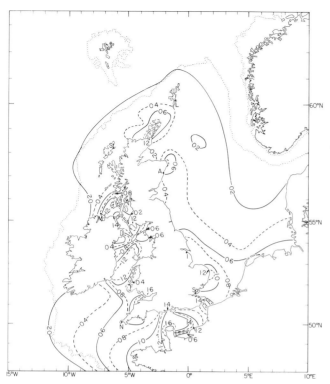

Fig. 3 Maximum depth-averaged current for an average spring tide (m s⁻¹). To obtain the maximum tidal current in the North Sea multiply by a factor of 1.4 (ref. 15)

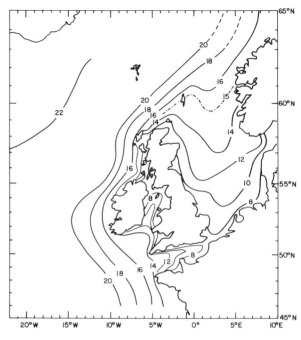

Fig. 5 Estimate of 50-year return significant wave height in m (ref. 15).

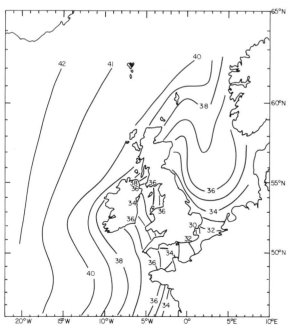

Fig. 4 Estimate of 50-year return omnidirectional hourly-mean wind speed in m s⁻¹ at 10 m above still water level (ref. 15).

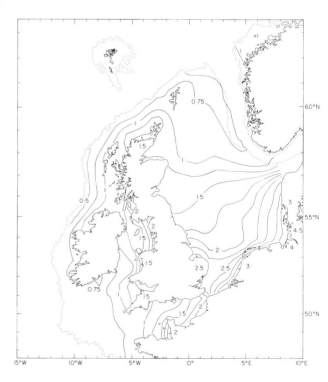

Fig. 6 Estimate of 50-year return positive storm surge elevations in m (ref. 15).

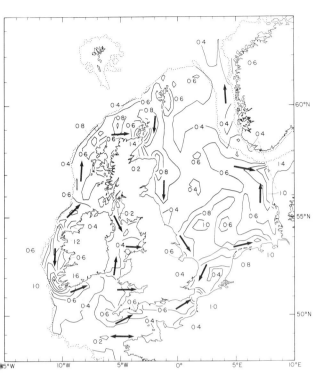

Fig. 7 Estimate of 50-year return depth-averaged hourly-mean storm surge currents in m s^{-1} (ref. 15)

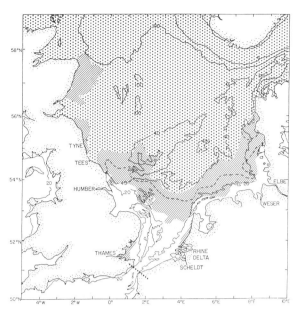

Fig. 9 Predicted position of summer stratified (dotted), homogeneous and frontal (hatched) regions in the North Sea, based on numerical model predictions (ref. 5).

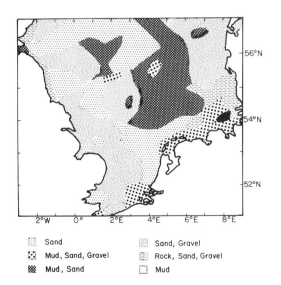

Sand

Mud, Sand, Gravel

Mud, Sand

Sand, Gravel

Rock, Sand, Gravel

Mud

Fig. 8. The large-scale distribution of sediment types in the North Sea (ref. 16). The local effects of estuaries, especially on fine sediments, is not shown.

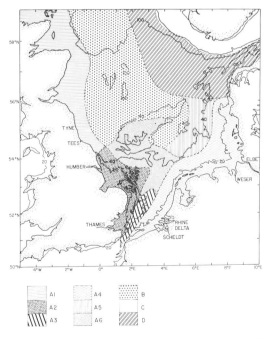

A1 A4 B
A2 A5 C
A3 A6 D

Fig. 10 Hydrographical regions of the North Sea. For a detailed account see ref. 9; broadly speaking, regions A1, A2 and A4 are dominated by coastal input, region A3 is oceanic water through the Channel and region B is the ocean water entering between Scotland and Norway. Regions A5 and A6 are areas of mixed water origin.

CONTINENTAL SHELF SEAS—PRESENT

Tides are the dominant dynamical processes in the continental shelf seas around the British Isles with their twice daily beat, fortnightly, spring–neap, cycle and longer period, monthly, six-monthly and annual cycles. The tides are basically long waves in shallow water and their dynamics is mainly linear and two-dimensional, so that on this basis the salient features of the spatial distribution of tidal elevations and currents can be predicted with numerical models. These models can be applied further to predict where homogeneous, summer stratified and frontal regions will occur (Fig. 9), since tides are the controlling source of energy for mixing down the potential energy generated by solar heat input at the sea surface during summer (refs 2 and 3). Non-linear effects are generally only significant near coasts, arising through friction, shallow water and large topographic variations. The waveform is distorted from sinusoidal by higher harmonics (with periods of six and four hours) and low frequencies generated at beat frequencies, primarily fortnightly. Models incorporating these dynamics have successfully predicted eddies and associated sandbanks near headlands and islands (ref. 4) and when allied with non-linear transport equations sediment transport paths (ref. 5).

The next most important forcing derives from winds and storms. Tidal models outlined above, with the addition of surface wind stress and atmospheric pressure gradient forcing, can predict storm surge levels in real time and extremes (refs 6 and 7). (Both tides and meteorological forcing together are necessary for the correct estimation of dissipation.) These models, when driven with mean wind stresses can successfully predict the long term movement and dispersion of a soluble contaminant, caesium 137, discharged from Sellafield (ref. 8).

Measurements of the distributions of caesium 137 in the seas around the British Isles have not contradicted traditional ideas of long term circulation patterns based on drifts and water mass studies (ref. 9). In broad outline, for the North Sea (Fig. 11), the largest inflow of oceanic water is in the vicinity of the Orkney and Shetland Islands. The majority of this does not directly penetrate to the southern North Sea but follows the 100 m isobath eastward across to the Skaggerak where it mixes with water flowing out of the Baltic and exits northward along the Norwegian coast. Some water flows southward along the Scottish and English coasts turning north-eastward off East Anglia to cross the central southern North Sea and join a northward flow past Denmark and into the Skaggerak. The other inflow of oceanic

water into the North Sea (less than 10% of the above) is through the Channel, where it is joined by freshwater discharges from continental estuaries, especially the Rhine and the Elbe. The volume of freshwater discharge is small (about 10% of the Channel inflow) but it is important dynamically because of its influence on the density field. These waters flow northeastward along the continental coast into the German Bight and then northward along the Danish coast. However, it must be emphasized that this picture of a weak mean circulation (currents generally less than 0.05 m s^{-1}) is periodically swamped by storm events when it could well be reversed. The sum of these events contributes largely to the circulation.

In most processes, apart from tides, the seasonal cycle predominates, arising from variations in meteorological forcing—in wind stress (winter

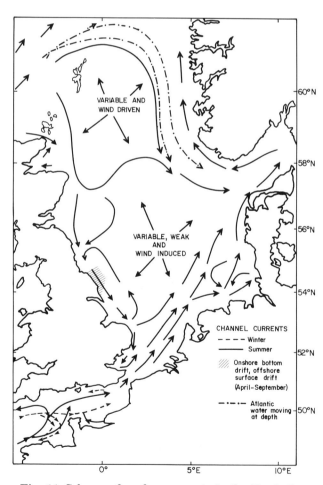

Fig. 11 Scheme of surface currents in the North Sea (ref. 9).

storms), in solar radiation (peaking in June), in river discharge (peaking in winter/spring), in evaporation and precipitation, in light intensity. This is reflected in seasonal cycles in temperature and salinity distributions (ref. 10), in stratification, in circulation, all of which impact on biological cycles, sediment movement and contaminant distribution and movement.

Improved understanding of physical processes will accrue from studies of interactions such as between tides, surges and waves where, for instance, waves significantly enhance bottom friction, and of small scale three-dimensional processes (including their incorporation into transport models), particularly the effects of fronts, both summer and river plume, on mixing and circulation. Are they barriers? How important to cross frontal transfer are ever present eddies? Are they conveyor belts, but also of eddies in general, of the vertical transfer of stress and of internal motion in stratified waters where, in addition, the surface well mixed layer becomes decoupled from the effects of bottom friction?

Although a transport model based on physical understanding is essential to the development of a water quality model, much insight into inter-related chemical, biological and sedimentological processes is also necessary. Contaminants are soluble in sea water, or buoyant, or particulate, usually with a marked affinity for fine sediments, or large particulate. The relationships between the suspended and sedimentary particulate and dissolved phases of contaminants is highly complex and variable, being affected by physical, chemical, biological and sedimentological conditions. The movement of conservative, soluble contaminants, such as caesium 137 and cadmium, can be predicted by a simple transport model alone. However, buoyant contaminants, for instance oil, plastics, fresh or warm water discharges, are moved by surface currents and the wind. Large particulate, heavy contaminants, for instance mining spoil and power station wastes, sink immediately to the sea floor under the action of gravity where only storms can provide sufficient energy for mobilization. Although only rarely mobile, they can still act as a continuous source as contaminants are leached out.

Particulate contaminants, for instance some forms of mercury, lead and plutonium, are adsorbed by chemical and physical processes dependent on their chemical state on to fine sediments and the dead remains of plankton—especially significant at the end of a bloom. Their movement is controlled by the dynamics of cohesive sediments, which are particularly complex, involving erosion, deposition and consolidation and where the stress needed to resuspend the sediment is difficult to predict, being generally greater than that allowing deposition (for the North Sea, see ref. 11). Resuspension events, often during storms, dominate fine sediment movement. Once in suspension the sediment will move with and can modify the near bed current. Within the seabed, movement is by deposition and erosion, by diffusion, especially via pore waters, and by bioturbation by worms and other benthic fauna, and moreover, the contaminant's chemical state and mobility may alter because of redox changes or the presence of organic material.

Sources of contaminants are estuaries, outfalls, dumping of sewage, industrial waste and dredge spoil, ships (oil and garbage), incineration at sea, the atmosphere, resuspension at the seabed and (possibly) the oceans. Much of the load is associated with man and estuaries and hence many contaminant levels are correlated inversely with salinity. Whilst the contaminant load at the landward end is frequently known, there are several major areas of uncertainty by the time the sea has been reached. In particular these relate to estuaries, quantification of air/sea fluxes caused by resuspension, as already mentioned. With estuaries, although the input of contaminants at the freshwater, river, end may be known, the output at the seaward end, where there will be the interface to the large scale water quality model, is not. Estuaries are a major source of fine sediments but are to some extent closed systems since seabed currents at their mouth tend to be shoreward. Estuaries act like buffers or reservoirs with output into the sea little related to changes in input. Moving seaward from an estuary, those particulate contaminants which are attracted to fine sediments show a proportionally greater reduction in concentration than those which are dissolved as the fine sediment settles out. Information on fluxes through the air/sea interface and their controlling processes is sparse but it is known that the atmosphere is an important source for some trace metals, such as lead (ref. 12). Contaminant fluxes are not the only interest since the ocean is the major sink for carbon in the atmosphere and hence plays an important role in the global carbondioxide budget and the 'greenhouse effect'. These are also significant fluxes out of the ocean, for instance near Sellafield bubbles carry plutonium ashore and in the North Sea plankton produce dimethyl sulphide in large quantities which is released into the atmosphere, contributing to its sulphur content and hence to acid rain.

A wide range of chemicals can act as contaminants —oil, pesticides, PCBs, anti-foulants like tributyl tin, heavy metals like cadmium, mercury and lead, mining and power station spoils, sewage sludge, fertilizers, radionuclides—all of which impose different strains on the system, especially on biological cycles. Some are poisonous; some physically harm the environment by reducing light levels or changing the seabed; some reduce available oxygen either chemically or biologically; some increase productivity leading to increased oxygen demand as the plankton die and decay. Even if not directly affecting the fish, the increase in stress can lead to increased disease amongst the fish.

The biological cycle itself is seasonal with primary productivity largely controlled by nutrient and light levels. Low activity in winter is followed by the spring bloom when nutrients and light are freely available, by a summer lull and possibly by a second bloom in the autumn as the autumnal overturn releases more nutrients into the well lit surface waters (ref. 13). This cycle is mirrored by nutrient levels which are highest at the end of winter and lowest in summer. Nutrient levels are important for the estimation of primary productivity and also appear to be significantly correlated with some trace metal levels (ref. 14). The primary source for nutrients is estuaries and rivers, both occurring naturally and through man's activities as nitrogen in fertilizers and phosphates in detergents.

Aim and objectives

The aim of the project is to develop a water quality model of the North Sea with the capability of predicting the long term effects of man's activities on the North Sea, particularly the dispersal of contaminants and eutrophication, so that the threats to the sea's well-being can be properly assessed as a guide to sensible management. As well as being of practical value and a stimulus to research during its development, such a model will be of great value to the shelf sea marine science community as an applied research tool needed in the pursuit of other goals. Within this broad aim three more specific objectives have been identified: the development of a physical transport model, improved understanding of non-conservative processes, and the definition of a seasonal cycle.

The physical transport model will be a three-dimensional hydrodynamic numerical model able to predict the evolving temperature and salinity fields and to model the advection and diffusion of conservative scalars, not just contaminants but properties like the movement of fish larvae. It will provide a framework which links existing process models and will form the foundation for subsequent applications including the development of models of non-conserva

tive processes. The model will require improved definition and parameterization of the physics, especially on sub-grid scales, and access to powerful computers, like the CRAY XMP at Rutherford.

To convert a transport model into a water quality model it is necessary to identify and quantify the non-conservative chemical, biological and sedimentological processes which determine the cycling and ultimate fate of contaminants. Many of the key mechanisms, such as the adsorption of contaminants on to particulates, have already been identified, but in most cases understanding of the rates of the processes and their seasonal variability needs significant improvement before these can be incorporated into models.

Definition of the seasonal cycle of the North Sea system by repeated surveys and moored instrumentation is necessary to improve understanding of the processes and to test theories and models, particularly for long term prediction. The database resulting from the survey period will form a unique reference for shelf sea studies.

NERC community projects

To tackle problems of this nature NERC Marine Sciences Directorate have set up Community Research Projects which bring together universities, polytechnics, NERC institutes and other laboratories. A community research project:

(1) covers a field in which there is widespread scientific and/or practical interest;
(2) is of a scale to require coordinated funding and facilities; and
(3) requires the collaboration of scientists in several laboratories.

A typical project may be planned to last five years at a total cost of the order of £10M. It may involve scientists from a dozen or more institutes including several NERC laboratories, in this case including University College North Wales (UCNW), Southampton, Liverpool, Cambridge, Essex and the East Anglia Universities, Plymouth Polytechnic and all NERC Marine Science Directorate Laboratories— especially IMER at Plymouth and POL—with the support of Research Vessel Services, Barry, and NERC computing services. The administration of each project is centred at a NERC laboratory (here the Proudman Oceanographic Laboratory (POL), Bidston Observatory) which provides day-to-day coordination and management through a scientific coordinator (J. M. Huthnance) and a project manager (M. W. L. Blackley). The science is guided by a steering group of the senior scientists involved, chaired by Professor J. H. Simpson (UCNW). NERC has approved the North Sea Project to run from 1987 to 1992 with an overall budget of about £8M of which £780 000 has been allocated to Special Topics. (This is not new money but a re-allocation of NERC's resources.) Included in this the Council has allocated sole use of RRS *Challenger* to the project for 15 months, from August 1988 to October 1989, half its time being dedicated to survey cruises running for 12 days every month and the remainder to process study experiments. A 15 month observational period was chosen for adequate determination of a seasonal cycle, including an estimate of longer period variability with the important autumnal overturn period duplicated. All data will be banked at the Marine Information Advisory Service. Close links are being maintained with other UK government laboratories, especially MAFF, Lowestoft, and with continental scientists through the International Council for the Exploration of the Sea.

The project has three components: modelling, the survey, and process studies.

Modelling

A scheme for a water quality model has been outlined by working group 25 of the Group of Experts on the Scientific Aspects of Marine Pollution (GESAMP) (Fig. 12). At present a box model approach is the only one to be able to incorporate all aspects of a water quality model, with the circulation and mixing of water given by exchanges between irregularly shaped boxes. Because these models rely on

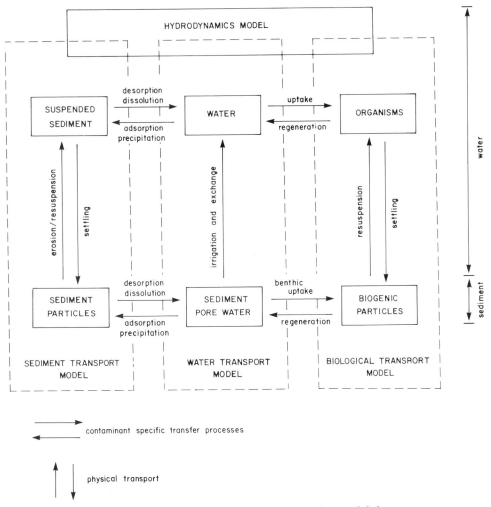

Fig. 12 Conceptual model for a water quality model, from GESAMP working group 25.

parameterization and are tuned to observations obtained from the relevant region, they are not easily transferred to other regions. In the GESAMP scheme the hydrographic water transport model forms the backbone to which are attached inter-related chemical, biological and sedimentological process models. Water quality models have been developed based on two- and three-dimensional transport models with imposed density fields and with contaminants treated as conservative soluble tracers. For the North Sea Project a three-dimensional transport model will be developed based on existing two-dimensional models and incorporating the vertical transfer of stress, stratification and internal dynamics, and some thermodynamics. The aim is for the model to be prognostic i.e. for it to predict the time evolution of the density (temperature and salinity) field.

The hydrodynamic model will provide both boundary and initial conditions for the localized models of specific processes which will in turn provide better parameterization and validation data for the hydrodynamic model. The development of process models need not await the availability of a full three-dimensional transport model but could usefully start with numerical experiments, particularly concerning sensitivity analysis, based on a simpler transport model. However, care is essential to ensure that the results are not considered out of context, in conditions which invalidate the basic assumptions. The process models will, in general, be of small scale phenomena and could be either point models (probably one-dimensional) or localized models, including fronts, patches, eddies. In both cases the models can be either stationary or drifting with the mean flow. Since the relative importance and balance of the processes controlling the movement of contaminants differ from contaminant to contaminant, each contaminant will need a specific model. Initial efforts will be concentrated on constituents whose involvement in chemical and biological processes is well specified.

Observations and models are inextricably linked: at the very least observations provide initial or boundary data for models and are needed to test the model predictions, whilst model calculations can help the design of experiments and the interpretation of results. Testing or validating a model is a vital stage in its development and is often part of an iterative loop. In particular testing of the transport model is not straightforward. Spatial scales are such and uncertainties in measuring the mean, which is weak in most areas, so great that it is impossible to obtain sufficiently detailed observations of the current field at fixed locations to provide a meaningful test of the model. However, an alternative approach which is well suited to long term transport studies, is provided by the movement and distribution of conservative tracers, like caesium 137, temperature and salinity. As mentioned above, an important, often the largest, component of long term signals is the annual, seasonal, cycle; this will provide a stringent test of a transport model's performance and is one objective of the survey.

Survey

The overall aim of the survey experiment is to define a particular seasonal cycle in the southern North Sea for various physical, chemical, biological and sedimentological parameters, with emphasis on variations in the vertical. The seasonal cycle is the largest component in the temporal variability of many of these parameters. The overall aim has been resolved into three specific objectives:

(1) to understand the horizontal and vertical spatial and temporal variations in the physical, chemical, biological and sedimentological observations and their inter-relationships and, by comparing the observations with long term mean seasonal data, where available, to determine the spatial distribution and movement of any anomalies;

(2) to provide a data set against which theories, prognostic transport models and water quality models can be strictly tested; and

(3) to provide boundary and initial data to force the models.

The survey will consist of 15 cruises each lasting 12 days repeated monthly (tied to the spring/neap cycle) and running from August 1988 to October 1989 preceded by a shakedown cruise in May 1988. During each cruise it is intended that the ship should cover the same track, with six categories of measurements being made: surface, water column, benthic, *in situ*, drifter, and ancillary (Table I). The experiments form nine groups: dynamics, density, suspended sediments, biology, nutrients, air/sea fluxes of trace metals and organics, trace metals, biogenic trace gases, and benthic fluxes (Table II).

The survey track (Fig. 13) is constrained by the 12 days allowed to be about 2000 miles long. This confines the survey area to the southern North Sea and poses some difficulty for the boundary prescription of oceanic inflow, particularly across the shelf edge Scotland–Shetland–Norway. However, it has the twin merits of restricting the coverage to the area most at pressure and of excluding the northern North sea, the Baltic outflow and the Norwegian Trench with their semi-enclosed circulation system. The

TABLE I
Survey scheme

Surface underway	Water column	Benthic	In situ	Drogues	Ancillary
					Meteorology
					Waves
					Tides
					Satellite IR
					CZCS
					River discharge
					Oil rigs
Temperature, conductivity	Conductivity, temperature		Thermistor chain	Temperature	CPR, UOR[a]
	Acoustic doppler profiler —underway		Current profiles	Currents	
Transmittance	Transmittance		Transmittance		
Fluorescence	Fluorescence		Fluorescence		UOR
Irradiance, masthead	Irradiance		Irradiance		UOR
Oxygen	Oxygen				
	Plankton species				CPR
Nutrients	Nutrients				
Air/sea fluxes					
Trace metals	Trace metals				
Biogenic trace gases	Biogenic trace gases				
		Benthic fluxes (cores)			
		Foraminifera			

[a]CPR = continuous plankton recorder; UOR = undulating oceanographic recorder

TABLE II
Survey: experimental groups

(1)	Density	Surface and water column profile measurements of conductivity and temperature. Thermosalinograph, CTD. Calibration samples Time series of temperature profiles–thermistor chains, April–October cf. experiments 4,5,7,8
(2)	Dynamics	Shipborne acoustic doppler current profiler (ADCP) Time series of current profiles—seabed ADCP, vector averaging current meter strings Satellite-tracked drogued buoys
(3)	Suspended sediments	Surface and water column profile measurements of transmittance, CTD Calibration samples Time series of transmittance?
(4)	Biology	Surface and water column profile measurements of fluorescence, irradiance, dissolved oxygen CTD plus pulsed O_2 sensor for surface oxygen Calibration samples. Masthead irradiance Calculate depth-integrated primary production Calibrate against on-deck carbon 14 incubations Time series of fluorescence (and irradiance?) Not winter months. Phytoplankton species composition and size distribution—50 to 100 samples per cruise, preserved, 99% not analysed. cf. experiments 1,5
(5)	Nutrients	Continuous surface and water column analysis with Autoanalyser. Nitrate, nitrite, ammonia, phosphate, silicate. cf. experiments 1,4,7,8
(6)	Air/sea fluxes	Wet and dry deposition. Trace metals and organics (hydrocarbons and PAH). cf. experiment 7
(7)	Trace metals	Surface? and water column samples. 150 samples per cruise. 6 cruises. Dissolved and to a lesser extent particulate. cf. experiments 1,3,5,6. Analysis on land
(8)	Biogenic trace gases	Continuous surface and water column samples Dimethyl sulphide and volatile halocarbons. cf. experiments 1, 4,5. February–October 1989
(9)	Benthic fluxes	6 cores. 6 cruises. Analysis at sea and on land

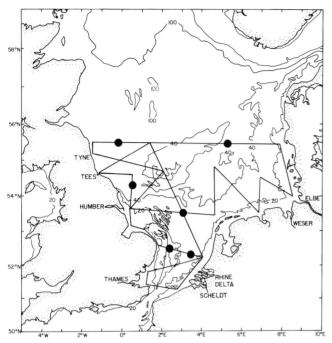

Fig. 13 Survey track. Dots denote *in situ* current and temperature profile sites.

Bight and northwards past Denmark and Norway. However, the significance of this is uncertain, both because this net movement is weak and frequently dominated and reversed by storms and also because the majority of contaminants do not behave as conservative soluble tracers.

It is confidently expected that more time will be lost through bad weather in winter than in summer. The situation is not entirely negative since the water column will also be well mixed then and offshore gradients weaker, necessitating a less dense coverage. The track has been devised so that up to 500 miles can be left out without seriously affecting the spatial coverage around the boundary; these 500 miles cover the Dogger Bank region which is frontal in summer but well mixed in winter.

Along the track will be made surface measurements, between 100 and 150 water column stations and, at about 6 sites, cores will be recovered. For the 12 day period a shipborne acoustic doppler current profiler (ADCP) will be continuously operated, providing important real time information on currents and current shears, particularly in summer stratified, fresh water stratified and frontal regions and on their seasonal variations. The measurement of vertical variations at the water column stations and with the ADCP is an important aspect of the survey. At the stations profiles will be measured with a CTD system fitted with O_2, transmittance, fluorescence and irradiance sensors and water samples will be collected with a rosette sampler fitted with twelve 10 litre Go-Flo water bottles. The stations will be about 40 miles apart in well mixed regions and 10–20 miles apart in frontal and stratified regions.

The survey observations, essentially once per month at each site with some sites being visited twice during each circuit, can only monitor low frequency motion although significant amounts of unpredictable energy exist at higher frequencies—in storms and blooms. This gap will be fitted to some extent by the process studies and also by *in situ* measurements and satellite-tracked drogued buoys. Current profile measurements will be recorded at six representative sites—two in homogeneous water (water depth 20–40 m), two in summer stratified water (water depth 60–80 m) and two in the separating frontal region (water depth 40–50 m)—either with seabed-mounted acoustic doppler current profilers or with strings of vector averaging current meters. Temperature profile measurements will be made with thermistor chains at the latter four sites for the period of stratification (April–October). These observations will be compared statistically with the transport model predictions. It is also hoped to develop *in situ*

track covers the different density regimes—summer stratified in the north, homogeneous in the Southern Bight, separated by a frontal band from Flamborough Head to the Dogger Bank to the Friesian Islands and the German Bight, and fresh water river discharge dominated especially along the Continental coast—and the different water masses—oceanic via the channel and the northern North Sea, British coastal and Continental coastal.

The major long term direct sources of most contaminants are associated with estuaries, each with an individual mix, and for a few contaminants, like lead, with the atmosphere. The survey track has been devised accordingly, paying particular regard to longshore and offshore gradients near the mouths of the Tyne, Tees, Humber, Wash, Thames, Scheldt, Rhine, Meuse, Lake Ijssel, Ems, Weser and Elbe estuaries and also covering the central southern North Sea. The accepted wisdom about the overall movement of contaminants, especially soluble conservative contaminants, was also considered during the design—the pattern being an anti-clockwise movement around the southern North Sea with British coastal water crossing the central southern North Sea and continental coastal water hugging the Continental coast, both passing through the German

TABLE III
Process studies

Title	Coordinator
(1) Nearshore fronts	J. Matthews (UCNW)
	D. Prandle (POL)
(2) Frontal mixing	E. Hill (UCNW)
	P. Linden (Cambridge)
(3) Frontal circulation	J. Howarth (POL)
	I. James (POL)
(4) Sand transport and drag over sand waves	D. Huntley (Plymouth Polytechnic)
(5) Bed shear stress, transports and form drag around a sandbank	M. Collins (Southampton)
(6) Estuarine plume	A. Morris (IMER)
(7) Bed stress induced resuspension	C. Reid (IMER)
(8) Air-sea exchange and initial dispersion by purposeful tracer release	A. Watson (MBA)
(9) Influence of plankton blooms on the chemistry of trace metals in the water column and an air-sea transfer of trace gases	D. Burton (Southampton)
(10) North Sea–Irish Sea comparative production	I. Joint (IMER)

instrumentation for measuring fluorescence and irradiance for time series estimation of phytoplankton biomass.

Process studies

In this short chapter it is impossible to do justice to each of the process studies so I will only briefly mention them (see also Table III). This should not be taken as implying they lack importance, indeed each studies a highly relevant process for a water quality model. The process study cruises will be in the 14 day periods interleaving the survey cruises. Some of the process studies are linked to the survey cruises: trace metals, biogenic trace gases, air/sea fluxes and gas exchange, suspended sediments. Of the others the estuarine plume; frontal dynamics, mixing, circulation and eddies; sediment resuspension; cohesive sediment dynamics experiments are all central to water quality studies.

CONCLUSIONS

The North Sea Project is an exciting opportunity to increase our understanding of shelf seas. The goal of a water quality model is ambitious but necessary for optimum benefits to be derived from the North Sea. The project is interdisciplinary and only tractable in the form attempted, bringing together the whole shelf seas community.

REFERENCES

1. Backhaus, J. O. (1985). A three-dimensional model of the simulation of shelf sea dynamics. *Deutsche Hydrographische Zeitschrift* **38**: 165–187.
2. Simpson, J. H. and Hunter, J. R. (1974). Fronts in the Irish Sea. *Nature* **250**: 404–406.
3. Pingree, R. D. and Griffiths, D. K. (1978). Tidal fronts on the shelf seas around the British Isles. *Journal of Geophysical Research* **83**: 4615–4622.
4. Pingree, R. D. and Maddock, L. (1979). The tidal physics of headland flows and offshore tidal bank formation. *Marine Geology* **32**: 269–289.
5. Pingree, R. D. and Griffiths, D. K. (1979). Sand transport paths around the British Isles resulting from M2 and M4 tidal interactions. *Journal of the Marine Biological Association of the UK* **59**: 497–513.
6. Flather, R. A. and Proctor, R. (1983). Prediction of North Sea storm surges using numerical models: recent developments in the UK. In: *North Sea Dynamics* (J. Sundermann and W. Lenz, eds). Springer-Verlag, Berlin, pp. 299–317.

7. Flather, R. A. (1987). Estimate of extreme conditions of tide and surge using a numerical model of the North-West European Continental Shelf. *Estuarine, Coastal and Shelf Science* **24**: 69–93.

8. Prandle, D. (1984). A modelling study of the mixing of 137 Cs in the seas of the European continental shelf. *Philosophical Transactions of the Royal Society A* **310**: 407–436.

9. Lee, A. J. (1980). North Sea: physical oceanography. In: *The North-West European Shelf Seas: The Sea Bed and the Sea in Motion*. Vol. II (F. T. Banner, M. B. Collins and K. S. Massie, eds). Elsevier, Amsterdam, pp. 467–493.

10. Maddock, L. and Pingree, R. D. (1982). Mean heat and salt budgets for the eastern English Channel and Southern Bight of the North Sea. *Journal of the Marine Biological Association of the UK* **62**: 559–575.

11. Eisma, D. (1981). Supply and deposition of suspended matter in the North Sea. *Special Publications, International Association of Sedimentologists* **5**: 415–428.

12. Cambray, R. S., Jefferies, D. F. and Topping, G. (1979). The atmosphere input of trace elements to the North Sea. *Marine Science Communications* **5**: 175–194.

13. Tett, P., Edwards, A. and Jones, K. J. (1986). A model for the growth of shelf-sea phytoplankton in summer. *Estuarine, Coastal and Shelf Science* **23**: 641–672.

14. Burton, J. D. and Statham, P. J. (1982). Occurrence, distribution and chemical speciation of some minor dissolved constituents in ocean waters. In: *Environmental Chemistry*. Vol. 2 (H. J. M. Bowen, ed). Royal Society of Chemistry Specialist Periodical Report, pp. 234–265.

15. Department of Energy (1987). Offshore Installations: Guidance on Design and Construction. Meteorological and Oceanographic Design Parameters. Draft version.

16. Lee, A. J. and Ramster, J. W. (ed.) (1981). *Atlas of the Seas around the British Isles*. Ministry of Agriculture Fisheries and Food, 100 pp.

30

Logistic Support for the Sandfill Operations in the Eastern Scheldt Tidal Basin

J. Vroon, A. van Berk, R. E. A. M. Boeters, R. 't Hart and *J. J. P. Lodder*

The final phase of the Delta Project undertaken in the south-west of the Netherlands consisted of erecting a closable storm surge barrier in the mouth of the Eastern Scheldt and constructing two compartmentalization dams, away from the mouth, at the eastern end of the basin. To limit the cost of this part of the project, it was decided to use sand instead of stone to effect the dam closures. However, in view of the high flow velocities that can occur during the final stage of sandfill closures, the success of this operation very much depended on the use of the newly built storm surge barrier to reduce the magnitude of the tidal motion. Due consideration was given to the best way in which the storm surge barrier could be employed. The main concerns involved were the technical implementation, environmental requirements and the protection of the barrier itself. A system was therefore set up to measure and predict such effects in order to ensure that the best possible timing, and closure settings were selected for the gates in the storm surge barrier. This chapter describes the system that was developed and discusses the results obtained.

COMPARTMENTALIZATION DAMS IN THE EASTERN SCHELDT

General

After large parts of the south-west Netherlands had been flooded during the catastrophic storm surge flood of February 1953, it was decided that action must be taken to improve the safety in the area. This resulted in the Delta plan being formulated. Basically, the plan envisaged the construction of a large number of enclosing dams at the sea inlets in the region, to minimize the risks of future flooding. The enclosure of the Eastern Scheldt was to be the final and largest single civil engineering operation in the whole project.

A summary of the salient features of the Delta Project and the dates at which the major engineering works were completed is given in Figure 1.

Originally, the enclosing dam for the Eastern Scheldt was to have been based on a fixed type of structure similar to that of the other dams. However, a growing realization of the high ecological value of

Advances in Underwater Technology, Ocean Science and Offshore Engineering, Volume 16: Oceanology '88

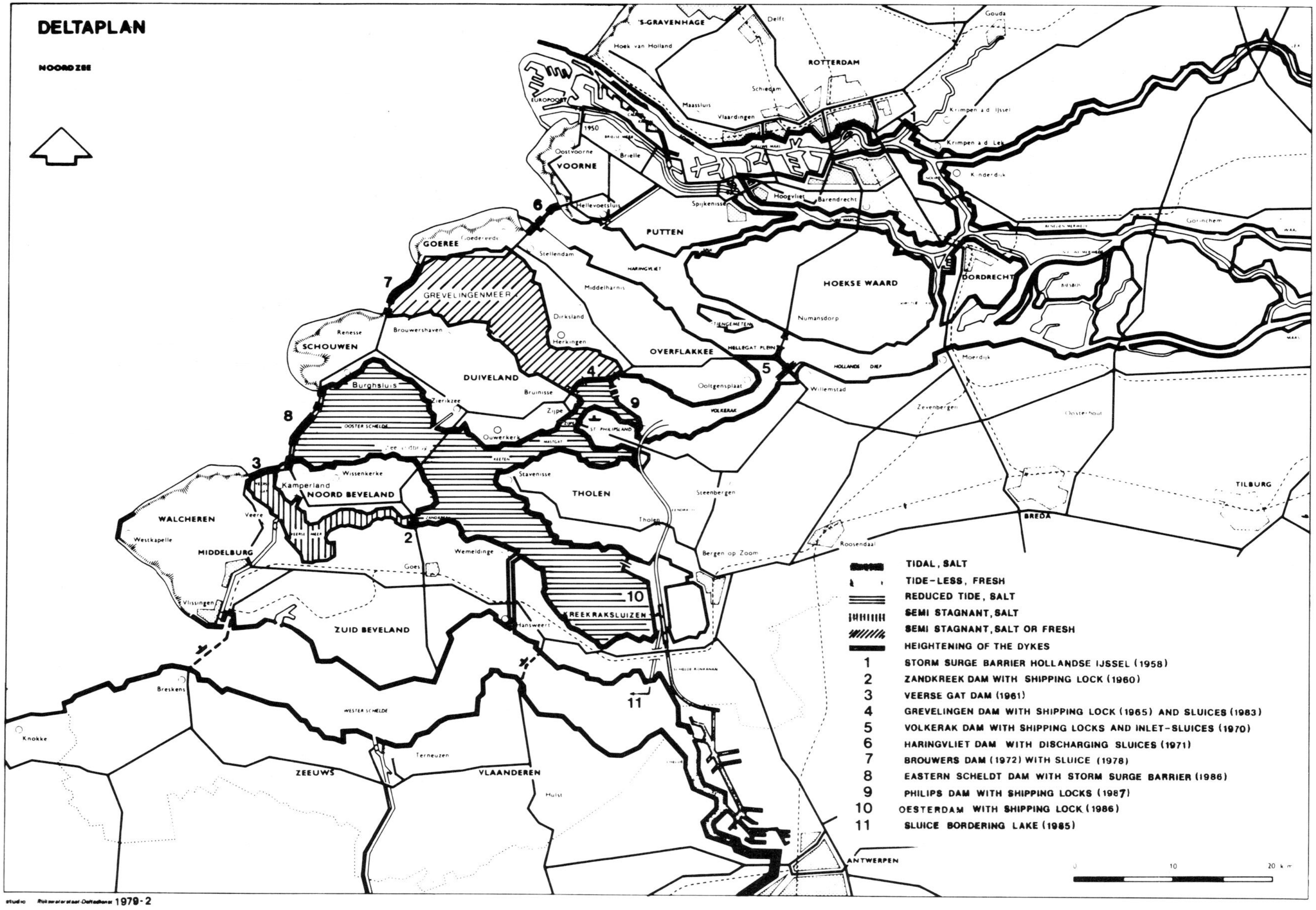
DELTAPLAN
NOORDZEE
'S-GRAVENHAGE
Hoek van Holland
Delft
ROTTERDAM
Schiedam
Maassluis
Vlaardingen
Gouda
Krimpen a.d. IJssel
Krimpen a.d. Lek
Kinderdijk
EUROPOORT
1950
Oostvoorne
VOORNE
Brielle
Spijkenisse
Hoogvliet
Barendrecht
Hellevoetsluis
PUTTEN
Gorinchem
Goedereede
GOEREE
Stellendam
HARINGVLIET
Middelharnis
HOEKSE WAARD
DORDRECHT
BIESBOS
GREVELINGENMEER
Dirksland
Herkingen
TIENGEMETEN
Numansdorp
HELLEGAT PLEIN
OVERFLAKKEE
HOLLANDS DIEP
Moerdijk
Renesse
SCHOUWEN
Brouwershaven
DUIVELAND
Bruinisse
Ooltgensplaat
Willemstad
VOLKERAK
Zevenbergen
Oosterhout
Burghsluis
Zierikzee
Zijpe
OOSTER SCHELDE
Ouwerkerk
MASTGAT
ST. PHILIPSLAND
KEETEN
Stavenisse
Wissenkerke
Kamperland
NOORD BEVELAND
THOLEN
Steenbergen
TILBURG
BREDA
WALCHEREN
Veere
VEERSE MEER
ZANDKREEK
Westkapelle
MIDDELBURG
Wemeldinge
Tholen
Roosendaal
Bergen op Zoom
Goes
Vlissingen
ZUID BEVELAND
KREEKRAKSLUIZEN
Hansweert
Breskens
WESTER SCHELDE
Knokke
Terneuzen
ZEEUWS
VLAANDEREN
Hulst
ANTWERPEN
1 2 3 4 5 6 7 8 9 10 11
TIDAL, SALT
TIDE-LESS, FRESH
REDUCED TIDE, SALT
SEMI STAGNANT, SALT
SEMI STAGNANT, SALT OR FRESH
HEIGHTENING OF THE DYKES
1 STORM SURGE BARRIER HOLLANDSE IJSSEL (1958)
2 ZANDKREEK DAM WITH SHIPPING LOCK (1960)
3 VEERSE GAT DAM (1961)
4 GREVELINGEN DAM WITH SHIPPING LOCK (1965) AND SLUICES (1983)
5 VOLKERAK DAM WITH SHIPPING LOCKS AND INLET-SLUICES (1970)
6 HARINGVLIET DAM WITH DISCHARGING SLUICES (1971)
7 BROUWERS DAM (1972) WITH SLUICE (1978)
8 EASTERN SCHELDT DAM WITH STORM SURGE BARRIER (1986)
9 PHILIPS DAM WITH SHIPPING LOCKS (1987)
10 OESTERDAM WITH SHIPPING LOCK (1986)
11 SLUICE BORDERING LAKE (1985)
0
10
20 km
1979-2

Fig. 1 Salient features of the Delta Project.

the Eastern Scheldt led to this decision being reconsidered. Many people thought that other technical solutions should be found to reduce the risk of flooding. In 1974 the Dutch parliament decided to close off the Eastern Scheldt by means of a closable dam which would minimize the impact on the environment. The design of the Eastern Scheldt storm surge barrier (SVKO) is such that, at times of high water, the barrier can be closed by lowering a series of gates.

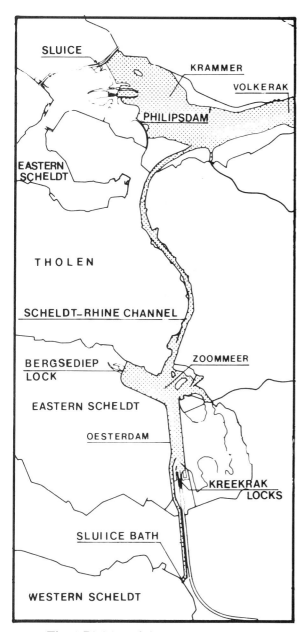

Fig. 2 Division of the Eastern Scheldt.

However, during normal tidal conditions the gates are kept open. This provides an effective area through which the water can flow of 17 900 m^2.

Because an opening of 17 900 m^2 is insufficient to maintain the same tidal range that existed before the storm surge barrier was constructed, it was necessary to reduce the capacity of the basin behind the barrier. This was accomplished by dividing the Eastern Scheldt into compartments (Fig. 2). The compartmentalization process also afforded an excellent opportunity to improve the freshwater facilities in the area. A further advantage was that the vital Antwerp–Rotterdam shipping route (Scheldt–Rhine connection) could be made non-tidal.

It was realized at an early stage that it would be necessary to provide technical assistance and guidance for the engineering work to be carried out in the Eastern Scheldt and, in particular, for the storm surge barrier and the two compartmentalization dams (C dams). A comprehensive system was therefore set up to collect relevant data and to formulate predictions. The total costs involved in the Eastern Scheldt project have amounted to Fl. 8000 million, of which Fl. 1000 million was required for the two compartmentalization dams (the Oester Dam and the Philips Dam) and for the associated engineering work. The cost of the comprehensive measuring and forecasting system was about Fl. 25 million.

Planning

According to the original planning schedule, it was envisaged that the compartmentalization dams and the storm surge barrier in the Eastern Scheldt would be completed at the same time. It would have been inadvisable to finish the dams first, since this would have increased the risk of flooding the dikes in the Eastern Scheldt in the period before the storm surge barrier was completed. A later completion date for the dams was also undesirable in view of environmental considerations. Reduced tidal motions would have occurred in the period between the completion of the storm surge barrier and the closure of the compartmentalization dams.

In order to effectively conclude the two projects simultaneously, it was necessary to work on the compartmentalization dams and the storm surge barrier independently. This meant that the closure gaps that were left in the Oester Dam and the Philips Dam (the Tholen Gap and the Krammer respectively) both needed to be closed off with a stone filling.

However, in the final plan, the proposal to use a stone filling for the closure operation was changed and a method employing sandfill was selected instead. To implement the new method successfully it was recognized that the flow velocity in the Eastern Scheldt would have to be reduced. It was therefore necessary to complete the storm surge barrier before the compartmentalization dams in order to make use of the barrier's ability to reduce the water flow by a partial closing of the gates. The time-table to be adhered to for the compartmentalization dams consequently became intrinsically linked to the completion date of the storm surge barrier. This resulted in the Eastern Scheldt project being delayed by six months compared with the original schedule. The fact that the tidal motion had to be reduced during this period obviously had unfavourable consequences for the environment. However, by using sand, which is much cheaper than stone, as the filler material, it was possible to make a considerable cost saving.

It was decided to implement the closure programme in a phased manner. First, the Tholen Gap was closed off in October 1986. This was then followed by the Krammer closure in April 1987. A phased approach was needed since it was not possible to operate the storm surge barrier in such a way as to accommodate work on both closure gaps at the same time. Furthermore, completing both dams simultaneously would have required an extremely high suction dredger capacity.

Design principles

Implementation

It was recognized that without the assistance of the storm surge barrier to control the water level, the flow velocities in the channels in the Krammer and Tholen Gap would have reached about 5 m s^{-1} during the final stages of the closure operations. In view of the experience gained with previous sandfill closures, involving flow velocities of up to 2.5 m s^{-1}, it was thought inadvisable to carry out such operations under these conditions. An upper limit of 2 m s^{-1} was therefore specified for the flow velocities in the closure gaps hence ensuring that the operations remained within the realms of previous experience.

Even when all the gates in the storm surge barrier have been lowered, a reduced level of tidal motion is still apparent because of leakage effects. Leak paths are formed in the gaps that exist between the stones under the sill beams and via the spaces between the gate stops and the concrete structure. This meant that during the final stage of the closure operation, when the gates in the storm surge barrier had been lowered, the flow velocities in the closure gap were determined by the amount of leakage that occurred. The extent of this phenomenon obviously had impli-

cations for the technical feasibility of the sandfill operation.

For design purposes the effective cross-sectional area of the opening through which leakage could occur was taken to be 1650 m^2. Immediately after the Tholen Gap had been closed this figure was revised on the basis of measurements taken at the time. The new information was used as the basis for planning the Krammer closure.

Environmental concerns

The use of the storm surge barrier placed an extra burden on the environment in the Eastern Scheldt area. It was therefore important to ensure that the period over which the barrier would operate should be kept to a minimum. Thus restrictions were placed both on the partial and complete lowering of the gates. It was finally decided to adopt scenarios for the Krammer closure in which the tidal periods were increased to twice the normal time span (TO). However, in the case of the closure operations in the Tholen Gap no such drastic measures were called for. An example of a '2T0 scenario' is given in Figure 3. The order of the planning scenarios was as follows:

- No tidal reduction (T0);
- Limited reduction T0;
- 2T0;
- 2T0 with a reduction at low water;
- 2T0 with a reduction at low water and closed at high water.

The scenarios also had to meet the following requirements:

- The storm surge barrier could not remain closed for more than two days, with allowance being made for a 10% chance of exceeding this limit;
- The reduced high water levels could not fall below AOD+1.10 m (AOD = Amsterdam ordnance datum, mean sea level as defined for Amsterdam) for a period longer than five days. Allowance had to be made for a 10–30% chance of exceeding this period;
- The Krammer closure had to take place in April which was the least vulnerable period for the environment. Only a 10% chance of exceeding the planned period was allowed.

It was necessary to specify the risks associated with the various operations in order to employ a probabilistic approach in planning the overall sandfill closure.

Protecting the storm surge barrier

Certain restrictions were placed on the use of the storm surge barrier as a means of reducing the tidal motion in the basin while the sandfill operations were being carried out. Such provisions were necessary to

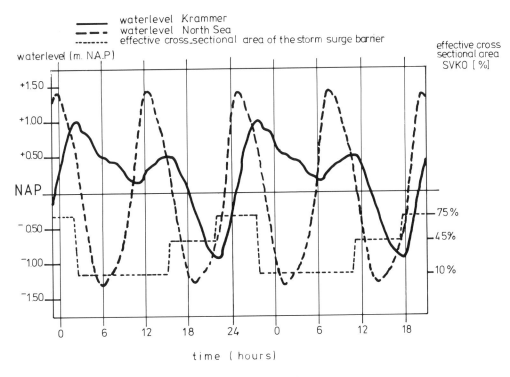

Fig. 3 2T0 scenario.

protect and maintain the structural integrity of the barrier.

Investigations carried out with a scale model have shown that problems can occur when the barrier is partially closed, especially if all the gates have been lowered to the same relative position. Under such conditions water tends to flow rapidly over the sill beam and is directed downwards in a definite flowstream. This can have detrimental effects on the part of the structure that is designed to protect the seabed at this point. It was therefore decided that for every 5% reduction in the total barrier opening specific gates would be closed according to a predetermined pattern. Alternate blocks of gates were designated either to be completely closed or to be only 25% closed. This combination of gate openings prevented the downward flow movement from occurring and hence minimized the risk of damage to the protective structure on the seabed. The required combination of opening settings was programmed into the barrier's central computer control system prior to the operation.

Normally, the hydraulic cylinders of the individual gates are automatically lubricated when the gates are lowered, but problems can arise if the gates have to be lowered for long periods of time. After more than 48 hours of being completely submerged in salt water, most of the lubricant will have dissolved, and this can cause considerable damage to the cylinders when the gates have to be raised. Maintenance considerations therefore dictate that the movable parts of the gates are not allowed to remain in the water for more than 48 hours. The gates are therefore fully raised at regular intervals to ensure sufficient lubrication.

During the actual project, the timing for this operation was especially critical at the final stage of the sandfill closure when the gates had to be fully closed. Had the gates been raised at the wrong time, translatory waves would have led to an unacceptable increase in the flow velocity in the closure gap. Computer simulations had indicated that the least number of translatory waves would be created in the basin by raising the gates just prior to the tide turning at high water. It was also shown that such wave effects could be further minimized by reducing the number of gates that were raised simultaneously.

Design methodology

To provide a balanced assessment of the suitability of the different constructional elements in the design, probabilistic design techniques were used. Fault trees were prepared based on the failure mechanisms associated with the various parts of the structure. This allowed the different components to be dimensioned after an estimate had been made of the various probabilities of failure of the remaining parts of the structure. By defining a generally accepted total risk of failure for the construction as a whole, acceptable probabilities of failure could thus be determined for the different constructional elements in respect to their specific failure modes.

Calculation models were used to determine the amounts of sand that would be lost during the closure operations. The Engelund-Hansen transport formula was selected for this purpose because of its good correlation with the results of model tests.

By using calculation models it was therefore possible to quantify the expected sand losses and to assess the associated uncertainties. It was not permitted for sand losses to influence the progress of the work in such a way as to prolong the use of the storm surge barrier. Assessments of the expected loss of sand and the required rate of progress determined the total production capacity necessary for the suction dredgers.

Construction method

Several different methods of closing off the flow channels were identified, a number of which are illustrated in Figure 4. The range of options offered by the various alternatives were all thoroughly investigated. Eventually, it was decided that a combined horizontal and vertical closing system would be the most suitable choice. By selecting a closure system employing a sill beam as the first stage, it was possible to ensure that the actual sandfill operations were kept within the realms of previous experience (i.e. operating in water depths of about 5 m). This meant that considerably less uncertainty existed regarding the length of the construction period—and hence the use that would need to be made of the storm surge barrier—than would have been the case with a horizontal closing system.

Sand collected by suction dredgers was sprayed into position in the closure gap from pressure pipes. The capacity of the suction dredgers needed to be large enough to ensure that the work progressed at a sufficient rate, as well as effectively compensating for the high sand loss experienced during the tidal phases. To satisfy these demands suction dredgers with a total gross capacity of 20 000 m^3 h^{-1} were required.

The sand needed for the closure operations could be extracted locally within a reasonable distance from the proposed courses of the dams. The large amounts of sand required (about 9 million m^3 in the case of the Krammer) meant that a relatively large area of land was involved. The stability of the slopes of the sand excavations (settlement flows) prevented the use of locations close to the actual dam construction sites. The D_{50} of the sand was determined on the basis of trial drillings in the extraction area and was found to be 180–200 μm.

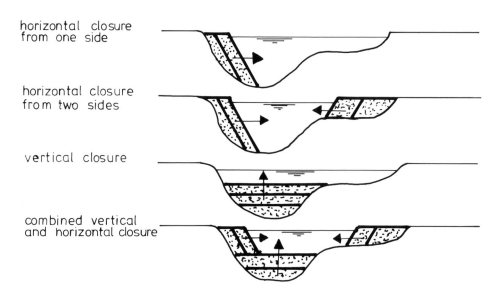

Fig. 4 Methods of gradual sandfill closures.

Use of the storm surge barrier

The principal concerns governing the use of the storm surge barrier, referred to above, were the technical implementation, the environmental requirements, and the protection of the barrier itself. These considerations necessitated a balanced approach to the way the barrier was operated. To achieve this, the various requirements were translated into unambiguous parameters. This meant that the flow velocities in the closure gaps and the safety requirements for the barrier were expressed in terms of a permissible drop in water level. The environmental concerns were translated into a required tidal range in the Eastern Scheldt basin. A predictive model was used during the construction period to make continuous prognoses about the drop in the water level and the tidal range. These results were then compared with preset standards, which varied for the different construction phases. Such comparisons allowed the barrier gate settings to be optimized using iterative methods. The relevant design standard is given in Figure 5.

It was also possible to check the predictions by means of continuous measurements of the water level and flow velocities made at special registration stations. Furthermore, an environmental monitoring system was set up to study the effects on the condition of benthic fauna, birds and mussels.

MEASURING AND FORECASTING SYSTEM

General

It was necessary to provide hydraulic and meteorological information on a daily basis during the construction of the storm surge barrier in the mouth of the Eastern Scheldt and the building of the compartmentalization dams at the other end of the basin. To supply these data it was decided to set up an instrumentation and information system. The system was primarily used to collect hydrological and meteorological data at various locations in the Delta area and to send these on to the Zierikzee Data Processing Centre (VCZ) where the data were checked and stored as 10-minute readings in a 24-hour master file of a central data acquisition computer. To be able to apply these data, the Hydro-meteo Centre (HMC) was established as part of a collaborative venture between the Department of Public Works and the Royal Netherlands Meteorological Institute (KNMI).

During the construction period, the HMC provided forecasts to help determine whether it was possible to operate the vessels involved in the construction work. In the final phase of the work on the compartmentalization dams, involving the actual closure of the remaining gaps in the dams, a second forecasting station, the Support Centre (SC), was set up on

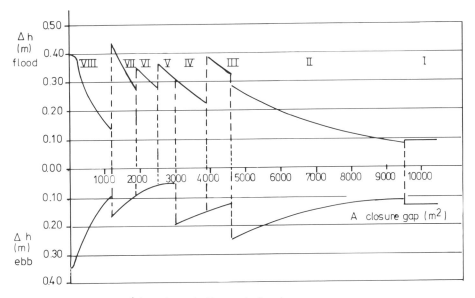

Fig. 5 Design standard.

site, to complement the activities of the HMC. The Support Centre's duties consisted of making recommendations about the use of the storm surge barrier on the basis of the requirements about technical feasibility, environmental concerns and the protection of the barrier. For this purpose, the Support Centre made use of a predictive system based on a water motion model coupled with a sandfill closure model. The system was used to predict the drops in water level for the closure gaps and for the storm surge barrier as well as the tidal ranges in the basin. The data generated were checked against predetermined norms. The standards applied were dependent on the stage the construction work had reached. If the predicted values did not meet the required standards, the calculations had to be repeated using a different discharge rate for the storm surge barrier. In the case of satisfactory agreement, the barrier gate setting used in the model was then recommended to the decision-making team. Whether this recommendation was implemented also depended on the information that had been received from the environmental monitoring system and from the situation at the construction site.

The boundary conditions relating to the natural environment that were needed for the predictive system were supplied by the HMC. Furthermore, up-to-date information about the closure gaps was also required to facilitate accurate calculations. Figure 6 outlines the measuring and forecasting system that was used during the closure operations of the compartmentalization dams.

A detailed description of the various components of the system is given below.

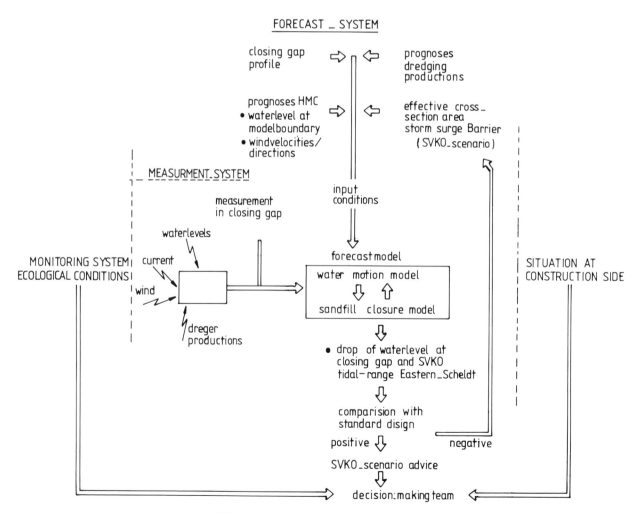

Fig. 6 Measurement and forecast system.

IMPLIC water motion model

IMPLIC is a one-dimensional numerical tidal model that was used to study certain aspects of the sandfill closure operations. The input data for this program consisted of a schematic representation of the seabed, describing the fore-delta of the Eastern Scheldt in about 170 sections (Fig. 7). In addition to requiring a prescribed outline of the seabed, IMPLIC also utilized the discharge characteristics of the closure gap. The discharge properties used in the IMPLIC model had been previously derived from a two-dimensional numerical analysis carried out with the WAQUA model and were known to be highly dependent on the size and shape of the closure gap.

The actual water movements predicted by the model were determined by the following time-dependent boundary conditions:

- Barrier gate settings;
- Natural boundary conditions consisting of:
  ○ water levels at the edge of the area encompassed by the model (North Sea),
  ○ wind forecasts for the Eastern Sheldt basin;
- Dimensions of the closure gap.

Barrier gate settings

The primary objective of the forecasting system was to determine the optimum gate settings for the barrier. Since this also formed one of the input parameters for the calculation procedure, an iterative approach was required. After calculations had been made with the tidal model, using a preliminary estimate of the barrier gate settings, estimates of the expected drop in the water level were arrived at. Comparison of these data with the specified requirements resulted in a particular barrier gate setting either being accepted or rejected. When the settings were found to be unacceptable, revised estimates were prepared and the iteration procedure repeated until suitable drops in the water level were found.

Natural boundary conditions

The boundary conditions defining the water levels used in the IMPLIC model were derived from two sources. Basic input information on water levels was taken from astronomical studies. This information was obtained by applying frequency analysis techniques to sets of water level readings that had been collected over many years. Further data were supplied regarding the expected deviation from current background levels that result from meteorological influences.

The procedure for forecasting deviations in water levels by the KNMI has become increasingly more automated. The predictions are based on wind data obtained from the fine-mesh atmospheric model used by the United Kingdom Meteorological Office. These data are then used as input variables for the imposed

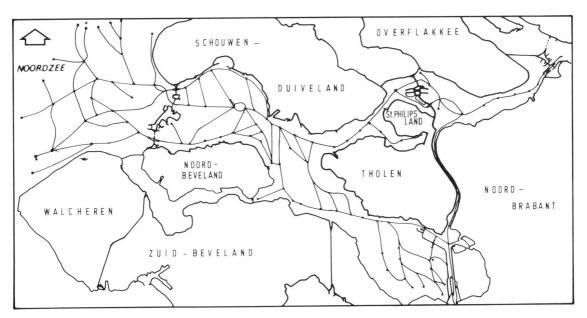

Fig. 7 The water motion model and IMPLIC for the
Eastern Scheldt.

deviation model. In spite of the significance of computer-based calculation methods, the contribution of the meteorologists and, in particular, their skill in interpreting the results remain of the utmost importance. The complex interrelation between the various meteorological effects that play a part in determining the imposed variations in water levels in the mouth of the Eastern Scheldt confirm the need for this type of expert advice. Apart from pressure distributions and wind patterns over the North Sea, pressure effects above deep water as well as dynamic effects may all influence the imposed variations. Pressure effects above deep water away from the North Sea area and continental shelf may be responsible for external surges. Dynamic effects can lead to situations in which resonances develop. For instance, a primary resonant condition can be induced when the speed of the imposed wave variation is equal to that of the wind causing the imposed deviation. A secondary resonant condition can occur when the intervening time between two successive storm surges is equal to the natural period of the North Sea.

Dimensions of the closure gap

To provide information about the dimensions of the closure gap which could be used in the IMPLIC model, a special sandfill model (MOZAS) was coupled to the basic hydraulic model in order to study the effect of sandfill closure operations.

The MOZAS sandfill closure model

The MOZAS model is a balanced calculation model which evaluates the progress of dam building activities and hence the size of closure gaps. Increases in the dam volume can be determined with the MOZAS model for a given sand production rate and expected sand loss. New closure gap profiles can be calculated once the time-dependent boundary conditions and the initial profile have been defined. An iterative procedure is applied at each individual stage in the process. An assessment is made of the progress of the dam construction based on the results obtained during the previous stages. From these estimates both the increase in the dam volume and the loss of sand can be determined. The estimated total sand loss and the increase in the dam volume must be equal to the amount of sand produced. If this is not the case, the calculation is repeated using a revised estimate of the progress made on the dam until a true equilibrium is reached between the increase in dam volume, sand losses and sand production.

The most important boundary conditions that must be specified in the model are:

- The average flow velocities in the closure gap;
- The sand production in $m^3 h^{-1}$ characterized by a D_{50} grain size distribution in μm;
- The profile of the dam including the as-poured slope and side slope.

The average flow velocities in the closure gap

To formulate consistent predictions, the IMPLIC hydraulic model and the MOZAS sandfill closure model were linked together to generate mutually complementary boundary conditions. The average flow velocities in the closure gap derived from the IMPLIC program were used as input data for the MOZAS model. Similarly, the dimensions of the closure gap calculated with the MOZAS program were subsequently used in the IMPLIC model.

Sand production

The predicted sand production rates for the individual suction dredgers were based on actual production figures that had been obtained in practice and on information supplied by the contractor about the down-time required for maintenance work, adjusting pipes etc. The D_{50} standard used in the calculations was based on information derived from soil drillings in the sand extraction areas and on the actual position of the various suction dredgers.

Dam profile

The original assumptions about the bank gradients were based on previous experience with sandfill closures. However, in the course of the actual closure operations these figures were revised on the basis of soundings taken in the area.

Calibration of the predictive model

To calibrate the predictive model information was required regarding the dimensions of the closure gap and the local hydraulic conditions. The following data were of fundamental importance.

The water levels on either side of the closure gap required to establish the drop in water level within the gap

Measuring poles were erected along the axis line on both sides of the closure gap to provide information about the current water level. The data obtained were automatically transmitted by means of a telemetric registration point to the processing centre in Zierikzee. Here the average values received were stored in the database once every ten minutes. This allowed the information required for predictive purposes to be easily retrieved from the system.

The closure gap profile needed to define the effective flow opening

When conditions permitted, gauging operations were carried out each time the tide turned to determine the water levels along five pre-defined rows: along the axis line and two rows either side. This enabled the average closure gap profile to be determined. The effective area of the gap was calculated by digitizing the plotted, average closure gap profile. This information was also stored in the database.

The sand production in each dredger in order to determine the total amount of sand extracted

Pre-calibrated sand meters were installed on each of the dredgers to determine the production rates achieved. This involved measuring both the velocity of the sand water mixture and the concentration of solid particles in the output line. Telemetric techniques were used to transmit the information to the data-processing centre in Zierikzee. The average values recorded were stored as 'ten-minute readings' in the database. The down-time of each dredger was also noted.

The overall volume of the dam construction which was used to monitor the progress achieved, by comparing the production data with the sand losses in a given period

Virtually every day extensive gauging operations were carried out in the area where the sand dam was actually under construction. For this purpose fixed gauging lines were laid out at 25-m intervals. The measured values were converted to volumetric data using the computer. Comparisons were drawn with the previous set of data as well as with those from the original gauging operations carried out at the start of the project.

In addition, regular flow rate measurements were taken in the closure gap, which were processed within an eight-hour period. These data were combined with the observed drops in the water level and the effective flow area and were used as the basis for monitoring a number of parameters in the hydraulic model that formed an integral part of the forecasting system. This was primarily associated with the first phase of the work as in the latter stages, progress was so fast that the flow rate data were of little use for the actual construction work.

Forecasting range and frequency

From the point at which the Krammer gap had been reduced to a size of about 8000 m², the Support Centre became involved in forecasting the sequence of events that would occur during the closure operation. Up until this time forecasts had been supplied directly from the HMC. The frequency of the forecasts generated increased as the closure operation progressed and as the storm surge barrier came to play a more significant role. At first, the Support Centre provided a twice-daily forecast, but from the time at which the closure gap had been reduced to about 5000 m², continuous forecasts were given. During this phase the organization was such that a special team was empowered to decide on the most suitable barrier gate settings four times a day. This meant that the predictive models needed to be capable of producing specific recommendations four times a day. However, whenever possible, this frequency was reduced to allow sufficient time to check the quality of the information being generated.

The forecasts produced were valid for a maximum period of about 30 hours. The inaccuracy of the predictions obviously became larger as the time period for which the forecasts were given extended further into the future. This was largely due to the increasing uncertainties associated with both the weather forecasts and the production rates of the suction dredgers. The start of the meetings of the decision-making team therefore had to be arranged as close as possible to the times at which particular forecasts came into force. These conditions were satisfied by organizing the meetings three hours after the input data had been received. The time needed to work out a forecast using the predictive models and for formulating recommendations was about 1 hour 45 minutes. The HMC used the remaining 1 hour 15 minutes to prepare the forecast.

Wind and water level data registered by the measuring system were validated and stored in the central database. A retrospective forecast was prepared from these data using the IMPLIC model to provide an accurate basis for the required predictions. Prognoses for the boundary conditions relating to the wind and water levels were formulated at the same time.

In addition to short-term forecasts extending over a 30-hour period, predictions were also made for the medium-term outlook. These forecasts referred specifically to a period of several days ahead. Moreover regular use was made of longer-term forecasts covering the period up until the dams were closed. Such forecasts were particularly useful in identifying potential bottlenecks well ahead of time and assessing the long-term consequences of specific barrier gate settings. Medium-term forecasts were normally prepared about once a day, depending on the changes in conditions.

Evaluation of the forecasting system

As indicated above, the main objective for setting up a forecasting system was to advise on how best to use the storm surge barrier. Where possible, attempts were made to adhere to the design standard laid down (see Fig. 5). The effectiveness of the forecasting system can be illustrated by comparing the original standard with the actual drops in water level that occurred. A comparison of this type has been prepared for the Krammer closure operation (Fig. 8).

A further evaluation of the sandfill closure operations will be carried out to analyse the part played by the various components of the forecasting system in introducing errors. This analysis will also be concerned with the accuracy of the forecast produced as a function of time.

Hydrological and meteorological information provided after the completion of the Delta Project

The need for accurate forecasts of hydrological and meteorological data has remained even though the Delta Project is now virtually completed. Examples are:

(a) Predictions that can assist with the management of the storm surge barrier. The barrier is generally closed when storm surges are expected. In view of the negative impact such closures have on the environment, unnecessary closures of the gates should be avoided. This emphasizes the importance of having accurate forecasts of expected water levels.

(b) Storm surge forecasts for the Western Scheldt. The construction of the storm surge barrier has resulted in the Eastern Scheldt basin being protected from the effects of storm surges. However, during such storms the influence of the tide can still affect the Western Scheldt. To be able to take prompt action, forecasts are needed about exceptionally high water levels.

(c) Forecasts supplied for dredging and shipping activities.

It is hoped that the above explanation has served to demonstrate that the work involved in preparing forecasts for the Delta Project only represents a part of the HMC's wide range of activities and to explain why the HMC will continue to function even though the Delta Project has almost been completed.

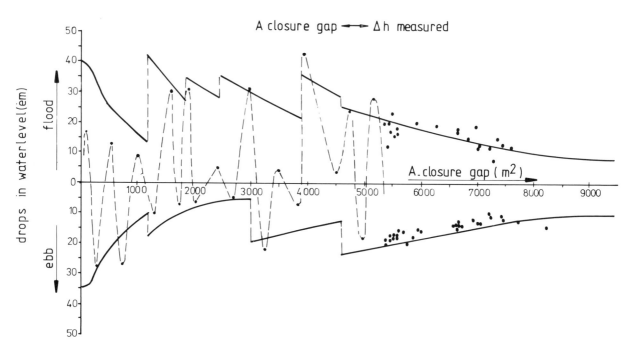

Fig. 8 Sandfill closure: design standard (——) and measured maximum drops (·····) in water level.

31

Limitations on Offshore Environmental Monitoring Imposed by Sea Bed Sampler Design

P. F. Kingston, Institute of Offshore Engineering, Heriot-Watt University, Edinburgh, UK

The pressures on coastal sea areas from industrial activity have been increasing in recent years, as have pressures from the use of the sea and sea bed as a site for the disposal of waste material. In parallel with this increased usage of the sea has been an increase in concern for the health of the marine environment by both public and, in reaction, by their governments.

UK offshore oil operators are now obliged by law to carry out environmental monitoring in the vicinity of their production platforms. Dredge spoil and sewage sludge disposal operations require monitoring programmes as a condition of licence to dump at sea. Prior to this many companies and other users of the sea conducted monitoring programmes regardless of their legal obligations.

The methods of monitoring the sea bed have changed little over the last 10 to 15 years. There are usually two approaches: one concerned with direct measurement of chemical contamination, the other with possible effects the particular operation is having on the biological environment.

Although the water column is sometimes sampled, the emphasis for offshore monitoring has invariably been the sea bed and its sediments. This is not unreasonable since the sediments, and the particles that make it up, are likely to accumulate any contaminants by adsorption.

Being for the most part sedentary, the animals that live in the sediment are unable to move away from the area of effect and so can be used as biological indicators of stressful conditions, being particularly of value for monitoring intermittent insult to the environment.

SAMPLING PROGRAMMES

Most offshore sea bed monitoring programmes require the biological and chemical analysis of sediment samples acquired remotely from surface operated sampler systems. Recently some attention has been turned to direct observation using underwater television. However, whilst this approach is satisfactory for so called 'skilled eye survey' for the detection of gross effects, accurate quantification of chemical and biological parameters is essential if incipient environmental changes are to be detected.

The sedimentary environment is theoretically one

Advances in Underwater Technology, Ocean Science and Offshore Engineering, Volume 16: Oceanology '88

of the easiest to sample quantitatively and has long been recognized as a potentially useful medium for sea bed environmental monitoring. However, in practice the reliable acquisition of samples of suitable integrity for chemical and faunal analysis has proved extremely difficult. This is largely a result of the design and mode of operation of existing equipment.

Most offshore monitoring surveys presently conducted by UK operators utilize a 0.1 m^2 grab sampler as their central piece of sampling equipment. These are used for both faunal samples, when the grab contents are retained in their entirety and then sieved to remove the biota from the sediment, and for chemical/physical samples, when a subsample is usually taken from the surface of the sediment obtained. In both cases the sampling programme is reliant on the grab sampler's taking consistent and relatively undisturbed sediment samples.

CURRENT GRAB SAMPLER TECHNOLOGY

Although there are a great many designs for benthic samplers, three types of bottom grab sampler are used almost exclusively by UK workers: the van Veen grab, Smith-McIntyre grab, and the Day grab (Fig. 1).

The Smith-McIntyre and Day grabs are similar in that the jaw buckets are supported in a frame. The buckets of both are semicircular in cross-section and produce a bite profile of similar shape. The van Veen grab consists of two jaws that are self supporting, the jaw buckets possessing a leading edge that is vertical as the instrument descends. Figure 1(b) shows the bite profiles of these samplers produced under test tank conditions using a hard-packed fine-sand substratum (Riddle, 1984). These profiles were

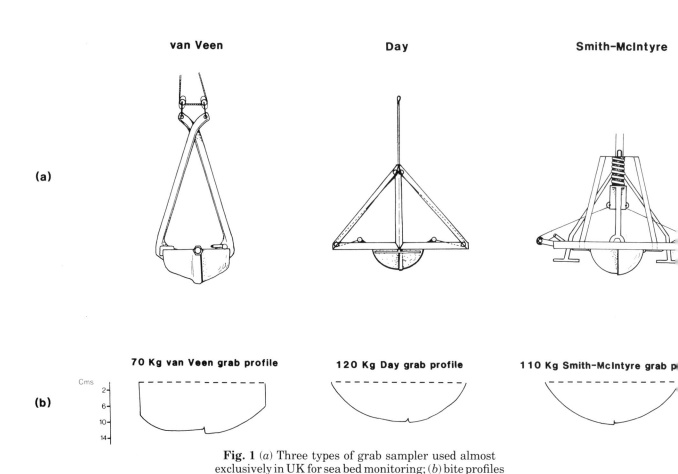

Fig. 1 (a) Three types of grab sampler used almost exclusively in UK for sea bed monitoring; (b) bite profiles of the three samplers obtained in a test tank using a fine sand substratum.

produced under ideal conditions and at a constant rate of hauling ($0.5\,\mathrm{m\,s^{-1}}$).

Under field conditions, in each case the sampler is lowered to the sea bed in a cocked (open) position. On reaching the bottom a release mechanism is triggered either by a bottom contact plate (Day and Smith-McIntyre) or by the weight of the sampler being taken off the warp. In either case the jaws are closed by the action of the warp as it hauls the sampler off the bottom.

WARP ACTIVATION

The use of the warp acting against the weight of the sampler to close the jaws has several disadvantages.

Warp heave

There is a tendency for the sampler to be pulled off the bottom as the jaws close, this tendency being related to the total weight of the sampler and the speed of hauling. For example, the theoretical maximum depth of bite of a 120 kg Day grab is 13 cm (based on direct measurements of the sampler); the bite profile given in Figure 1(b), however, shows a maximum depth achieved of only 8 cm. The influence of warp action on the digging efficiency of a grab sampler can be further demonstrated by comparing the results obtained by Riddle (1984) for chain and warp rigged instruments of similar weight (approximately 30 kg). Figure 2(a) shows the bite profile of the chain-rigged sampler, in which the end of each arm is directly connected to the warp end by a chain. The vertical sides of the profile represents the initial penetration of the grab and the central rise the upward movement of the grab as the jaws close. Figure 2(b) shows the bite profile of the van Veen rigged with an endless warp in which the arms are

closed by a loop of wire passing through a block on the end of each arm (as in Fig. 1(a)). The vertical profile of the initial penetration is again apparent, but in this case the overall depth of the sampler in the sediment is maintained as the jaws close.

The endless warp rig increases the mechanical advantage of the pull of the warp whilst decreasing the speed at which the jaws are closed. The result is that the sampler is 'insulated' from surface conditions to a greater extent than when chain rigged, giving a better digging efficiency.

Grab 'bounce'

In calm sea conditions it is relatively easy to control the rate of warp heave and obtain at least some consistency in the volume of sediment secured. However, such conditions are seldom experienced in the open sea, where it is more usual to encounter wave action. Few ships used in environmental monitoring studies are fitted with winches with heave compensators so that the effect of ship's roll is to introduce an erratic motion to the warp. This may result in the grab 'bouncing' off the bottom where the ship rises just as bottom contact is made or in the grab being snatched off the bottom where the ship rises just as hauling begins. In the former instance it is unlikely that any sediment is secured; in the latter the amount of material and its integrity as a sample will vary considerably depending on the exact circumstances of its retrieval.

The intensity of this effect will be dependent upon the severity of the weather conditions. Figure 3 shows the relationship between wind speed and grab failure rate which is over 60% of hauls at wind force 8. What

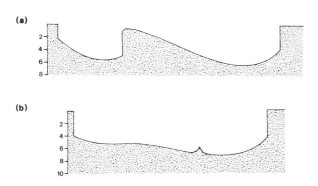

Fig. 2 (a) Bite profile of a chain-rigged van Veen grab sampler weighted to 35 kg in fine sand. (b) bite profile of a similarly designed and weighted van Veen, but rigged with an endless warp.

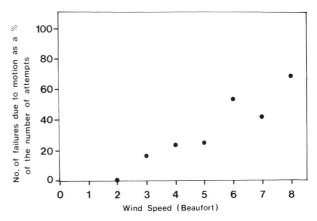

Fig. 3 Relationship between wind speed and grab failure rate. Data obtained from an offshore survey in the central northern North Sea using a 70 kg endless warp rigged van Veen and a 60 metre vessel (from Riddle, 1984).

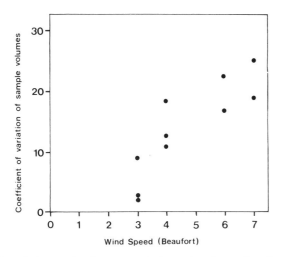

Fig. 4 Relationship between wind speed and sample volume variability. (Data from same source as Fig 3.)

is of more concern to the scientist attempting to obtain quantitative samples is the dramatic increase in variability with increase in wind speed with a coefficient of variation between 20 and 30 at force 7 (Fig. 4). The high cost of ship time places considerable pressure to operate in as severe weather conditions as possible, and it is not unusual for sampling to continue in wind force 7 conditions with all its disadvantages.

Drift

For a warp-activated grab sampler to operate efficiently it should be hauled with the warp positioned vertically above. If there is a strong wind or current, these conditions may be difficult to achieve. The result is that the grab samplers are pulled on to their sides. This is a particular problem with the van Veen grab, which does not have a stabilizing frame. Diver observations, however, have shown that at least in shallow water where the drift effect is at its greatest on the bottom, even the framed heavily weighted Day and Smith-McIntyre grabs can be toppled.

Initial penetration

It is clear that the weight of the sampler is an important element in determining the volume of the sample secured. Ankar (1977) doubled the volume of the sample taken by a 25 kg van Veen grab with the addition of an extra 20 kg of lead. This was largely the result of increasing the initial penetration of the sampler on contact with the sediment surface.

Initial penetration is one of the most important factors in the sequence of events in grab operation

determining the final volume of sediment secured. Figure 5 shows the relationship between initial penetration and final sample volume obtained for a van Veen grab (Ankar, 1977). Over 70% of the final volume is determined by the initial penetration. Subsequent digging of the sampler is hampered, as already shown, by the pull of the warp.

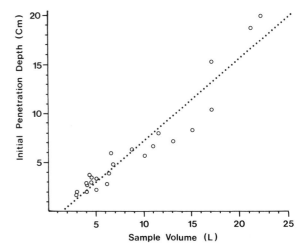

Fig. 5 Relationship between initial penetration of a 25 and 45 kg van Veen grab and sample volume secured (adapted from Ankar, 1977).

For benthic faunal studies it is important for the sampler to penetrate at least 5 cm into the sediment (for a $0.1 m^2$ surface area sample this gives 5 litres of sample). In terms of number of species and individuals, over 90% of benthic macrofauna is found in the top 4–5 cm of sediment (Ankar, 1977). Figure 6 shows how the number of individuals relates to average sample volume for eighteen stations in the southern North Sea (each litre recorded represents 1 cm of penetration). Although there is considerable variation in the numbers of individuals between stations, there is no significant trend linking increased abundance with increased sample volume penetration. In no case were sample volumes of less than 4.5 litres taken, indicating that at that level of penetration most of the fauna were being captured.

Samplers in which the jaws are held rigidly in a frame have no initial penetration if the edge of the jaw buckets, when held in the open position, are on a level with the base of the frame. The lack of any initial penetration in such instruments has the added disadvantage in benthic fauna work of under sampling at the edges of the bite profile (see Figure 1(b)) although the addition of weight will usually increase sample volume obtained.

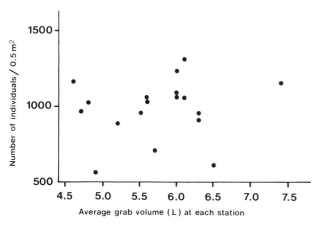

Fig. 6 Relationship between the number of benthic faunal individuals captured and sample volume for a sandy substratum in the southern North Sea.

Pressure wave effect

The descent of the grab necessarily creates a bow wave. Under field conditions it is usually impracticable to lower the grab at a rate that will eliminate a preceding bow wave, even if the sea were flat calm. There have been several investigations of the effects of 'down wash', both theoretical, using artificially placed surface objects (Wigley, 1967), and *in situ* (Andersin and Sandler, 1981; Riddle, 1984). The effects of down wash can be reduced by replacing the upper surface of the buckets with an open mesh. Although there is still a considerable effect on the surface flock layer (rendering the samples of dubious value for chemical contamination studies), the effect on the numbers of benthic fauna is generally very small (Lie and Pamatmat, 1965; Riddle, 1984).

ALTERNATIVES TO GRAB SAMPLERS

Ideally a benthic sediment sample for faunal and in particular for chemical/physical studies should be straight sided to the maximum depth of the sample and should retain the original stratification of the sediment. Grab samplers, by the very nature of their action will never achieve this end.

One solution is to employ some sort of corer designed to take surficial samples of sufficient surface area to satisfy the present approaches to benthic monitoring studies. Such devices do exist and are used widely, particularly in deep sea studies.

Spade box corers

The most widespread design is that of the spade box corer, first described by Reineck (1958) and later subjected to various modifications (Fig. 7). The sampler consists of a removable steel box open at both ends and driven into the sediment by weights. The lower end of the box is closed by a shutter supported on an arm pivoted in such a way as to cause it to slide through the sediment and across the mouth of the box. As with the grab samplers previously described, the shutter is driven by the act of hauling on the warp with all the attendant disadvantages. Nevertheless, box corers are very successful and are used widely for obtaining relatively undisturbed samples of up to $0.25\,\text{m}^2$ surface area. One great advantage of the box corer is that the box can usually be removed with the sample and its overlying water intact, allowing detailed studies of the sediment surface. Furthermore it is possible to subsample using small diameter corers for chemical and physical characteristic studies. Despite their potential of securing the 'ideal' sediment sample, box corers are rarely used for routine benthic monitoring work. This is largely because of their size (a box corer capable of taking a $0.1\,\text{m}^2$ sample weighs over $\frac{3}{4}$ tonne and stands $2\,\text{m}$ high) and the difficulty in deployment and recovery in heavy seas.

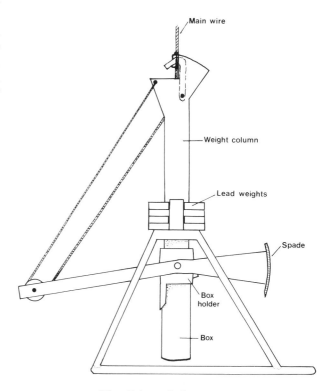

Fig. 7 A spade box-corer.

Precision corers

For chemical monitoring it is important that the sediment/water interface is maintained intact, for it is the surface flock layer that will contain the most recently deposited material. Unfortunately such undisturbed samples are rarely obtained, corers such as the craib corer (Craib, 1965), which are capable of securing undisturbed surface sediment cores, are unsuitable for routine offshore work because of the time taken to secure a sample on the sea bed and dependence on warp-activated closure. Large multiple precision corers have been constructed but again are too large and difficult to deploy for routine monitoring application. At present there is no instrument that fulfils the requirements of a quick turn-round precision multiple corer for offshore sediment monitoring.

SELF-ACTIVATED BOTTOM SAMPLES

There can be little doubt that one of the most important factors responsible for sampler failure or sample variability in heavy seas is the reliance of most presently used instruments on warp activated closure. The most immediate and obvious answer to this problem is to make the closing action independent of the warp by incorporating a self-powering mechanism.

Spring-powered samplers

Spring-powered samplers are in existence, possibly the most widely used being the Shipek grab, a small ($0.04\,\mathrm{m}^2$) sampler consisting of a spring-loaded scoop. This instrument is widely used where small surficial sediment samples are required for physical or chemical analysis. The use of a pretensioned spring unfortunately sets practical limits on the size of the sampler, since to cock a spring to operate a sampler capable of taking a $0.1\,\mathrm{m}^2$ sample would require a force that would be impracticable to apply routinely on deck. In addition, in rough weather conditions a loaded sampler of this size would be very hazardous to deploy.

Compressed-air-powered samplers

Another approach has been to use compressed air power. Flury (1963) fitted a compressed air ram to a modified Petersen grab with success. However, the restricted depth range of the instrument and the inconvenience of having to recharge the air reservoir for each haul limit its potential for routine offshore monitoring work.

Hydrostatically powered samplers

Hydrostatically powered samplers use the potential energy of the difference in hydrostatic pressure at the surface and the sea bed. The idea of using this power source is not new. Joly (1914) described a 'hydraulic engine' using hydrostatic pressure to drive a rock drill. Hydrostatic power has been widely used to drive corers largely for geological studies (Rosfelder and Marshall, 1967; McCoy and Selwyn, 1984). However, these instruments were principally concerned with deep sediment corers and were not designed to collect the flock layer at the sediment/water interface.

The potential for use of hydrostatic power for driving bottom samplers has not been fully realized, and its application is only now being seriously addressed.

A new hydrostatically powered grab sampler

In an attempt to find a solution to the problem of obtaining reliable and consistent sea bed samples for offshore monitoring studies a hydrostatically powered grab sampler has been developed at the Institute of Offshore Engineering.

Two approaches to the problem were adopted. The first was to use hydrostatic pressure to cock a powerful spring as the grab descends, the energy stored by the spring being released when the sampler touches bottom. The second was to use the hydrostatic power to directly drive the jaws shut on the sea bed.

The advantage of the first approach is that the spring-driven action ensures a consistent driving force to close the jaws. Its disadvantage was the limitation placed on the power of the instrument by the need to contain springs of sufficient strength to utilize the full potential of the hydrostatic pressure. In addition the system was not suited to shallow water operation.

The use of a direct system avoided the problem of spring containment and simplified the design of the ram system. Figure 8 shows the general layout of the instrument. Water enters the upper chamber of the cylinder when the sampler is on the sea bed, forcing down a piston which is connected to a system of levers which close the jaws.

The activating valve is held shut by the weight of the sampler and there is a delay mechanism to prevent premature closure resulting from 'bounce'. On the surface a one-way valve prevents water venting from the actuating valve. Instead a water-way through the connecting rod leading to a collar valve immediately above the jaws is used to drain the upper chamber, the piston being driven up by a light spring (not shown in Fig. 8) and the air compressed by the action simultaneously drawing open the jaws and washing out the sample. The closing action is independent of the warp and, provided sufficient excess wire is deployed, will be unaffected by surface weather conditions.

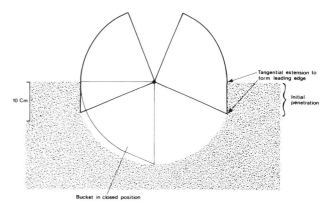

Fig. 9 Bucket design and bite profile of the hydrostatically powered sampler showing the initial penetration and sweep of the leading edge.

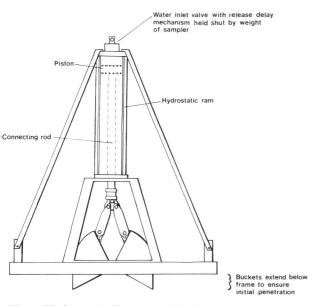

Fig. 8 Hydrostatically powered bottom grab sampler.

In designing a new sampler, opportunity was also taken to address the problem of initial penetration. Earlier studies have shown that the static weight of a sampler of 70 kg or more is adequate to ensure on initial penetration of 5–8 cm in even the hardest-packed sand substrata. Bucket design and configuration, however, are important if this is to be achieved. In the new sampler the leading edge of buckets have been tangentially extended by 8 cm so that they protrude below the level of the supporting frame. The result is that not only do the leading edges of the jaws begin closing 8 cm below the sediment surface they also sweep an arc of a radius greater than the body of the buckets, effectively digging themselves into the sediment (Fig. 9). The upper parts of the buckets are fitted with mesh-covered vents with a rubber flap which opens on descent and stretches across the vents sealing them off on ascent. The sampler has been built and tested and it performs as expected.

DISCUSSION

The main factors limiting the use of currently employed benthic samplers centre on their reliance on warp-driven closing mechanisms. This has the effect of severely restricting the 'weather window' in which they may be successfully used and has considerable influence on the overall sampling strategies adopted in most offshore monitoring programmes. The most important of these are the number of hauls made at each site and (partly as a consequence) the size of the unit sample. Where benthic faunal studies are involved, the standard sample size is usually 0.1 m². This is not based on any statistical consideration, as it would be in, for example, terrestrial work, but on what is the most practical unit size for routine offshore sampler operation.

Benthic faunal monitoring

Benthic fauna are contagiously distributed, that is, they tend to be aggregated or patchy in their distribution. There is also considerable temporal variation in the species composition and abundance of benthic communities. To sample these communities with a precision that would enable distinction between temporal variation and incipient change resulting from pollution effects, it is generally accepted that full replicate 0.1 m² hauls from each station are necessary (giving a precision at which the standard error is no more than 20% of the mean).

This frequency of sampling requires approximately 10–15 man days of sediment faunal analysis per site. The high cost of analysing samples and the lack of statutory requirement to carry out faunal monitoring studies have led offshore operators to reduce sampling frequency to two replicate hauls per station. This reduces the precision with which faunal abundance can be estimated to a level at which only gross change can be demonstrated (usually under conditions that do not require detailed analysis for their detection anyway).

However, sampling to an acceptable precision may be achieved from an area equivalent to one 0.1 m² grab sample if a smaller unit sample is used. For example, fifty 5 cm diameter core samples (with a total area of about 0.1 m²) will give a similar precision to that of five 0.1 m² grab samples (Riddle, 1984). Thus a similar degree of precision may be obtained for around one-fifth the analytical costs of using a conventional sampler.

The problem is to be able to secure 50 core samples per site in a timescale that would be realistic offshore. At present there is no instrument capable of supporting such a sampling demand. Clearly a single core sampler would be impracticable, and one must look to some sort of multiple corer that is capable of flexibility in its operation and will allow a quick on-deck turnaround between hauls. The huge saving in work-up costs will allow some increase in sampling time while retaining cost effectiveness. The smaller size of the unit sample would simplify automation of sieving on deck, if sieving would be necessary at all in the field.

Chemical monitoring

The present practice of subsampling a grab for chemical analysis is far from satisfactory. Not only is there a massive disturbance of the sediment surface layer as the grab sampler makes contact with the sea bed, but the stratification of the sediments is also disturbed by the action of the grab jaws as they close to secure the sample.

Since recent inputs into the sedimentary environment are likely to accumulate initially in the surface flock layer, it is vital that this should be adequately sampled in any monitoring programme designed to detect chemical environmental contamination.

The only way this can be effectively achieved is by precision corer. There are, however no instruments presently available that work satisfactorily offshore, which is why the use of grab subsamples is so widespread.

Non-random sampling

In order to determine the spatial distribution and abundance of benthic fauna it is usual to sample the sea bed at random. Such an approach is appropriate in a homogeneous environment such as prevails in offshore sedimentary habitats and has been effectively used in establishing the status of the sea bed fauna and sediment characteristics prior to any offshore operations.

In monitoring a point source of contamination, such

as the discharge of oily drilling cuttings, a random sampling programme may, however, be very wasteful of effort, particularly where the contamination is very localized and patchy in its distribution.

A more cost-effective approach would be to sample the contaminated and uncontaminated areas separately (stratified random sampling). This would take fewer samples overall and would give considerably more accurate information. The snag is that present sampling equipment is designed for random sampling with no facility for the operator to choose where it lands. It is theoretically possible to fit an underwater television camera to one of the currently used grabs, and use this to position it; however, the 'one shot' capability of such a configuration would not make its routine employment a viable proposition.

New approach to benthic sampler design

From the discussion above it is clear that there is a requirement for a benthic sampler capable of performing a combination of tasks which in many cases are individually performed by instruments already in existence but never together in a single piece of apparatus. The requirements of such a sampler may be summarized as follows:

(1) Sampler operation should be self powered, independent of warp action.
(2) Sampler should take a relatively small (e.g. 5 cm diameter) unit sample but be capable of a multiple function, each unit able to be activated independently.
(3) The sample should be secured with no disturbance of the surface sediment layers or stratification of the sediment.
(4) The sampler should be fitted with a means of direct observation to enable active positioning and repositioning between individual sample acquisitions without the need to return to the surface.
(5) The sampler units should be designed in such a way to facilitate quick on-deck turnaround.

All of these requirements could be met by a multiple corer made up of self-powered (hydrostatically driven) core units capable of surface controlled individual operation using underwater television to select the sampler position. Such a device would allow the rapid and routine core sampling required for benthic faunal monitoring, enabling a reduction in total sediment requiring faunal analysis while increasing the number of unit samples. It would also enable *in situ* site selection and the capability to obtain relatively undisturbed sediment cores for chemical and physical studies.

CONCLUSIONS

Present sampling practice in offshore benthic monitoring is far from satisfactory. The limitations on the quality of samples obtained casts doubt on the validity of many of the measurements made, particularly of levels of chemical contamination in the surface sediment layers. Yet there are no practical alternatives available. Present monitoring methods cannot be cost effective while the present approaches continue. It is an illusion to believe that the present monitoring programmes give good value for money when serious doubts exist as, in the case of chemical analysis, to the validity of the levels of contamination determined. In the case of faunal monitoring where surface disturbance is not so critical, an equal amount of information, if not more, can be obtained from a fraction of the amount of sediment currently used as the 'standard' sample size, by adopting an approach of much more frequent sampling using a much smaller sampling unit.

Benthic monitoring is largely bound by traditional approaches to studying the sea bed. Offshore operators are understandably conservative about using new approaches to a subject that is politically sensitive. Yet the adherence to these traditional (yet apparently acceptable) methods is resulting in a loss of important data and poor value for money in terms of return for effort and quality of information.

REFERENCES

Andersin, A. and Sandler, A. (1981). Comparison of the sampling efficiency of two van Veen grabs. *Finnish Mar. Res.* **248:** 137–142.

Ankar, S. (1977). Digging profile and penetration of the van Veen grab in different sediment types. *Contributions from the Asko Laboratory,* University of Stockholm, Sweden. No. 16, 22 pp.

Craib, J. S. (1965). A sampler for taking short undisturbed cores. *Journal du Conseil Perm Int par l'Exploration de la Mer.* **30:** 34–39.

Flury, J. (1963). A modified Petersen Grab. *J. Fish. Res. Bd Canada* **20:** 1549–1550.

Joly, J. (1914). On the investigation of the deep sea deposits. *Sci. Proc. R. Dublin Soc. NS* **14:** 256–267.

Lie, V. and Pamatmat, M. M. (1965). Digging characteristics and sampling efficiency of the 0.1 m² van Veen grab. *Limnol. Oceanogr.* **10:** 379–385.

McCoy, F. W. and Selwyn, S. (1984). The hydrostatic corer. *Marine Geology* **54:** 33–41.

Reineck, H.-E. (1958). Kastengreifer und Lotrohre 'Schnepfe'. *Senckenbergiana Lethaea* **39:** 42–48 and 54–56.

Riddle, M. R. (1984). Offshore Benthic Monitoring Strategies. PhD Thesis, Heriot-Watt University, Edinburgh, 335 pp.

Rosfelder, A. M. and Marshall, F. (1967). Obtaining large, undisturbed and orientated samples in deep water. In: *Marine Geotechnique* (A. F. Richards ed.). University of Illinois Press. pp. 243–263.

Wigley, R. L. (1967). Comparative efficiencies of van Veen and Smith-McIntyre Grab samples as revealed by motion pictures. *Ecology* **48:** 168–169.

32

Design and Use of a Clean Shipboard Handling System for Seawater Samples

N. H. Morley, Department of Oceanography, The University, Southampton, SO9 5NH, *C. W. Fay,* Natural Environment Research Council, Research Vessel Services, No. 1 Dock, Barry, South Glamorgan, CF6 6UZ *and P. J. Statham,* Department of Oceanography, The University, Southampton, SO9 5NH

This chapter describes the design, construction and deployment of a shipborne facility intended to minimize trace metal contamination during the sampling and initial processing of open ocean water samples. Results obtained to date confirm the overall success of the approach, both in absolute values for trace metal concentrations and in relation to the nutrient and other data obtained from the same samples.

HISTORICAL BACKGROUND

During the past 100 years the apparent concentrations of many trace elements in sea water have shown a considerable decrease, with current values, which are considered to be accurate, typically at nano- and picomole l^{-1} levels. Part of this decrease can be accounted for by improvements in analytical techniques, particularly the advent of graphite furnace atomic absorption spectrophotometry and electrochemical techniques (compare, for example, methods in refs 1 to 5). With these advances in analytical capability it has become increasingly obvious that contamination from reagents and containers used in analyses may be an important factor in data obtained and that the scrupulous use of clean techniques is required to ensure adequate precision, accuracy, and detection limits (refs 6 to 8).

As the improvements in analytical technique cited above took place, two further major sources of errors in data became apparent: adventitious contamination from the use of inappropriate sampling materials and techniques, and loss of metals from samples by adsorption on to container walls. The latter problem can generally be controlled, for filtered samples, by acidifying to pH < 2 prior to storage. To avoid contamination sample collection has to be via a clean and metal-free system. The use of plastic or rubber components containing metal fillers must be strictly avoided. Teflon is potentially the cleanest constructional material; however, it is expensive and poses mechanical problems (ref. 9). Generally acceptable substitutes include polyethylene, polypropylene and polymethylmethacrylate (Perspex) (ref. 6).

Material from steel hydrowires, associated lubricants and surface slicks, often derived from the

Advances in Underwater Technology, Ocean Science and Offshore Engineering, Volume 16: Oceanology '88

sampling ship, can contaminate sampling bottles. On recovery of a representative water sample, its initial processing (e.g. filtration, transfer to storage containers, and preservation) on board a ship not specially prepared is carried out in an environment likely to be rich in potential sources of metal contamination: rusty iron, metallic paints, stack fumes etc. To avoid the problem of shipboard contamination it is necessary to isolate the samples from the local environment and to provide a clean area in which to work. At its most basic this may be obtained by use of a plastic tent, and this has proved adequate for work on some metals in coastal and estuarine areas. The next stage is to provide a class 100 laminar flow hood for critical operations (e.g. filter changing and acidification of samples), while keeping less contamination-prone steps within a tent. This extends the range of sample types that can be usefully handled, but is still not ideal. A more rigorous approach is to provide a self-contained clean environment in which all operations may be carried out. A particularly appropriate and convenient solution is the use of a containerized clean laboratory.

THE CLEAN SHIPBOARD SAMPLE HANDLING SYSTEM

Early in 1985, the Natural Environment Research Council (NERC) took delivery of their newest research ship—the RRS *Charles Darwin*. Unlike her older sister ship, RRS *Discovery*, *Darwin* does not have laboratories fitted out for specific scientific disciplines. Instead she has a series of large scientific working areas and deck spaces which are designed to be adapted by the temporary installation of laboratory furniture and equipment for an individual cruise.

In 1984, while the RRS *Charles Darwin* was under construction, the Research Vessel Services (RVS) embarked on the creation of a new marine chemistry facility. This was the result of the emergence of marine chemistry as a major research area requiring logistic support. Also, NERC had declared an intention to mount an oceanographic campaign in the Indian Ocean which was to involve a significant content of deep-ocean marine chemistry, commencing with cruise RRS *Charles Darwin* 15 (CD 15). In consultation with potential users from universities and NERC institutes, a programme of development and equipment procurement was established. This programme is now largely complete, but it has had to span over a number of years for financial reasons.

Among a number of facilities required by the research community was a clean sampling and handling system specifically designed for trace-metal work. The sampling problem could be solved by the purchase and modification of a suitable system, but no such solution to the handling problem was available so some method of clean sample processing had to be designed and manufactured. A major consideration was the provision of a contamination-free space in which shipboard processing could be carried out. As indicated previously, and pointed out by other workers (refs 10 and 11), containerized laboratories offer a convenient and efficient means of providing a clean facility on a ship. The principal criteria used in the design of the container laboratory described here relate to the contamination problems mentioned earlier. Particular emphasis has been placed on controlling contamination at the most critical stages of sample handling.

SAMPLE COLLECTION

Sample collection is performed using 10-litre GO-FLO water sampling bottles attached to a 12-bottle rosette with an integral CTD instrument package. Fluorescence and dissolved oxygen probes can also be fitted to this unit. The GO-FLO sampling bottles are manufactured from Teflon-lined PVC with PVC end valves. Closure is by means of an external rubber band. The bottles are designed to penetrate the water surface closed and to open at a depth of about 5 m, thus preventing internal contamination from surface slicks. Prior to use the bottles required the replacement of the existing taps (which included a stainless steel pin) with all PTFE units. While the rosette system is suitable for most purposes, a totally non-metallic system was also considered desirable for critical sampling. This was provided by deploying the GO-FLO bottles from 6-mm KEVLAR hydrographic line (with an epoxy-resin-coated weight) and using solid PTFE messengers. The 200-m depth limit of this system could be extended by linking the KEVLAR line to the ship's hydrographic wire.

CONSTRUCTION OF THE CLEAN CONTAINER LABORATORY

The clean container laboratory (see Fig. 1 for plans) was constructed from a standard hard-topped 20′ × 8′ × 8′ (6058 mm (length) × 2438 mm (width) × 2950 mm (height)) transit container conforming to Lloyd's, TIR and IMCO specifications. The outside skin is of 6-mm mild steel plate. The inside bulkheads and deckheads are of epoxy-painted 11-mm marine-grade plywood with 60 mm of fibreglass wool insulation between the inner and outer

SECTION SHOWING WET BENCH

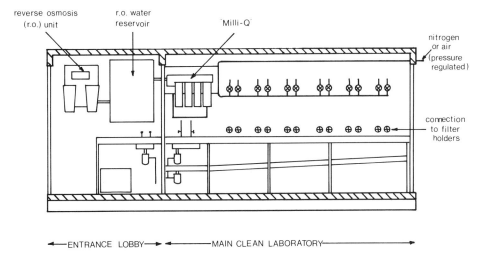

PLAN VIEW

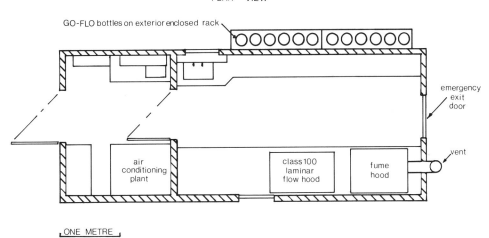

Fig. 1 The clean laboratory container.

skins. At each end is a single width door: one is the normal extrance; the other is an emergency exit from the laboratory. Inside, there are two compartments: the clean laboratory, measuring 4030 mm × 2200 mm, and an entrance lobby, measuring 1700 mm × 2200 mm. In both compartments the floor is finished with 5-mm-thick studded rubber floor tiles. The services fitted into the unit include air conditioning, single-phase 240 VAC electricity with 13-amp socket outlets, gas lines for low pressure nitrogen or air, fresh water and non-toxic seawater supplies, drainage, smoke detectors, fire alarm, telephone, and a loudspeaker extension from the ship's public address system. The lighting throughout is provided by 240 V AC fluorescent tubes in plastic housings. The emergency lighting, which switches on automatically with loss of mains power, is driven from internal 24-V batteries which are on continuous trickle charge. Connection of the 240 V AC power supply from the ship to the clean container laboratory is via a splash-proof plastic 63-amp Duraplug conforming to BS 4343.

The lobby houses the air conditioning equipment, the electrical service terminations (fuse boxes, fire detection control panel, and intercom), a sink, a small refrigerator, a Nitrox NG300 portable nitrogen generator, and a Millipore water purifier capable of producing 20 litres per hour of Grade 1 analytical water. A Halon fire extinguisher is installed. The only normal access to the clean laboratory is via the lobby, which thus serves as an 'airlock' and changing room to reduce the transfer of contaminants from the outside into the clean laboratory.

The internal surfaces of the clean laboratory need to be non-metallic and non-particle-generating. Where necessary, surfaces have been coated with an epoxy resin paint in order to achieve this. One side of the laboratory has a specially designed wet bench with N₂/air, water and drainage services. On the other side is a simple dry bench with a cupboard and drawers below. Ten wall-mounted splash-proof 13-amp electrical outlet sockets provide power for instrumentation. On the dry bench is a fume hood which has been adapted both for direct venting to the atmosphere and for use in a recirculating mode; the latter feature was included because the flow of exhaust gases from a vent on a ship is very indeterminate owing to wind variation and could cause contamination or a safety hazard. Also provided on the dry bench is a class 100 HEPA filter laminar flow hood.

On the outside of the container, adjacent to the laboratory's wet bench, is a demountable water bottle rack capable of holding 12×10-litre GO-FLO water sampling bottles. Samples are fed to the wet bench area via through-bulkhead connections of FEP tubing fitted with PTFE taps. Pressurized (1 bar) gas may be fed to the sample bottles during sample filtration. Waste seawater flows from the bench to a trough below and thence into the drain service.

FILTRATION OF SEAWATER SAMPLES

Removal of particulate material from seawater is necessary because its presence may affect the concentration of the dissolved metals. Also analysis of the particulate material obtained may yield valuable geochemical information. However, this is a critical sample handling stage, and as most commercially available filter units have potential for metal contamination at the ultra trace level, the unit shown in Figure 2 was designed and built especially for use with open ocean seawater samples. The entire unit, with the exception of a polyethylene frit upon which the 0.4 μm membrane filter rests, is machined from solid PTFE stock. When the unit is assembled the filter is sealed in position by a fluorinated ethylene propylene (FEP) coated 'O' ring. The cap is machined to accept the ⅝″ OD FEP tubing used with the GO-FLO bottles and is thus compatible with the rest of the

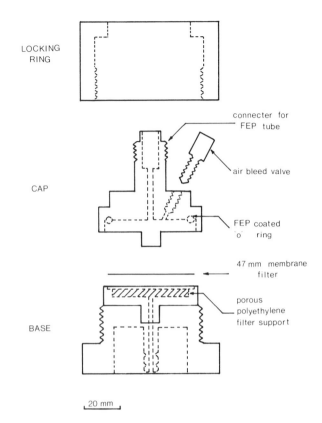

Fig. 2 The filter holder. The unit is constructed from 75 mm diameter PTFE rod. The tightening nut for the FEP tube connector is not shown.

clean system. The seawater sample passes directly from the GO-FLO bottle to the filter unit and then to the storage vessel, only coming into contact with FEP, PTFE and polyethylene components, and the filter itself.

The filter unit can be used at pressures of up to 1 bar, which is the maximum recommended internal pressure for the GO-FLO bottles. Flow rates of about 200 to 500 ml min⁻¹ for clean, open ocean water are typical; however, filtration time will rapidly increase in waters with high particulate loadings.

PRE-CRUISE PREPARATION

Prior to cruise CD 15 all equipment to be used for trace-metal sampling or storage was subjected to rigorous cleaning. After checking for flaws or inclusions the 1 l polyethylene sample storage bottles used were soaked in 2% v/v Micro solution then 6 M HCl followed by 6 M HNO₃ for 7 days in each. The bottles were rinsed with distilled water followed by quartz grade sub-boiling distilled water and dried overnight in a clean air cabinet (ref. 12). They were then individually sealed into plastic bags and boxed ready for shipping. The filter units were cleaned in the same way as the storage bottles. The GO-FLO bottles were cleaned by filling with 2% Micro followed by 2 M HCl for 24 h each. No nitric acid cleaning stage was employed, because of the risk of nitrate contamination of subsequent nutrient samples. The membrane filters were cleaned by soaking overnight in 3 M HCl and stored in sub-boiling distilled water. All other plastic ware that would come into contact with the samples was cleaned in the same way as the storage bottles, with the omission of the nitric acid soak.

COMMISSIONING THE CLEAN CONTAINER LABORATORY

Final preparation of the clean container laboratory was carried out aboard RRS *Charles Darwin* while on passage from Jeddah to the Seychelles (cruise CD 14a). Work included the making of connections to the ship's services and testing certain aspects of the container's functioning. For example, the nitrogen generator fitted was found to be inadequate for our pressure filtration requirements and a compressed air system was substituted. The de-ionized water supply was brought on-line, and those fittings which had been retained at Southampton for cleaning were also installed. The inside of the container was thoroughly cleaned at this stage.

USING THE SYSTEM

During the 21 days of cruise CD 15 (in the south-west Indian Ocean), seven stations were occupied (on a north-south transect between 5°S and 27°S). About 200 seawater samples were obtained, from throughout the water column, and processed through the clean container. The full suite of analyses intended to be carried out on these samples is given in Table I. To obtain suitable sub-samples for these analyses entailed (for each GO-FLO sample) the rinsing and filling of up to seven bottles with unfiltered and eight bottles with filtered water. A full rosette of 12 GO-FLOs took about four hours to process completely when two people were working in the container. Preparation of the container laboratory to receive the next batch of samples (e.g. bottle collection and numbering, filter preparation) took a further hour. On two stations over 40 GO-FLO samples were obtained and these required some 14 hours' occupancy of the container with a further six hours of associated work.

TABLE I

List of analytes to be determined on samples processed through the clean container during cruise CD 15

(1) Non-filtered samples
  (a) Shipboard analysis
    Salinity
    Dissolved oxygen
    Silicate
    Nitrate
    Phosphate
    Alkalinity
  (b) Stored
    Iodine
    Oxygen-18
    Ge
    Bacteria

(2) Filtered samples
  (a) Shipboard analysis
    Ni
    F^-
  (b) Stored
    Al, Cd, Cu, Mn, Ni, Zn, Ag, Co, Fe, Pb
    Rare earth isotopes
    As, Se

This listing is not exhaustive as stored samples will be available for the determination of further analytes.

During protracted processing periods, further water sampling was being carried out using alternative systems or the ship was on passage and thus processing time was not a limiting factor on station time. Sample loss rates were about 10%, due to system malfunction, with no sample obtained (sampler failing to trip due to rosette defects or end valves not fully closing were the most common of these) and a further 3% where samples were unable to be filtered successfully owing to tubing blowing out of taps under pressure or to leaks from the sampling bottles.

QUALITY OF DATA OBTAINED

Currently, few reliable data sets are available concerning the trace-metal regime of the Indian Ocean waters, hence no direct comparison of results is possible. Also it is not feasible to obtain system, as opposed to analytical, blanks. However data obtained using the system described are in close agreement with recently reported Atlantic and Pacific Ocean trace metal concentrations (refs 13, 14 and 15). For cadmium (at stn CD 15/4 27°S 56°E) the expected low surface concentrations (< 20 pmoles l^{-1}) and increase

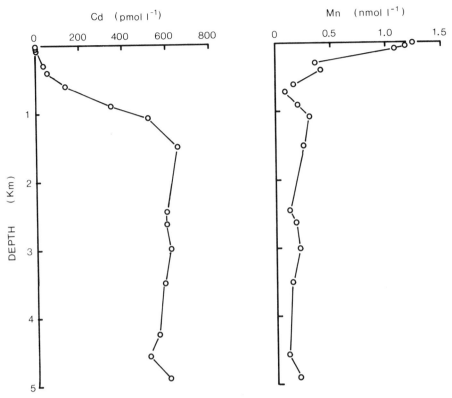

Fig. 3 Vertical distributions of dissolved metals at Station Cd 15/4 (27°S, 56°E): (*left*) cadmium, (*right*) manganese.

with depth (Fig. 3(*a*)), closely correlate with phosphate in the same samples (ref. 16). The dissolved manganese profiles at the same station exhibit the expected surface maxima with rapid depletion to close to the analytical detection limits (0.1 nmoles l^{-1}) in the deeper waters (Fig. 3(*b*)). In general the data obtained conform to the criteria of oceanographic consistency, with smooth profiles that show no major deviations from those found in other oceanic areas.

CONCLUDING COMMENTS

Research into techniques which will allow real-time or close to real-time analysis of a range of trace constituents in seawater, often *in situ*, are advancing rapidly. In these methods sample handling and thus the opportunities for contamination are kept to a minimum. However, these measurement techniques still require further development and do not cover the full range of trace metals of interest. As long as this is so, there will be a requirement for the type of clean facility described in this paper. The containerization of this facility allows its deployment on any vessel equipped to receive it, thus increasing the cost effectiveness of the clean system, while not wasting valuable shipboard space when other research work is being carried out.

ACKNOWLEDGEMENTS

The authors wish to acknowledge the valuable advice of Dr H. Elderfield and help of Mr M. Greaves (Cambridge University), the assistance with design and construction of the container laboratory given by Mr R. Griffiths and the RVS technical staff, and the construction of the filter holders by Mr J. Nevill (Southampton University).

REFERENCES

1. Riley, J. P. (1965). Analytical Chemistry of Seawater. In: *Chemical Oceanography* Vol. 2 (Riley and Skirrow eds). Academic Press, London and New York, pp. 295–424.
2. Riley, J. P. (1975). Analytical Chemistry of Seawater. In: *Chemical Oceanography, 2nd Edn.* Vol. 3 (Riley and Skirrow eds). Academic Press, London, pp. 267–407.
3. Bruland, K. W., Franks, R. P., Knauer, G. A. and Martin, J. H. (1979). Sampling and Analytical Methods for the Determination of Copper, Cadmium, Zinc and Nickel at the Nanogram per Liter Level in Seawater. *Anal. Chim. Acta.* **105**: 233–245.
4. Statham, P. J. (1985). The Determination of Dissolved Manganese and Cadmium in Seawater at Low nmol l^{-1} Concentrations by Chelation and Extraction Followed by Electrothermal Atomic Absorption Spectrometry. *Anal. Chim. Acta.* **169**: 149–159.
5. Bruland, K. W., Coale, K. H. and Mart, L. (1985). The Analysis of Seawater for Dissolved Cd, Cu and Pb, an Intercomparison of Voltammetric and Atomic Absorption Methods. *Mar. Chem.* **17**: 285–300.
6. Robertson, D. E. (1968). The Role of Contamination in the Trace Element Analysis of Seawater. *Anal. Chem.* **40**: 1067–1072.
7. Mitchell, J. W. (1982). Purification of Analytical Reagents. *Talanta.* **29**: 993–1002.
8. Moody, J. R. and Beary, E. S. (1982). Purified Reagents for Trace Metal Analysis. *Talanta.* **29**: 1003–1010.
9. Freimann, P., Schmidt, D. and Schomaker, K. (1983). Mercos, a Simple Teflon Sampler for Ultratrace Metal Analysis in Seawater. *Mar. Chem.* **14**: 43–48.
10. Wong, C. S., Cretney, W. J., Piuze, J., Christensen, P. and Berrang, P. G. (1977). Clean Laboratory Methods to Achieve Contaminant-free Processing and Determination of Ultra-trace Samples in Marine Environmental Studies, NBS Spec. Publ. 464, 249–258.
11. Danielsson, L. G. and Westerlund, S. (1983). Clean Room Container Laboratory For YMER 80. *Ocean Science and Engineering* **8**: 53–62.
12. Moody, J. R. and Lindstrom, R. M. (1978). The Cleaning, Analysis and Selection of Containers for Trace Element Samples. NBS Spec. Publ. 501, 19–32.
13. Bruland, K. W. and Franks, R. P. (1983). Mn, Ni, Cu, Zn and Cd in the Western North Atlantic. In: *Trace Metals in Seawater* (Wong, Boyle, Bruland, Burton and Goldberg eds). NATO Conf. Series. Series IV, Mar. Sciences, Vol. 9, pp. 395–414. Plenum Press, New York and London.
14. Bruland, K. W. (1980). Oceanographic Distribution of Cadmium, Zinc, Nickel and Copper in the North Pacific. *Earth Planet Sci. Lett.* **47**: 176–198.
15. Statham, P. J., Burton, J. D. and Hydes, D. J. (1985). Cadmium and Manganese in the Alboran Sea and Adjacent North Atlantic: Geochemical Implications for the Mediterranean Sea. *Nature* **313**: 565–567.
16. Statham, P. J. (1983). Analytical and Environmental Studies on Some Dissolved Trace Metals in Seawater. PhD Thesis, University of Southampton, 209pp.

33

Some Ecological Consequences of the Use of Anti-fouling Paints Containing Tributyltin

Andy R. Beaumont, School of Ocean Sciences,
University College of North Wales, Marine Science Laboratories,
Menai Bridge, Gwynedd LL59 5EY, UK

Tributyltin (TBT) has been extensively used over the last decade as a biocide in anti-fouling paints. Although TBT is extremely effective in preventing the settlement of plants and animals on ship hulls, it is now clear that other marine organisms besides those targeted are significantly affected by the TBT leaching from anti-fouling paints.

In estuaries and marinas in the UK, where large numbers of pleasure craft are moored, levels of TBT during 1986 ranged from less than 1 nanogram (ng) l^{-1} in the winter up to 1500 ng l^{-1} in the summer (ref. 1). Sensitivity to TBT varies among aquatic species, with gastropods and bivalves being the most susceptible (20–140 ng l^{-1}) followed by crustaceans (90–140 ng l^{-1}), algae (100–250 ng l^{-1}) and fish (200 ng l^{-1}) (for reviews see refs 1–3).

Most TBT toxicity tests have been carried out on single species in the laboratory, and although these tests provide useful information, they can be criticized because they may not be relevant to the undoubtedly more complex situation in the wild. Field trials, such as the transfer of animals between clean and polluted areas, can also be criticized for their lack of controlled conditions. This chapter reviews laboratory and field data and investigates evidence of the ecological effects of TBT in a microcosm toxicity study involving a range of species in a muddy-sand substrate (refs 4,5). Such microcosm studies are mostly under laboratory control, but exhibit much of the ecological complexity found in the wild.

The microcosm study confirms the high sensitivity of molluscs to TBT and reveals the significant ecological damage which can accrue from the use of TBT-based anti-fouling paints on pleasure craft in enclosed waters.

LABORATORY AND FIELD STUDIES

Walsh *et al.* (ref. 6) and Beaumont and Newman (ref. 7) have demonstrated that 100–300 ng l^{-1} TBT reduces the growth of certain microalgal species. The most sensitive so far tested is the diatom *Skeletonema costatum*, which shows no growth at all at 100 ng l^{-1}. Preliminary trials investigating respiration and photosynthesis in the microalga *Pavlova lutheri* and the macroalga *Ulva lactuca* (ref. 4) indi-

Advances in Underwater Technology, Ocean Science and Offshore Engineering, Volume 16: Oceanology '88

cate that 100 ng l^{-1} TBT can cause a significant increase in algal respiration. This may be due to damage caused to the mitochondria since there is evidence of mitochondrial damage due to TBT in animals (ref. 8). The effects of TBT on photosynthesis could not be resolved in these trials owing to interference by the acetic acid used as a carrier to introduce TBT into solution.

Microalgae are primary producers in the aquatic environment, and therefore reduced growth or mortality of microalgae could have far-reaching effects on life at higher levels in the food chain.

A number of crustaceans have been tested for the effects of TBT. Levels of between 200 and 1000 ng l^{-1} significantly reduce growth of larvae of *Homarus americanus,* the American lobster (ref. 9), and juveniles of the mysid shrimp *Acanthomysis sculpta* (ref. 10). Effects on reproduction were also noted in adult female *A. sculpta,* and behavioural effects have been demonstrated in the freshwater flea, *Daphia magna* (ref. 11). The copepod *Acartia tonsa* is also significantly affected at these levels of TBT (ref. 12).

Molluscs are particularly susceptible to TBT. Bryan *et al.* (refs 13,14) and Gibbs *et al.* (ref. 15) have demonstrated that the phenomenon of 'imposex', the imposition of male sexual characteristics onto the female, in the dogwhelk, *Nucella lapillus* is caused by very low levels of TBT (< 20 ng l^{-1}). Female dogwhelks develop a penis which may grow to the same size as the true penis of the males, and, associated with the penis, a vas deferens develops. The growth of the vas deferens eventually occludes the opening of the oviduct and prevents the release of egg capsules, thus rendering the female sterile (ref. 16). Since the early 1970s sufficiently high levels of TBT have been present in many estuaries in southwest England to cause a very significant decline, and in some cases virtual extinction, of dogwhelk populations (ref. 13).

A number of studies have been carried out on the incidence of shell thickening and malformation in the Pacific oyster, *Crassostrea gigas,* as a result of exposure to low concentrations of TBT. This thickening, commonly called 'balling', occurs as the result of the deposition of new nacreous layers separated by gel-filled cavities. Ten or more layers produce malformation of the shell, making the oysters unsuitable for marketing. The phenomenon of balling was first noted in the UK by Key *et al.* (ref. 17), but the French were the first to link balling in commercial oysters to the proximity of pleasure craft activity (refs 18, 19). As a result of Alzieu's work, the French government banned the use of TBT-based anti-fouling

paints on vessels under 25 metres in 1982. Subsequent studies (refs 20–22) have confirmed that TBT is the cause of shell thickening and malformation in *C. gigas.* Interestingly, this phenomenon does not appear to occur in other oyster species such as the American oyster, *Crassostrea virginica,* and the European flat oyster, *Ostrea edulis,* but the reasons for this are not yet understood (ref. 23).

Although imposex and balling are important and easily recognized effects of TBT on adult molluscs, reproduction and larval development of molluscs are also affected. Beaumont and Budd (ref. 24) have shown that veliger larvae of the mussel, *Mytilus edulis,* suffer high mortality after being exposed to 100 ng l^{-1} TBT for 15 days. Surviving larvae were significantly smaller and weaker than controls, and Laughlin *et al.* (ref. 25) have recently demonstrated a similar significant reduction in growth of hard shell clam, *Mercenaria mercenaria,* larvae after 14 days at 25 ng l^{-1} TBT. Significant effects have been demonstrated at 100 ng l^{-1} TBT during the early development of mussel embryos (ref. 26). Furthermore, apparently normal veligers derived from mussel embryos developing in 100 ng l^{-1} TBT, grew significantly slower than control veligers (ref. 27). Reproductive performance of *O. edulis* during long-term exposure to 240 ng l^{-1} TBT has been assessed by Thain (ref. 23). *O. edulis* is a sequential hermaphrodite, that is, normal sexual development involves the initial production of male gonad with subsequent development of the female gonad and loss of the male gonad. In Thain's experiments, contrary to controls where females developed normally and produced larvae, males predominated and no larvae were released in TBT-treated stocks. Growth of post-larvae (spat) of several bivalve species was also significantly inhibited by treatment with 240 ng l^{-1} TBT over a 6-week period. Spat of some species, however, such as *O. edulis* and the manilla clam *Tapes semideccusatus,* showed no significant growth effects at this low level of TBT (ref. 23).

Sublethal physiological effects of TBT on *C. gigas* have been investigated by Lawler and Aldrich (ref. 28). Significant effects on oxygen consumption and feeding rate were noted at 50 ng l^{-1} TBT, growth was affected at 20 ng l^{-1} TBT, and the ability to compensate for hypoxia was reduced at 10 ng l^{-1} TBT. Lawler and Aldrich's study shows the value of investigating the physiological effects of toxins since it is clear from their work that subtle effects are occurring at very low levels of TBT.

Studies on fish have indicated that behavioural effects only occur at quite high levels of TBT (2500 ng l^{-1}) and no significant growth reduction or mortalities have been demonstrated at low levels of TBT (ref. 3). Nevertheless, high bioaccumulation of butyltins has been demonstrated in Chinook salmon (*Onchorhyncus tshawytscha*) reared in sea pens treated with TBT anti-fouling paints (ref. 29), and this provides evidence of another method of entry of butyltins into the human diet besides the ingestion of shellfish which also bioaccumulate TBT.

MICROCOSM STUDY

To overcome some of the criticisms levelled at single-species laboratory experiments and uncontrolled field studies, a trial was undertaken using microcosms which effectively encapsulate a small portion of the wild but retain laboratory control. Each microcosm used in the trial consisted of a 1.5×0.5×0.5 m tank containing a muddy-sand substrate 10 cm deep over-laid by flowing unfiltered seawater to a depth of approximately 15 cm (Fig. 1). A total of nine micro-cosms were used; three were subjected to high levels of TBT (1000–2000 ng l^{-1}), three to low levels (100–200 ng l^{-1}), and three remained untreated. TBT was provided by painting a TBT-containing anti-fouling paint on to slate panels suspended in the mixing-header tank for each system. Known numbers of a range of invertebrates such as the bivalves, *Cerasto-derma edule* and *Scrobicularia plana,* the ragworm, *Neries diversicolor,* another polychaete, *Cirratulus cirratus,* and the winkle, *Littorina littorea,* were introduced into the tanks at the beginning of the

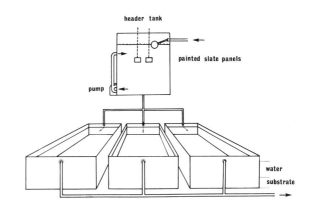

Fig. 1 Diagram of microcosm system for studying the effects of TBT on muddy-sand substrate fauna and microflora. Three such systems were employed in the trial: one control, one with low TBT (100 ng l^{-1}) and one with high TBT (1000–2000 ng l^{-1}) levels.

trial. Full details of the methodology are presented elsewhere (refs 4,5).

RESULTS

Table I indicates the percentage mortalities of the introduced invertebrate species in the microcosms. Owing to the practical problems associated with analysing all the microcosms at the same time, one microcosm from each system was analysed after 11 weeks, a further one from each system after 14

TABLE I
Percentage mortalities of introduced species in TBT treated and untreated microcosms. Duration of exposure in weeks is given in parentheses

Species introduced	Number per tank	Treatment								
		High level TBT			Low level TBT			Control		
		1	2	3	4	5	6	7	8	9
Cerastoderma edule	50	100(2)	100(2)	100(2)	20(11)	38(14)	80(17)	0(11)	6(14)	14(17)
Scrobicularia plana	26	27(5)	81(7)	100(10)	11.5(5)	11.5(8)	11.5(11)	0(5)	4(8)	0(11)
Neries diversicolor	50	90(11)	88(13)	74(16)	42(11)	80(14)	78(17)	64(11)	88(14)	84(17)
Cirratulus cirratus	50	10(9)	14(11)	2(14)	0(9)	8(12)	0(15)	4(9)	9(12)	0(15)
Littorina littorea	12	0(7)	0(9)	8(12)	0(7)	0(10)	0(13)	0(7)	0(10)	0(13)

[a]Microcosm

weeks, and the final three microcosms at 17 weeks. The duration of exposure to TBT is given in Table I because, apart from variations in the time of final analysis, one species suffered 100% mortality within two weeks and some species were added to the microcosms a few weeks after the start of the trial. It is clear from Table I that neither of the bivalve species survived the high level of TBT and that significant mortalities of *C. edule* were evident after 17 weeks in the low level TBT microcosms. Few *N. diversicolor* survived in any microcosms, but *C. cirratus* tolerated high levels of TBT without significant mortality. The winkle *L. littorea* was also apparently resistant to TBT but possible reproductive abnormalities were not investigated in these animals.

Bivalve spat and small juveniles of other invertebrates were undoubtedly present initially in the substrate, which was sieved down to 2 mm, and larval invertebrates will also have entered the microcosms via the unfiltered seawater supply during the trial. These non-introduced animals were collected, identified, and counted from each microcosm at the end of

the trial, and the results are presented in Table II. It is clear that virtually no recruitment from this source occurred in the high TBT microcosms and that fewer non-introduced animals found their way into, or survived in, the low level TBT microcosms compared with controls. Details of the numbers and sizes of the non-introduced bivalve species collected at the end of the trial reveal that for certain species, notably the mussel, *Mytilus edulis,* both numbers and mean size (measured as shell length) are significantly greater in the controls compared to the low TBT microcosms (ref. 5).

Sediment core samples were taken from the microcosms at the beginning and end of the trials and the upper 0.5 cm of these samples were analysed for chlorophyll-a using spectrophotometry. These measurements of chlorophyll-a provide an indication of the amount of living photosynthetic tissue present in the benthic microalgae at the sediment surface. Analysis of the chlorophyll-a data (ref. 4) revealed that the highest final levels occurred in the high TBT microcosms where virtually no grazing invertebrates

TABLE II
Total numbers of naturally occurring juvenile bivalves and other invertebrates in TBT-treated and untreated microcosms (from ref. 5)

| | Treatment | | | | | | | | |
| | High TBT | | | Low TBT | | | Control | | |
	1[a]	2	3	4	5	6	7	8	9
Bivalves	0	1	1	88	66	109	162	137	179
Neries diversicolor	0	0	0	29	5	38	17	113	22
Arenicola marina	0	0	0	5	3	17	5	11	23
Phyllodoce mucosa	0	0	0	0	2	1	8	2	4
Eteone sp.	0	0	0	0	0	3	1	4	2
Pholoe sp.	0	0	0	0	4	9	4	2	6
Scoloplos armiger	0	0	0	0	0	22	0	2	1
Hydrobia ulva	0	0	0	2	1	5	3	10	6
Nemertine spp.	0	0	0	0	0	0	1	2	2
Carcinus maenas	0	0	0	0	0	0	0	3	0
Tunicate sp.	0	0	0	0	0	0	2	0	0
Ophiuroid sp.	0	0	0	1	0	0	0	0	0
Anemone sp.	0	0	0	0	0	0	0	1	0
Totals	0	1	1	125	81	204	203	287	245

[a]Microcosm

were present. Lowest final levels were found in the low level TBT, and because fewer grazing invertebrates were present in the low level TBT microcosms than in the control microcosms (Table II), this strongly suggests that the low level of TBT may have depressed the growth of benthic microalgae compared to the controls.

DISCUSSION

Laboratory toxicity trials have demonstrated that bivalves are especially susceptible to TBT (refs 20,23,24). Furthermore, data from field trials (refs 30,31) and anecdotal reports in numbers of bivalves in certain UK estuaries provide further evidence of the high toxicity of TBT to bivalves. The microcosm study reviewed here confirms the validity of the laboratory and field data. For example, levels of 1000–2000 ng l^{-1} rapidly kill adult cockles which are living in substrate similar to their natural habitat. After 17 weeks in such conditions at a concentration of 100 ng l^{-1} TBT there was an 80% mortality of cockles. Of particular interest is the demonstration that recruitment of certain bivalves was significantly affected at low levels of TBT and that there was virtually no recruitment of bivalve species at higher levels of TBT. The adult polychaetes introduced into the microcosms were not significantly affected by high or low levels of TBT compared to the controls, but at the high level no young polychaetes were retrieved from the substrate. This is the first evidence that the life cycle of polychaetes can be affected by TBT. However, it seems that relatively high levels (1000 ng l^{-1}) are necessary for that effect to be demonstrated.

Laboratory algal toxicity tests have shown effects on growth of certain algae at 100 ng l^{-1} TBT (refs 6, 7). The microcosm study, whilst not providing incontrovertible evidence, does strongly suggest that growth of benthic microalgae may be similarly reduced at 100 ng l^{-1} TBT.

Although we now have a little knowledge of TBT toxicity to marine life, a number of important areas of uncertainty remain. For example TBT normally breaks down by photolysis and biological degradation through dibutyltin (DBT) and monobutyltin (MBT) to inorganic tin. The half-life of TBT is variously estimated at from 6 days in freshwater at 20°C to 90 days in seawater at 5°C (ref. 32). However, the persistence of TBT in the water column and the sediments is very difficult to predict since a number of factors such as the presence of microorganisms and the degree of adsorption of TBT on to sediments will affect the rate of degradation. Preliminary studies on DBT

suggest that although it has a similar half-life, it is considerably less toxic than TBT.

Another area of uncertainty concerns the accumulation of TBT along marine food chains. First, evidence presented by Lee et al. (ref. 33) suggests that microalgal cells take up TBT but then may rapidly degrade the TBT to DBT within the cell. The presence of microalgae may therefore reduce the background levels of TBT and microalgae may also not be a major source of input of TBT into the food chain. Secondly, Beaumont et al. (ref. 27) have shown that algae grown in TBT and then fed to bivalve larvae can significantly reduce the growth of the larvae. However, in these trials, the direct effect of TBT itself on the larvae seemed to be far greater than the effect due to ingestion of TBT contaminated algae.

It can be seen from the information so far presented that the localized ecological consequences of TBT may be very extensive. For example, if levels of around 100 ng l^{-1} TBT are maintained for the summer months—a situation that has commonly occurred in certain UK estuaries and marinas (ref. 34)—a number of organisms are threatened. Overall algal growth may be reduced, and certainly the species composition of the phytoplankton may be changed. Algal species such as the diatom, *Skeletonema costatum*, which has been demonstrated to suffer very reduced growth in TBT, are likely to be less well represented in the plankton than other, more tolerant species. This may reduce species diversity and therefore also reduce the range of foods available for the zooplankton. Furthermore, TBT taken up by microalgae will provide a source of TBT, albeit perhaps a minor one, for zooplankton. Organisms such as copepods and the larvae of crustacea and bivalves, which form a part of the zooplankton, are directly affected by TBT. It is likely therefore that the overall health of the zooplankton will suffer from both these causes. Reductions in numbers, health and diversity of zooplankton organisms will further affect fish and other organisms which feed on the zooplankton.

Apart from these general ecological considerations there are specific organisms which would be seriously effected by levels of TBT in the 10–100 ng l^{-1} range. Dogwhelk populations would be threatened by imposex, farmed Pacific oysters would suffer 'balling' and become unmarketable, and common bivalves such as the cockle and the mussel would suffer considerable problems with mortality and recruitment.

UK LEGISLATION

In February 1985, in response to the evidence building up on the high toxicity of TBT, the UK Government prepared a consultation paper which proposed that the levels of tin in organotin anti-fouling paints should be considerably reduced, at least for vessels less than 12 metres in length. This would have effectively ruled out the use of the then current commercial organotin paints on yachts and pleasure craft. As a result of a strong parliamentary lobby financed by a paint company and organized by a private lobby company (ref. 35) the Government introduced, in July 1985, a series of much reduced measures which were designed to lower environmental TBT concentrations while allowing the paint industry and their customers time to adjust. The package proposed an environmental quality target (EQT) of 20 ng l^{-1} TBT (ref. 36), but it soon became clear as a result of routine monitoring and further toxicity studies that, first, TBT levels in estuaries and marinas were often considerably higher than the EQT and, secondly, effects had been noted in dogwhelks and oysters at levels of TBT lower than the EQT.

After further legislation, under the Control of Pollution Act (COPA), which progressively reduced the amount of TBT allowed in anti-fouling paints for small boat users, the Government finally banned the retail sale of all anti-fouling paints containing TBT in May 1987. This also included a ban on salmon sea pen net treatments containing TBT. Furthermore, in future, anti-fouling agents of all kinds will have to be screened under the Food and Environment Protection Act (FEPA) before being introduced on to the market.

Outside the UK, the French ban on TBT remains, Germany and Switzerland have banned TBT in anti-fouling paints for use in freshwater, and the American Environmental Protection Agency (EPA) has recently announced a ban on TBT paints for vessels less than 25 metres in length.

CONCLUSION

In the last eight years, considerable TBT toxicity data have been collected on marine organisms. Also during this period, water in estuaries, marinas and freshwater broads has been sampled and analysed for the presence of TBT. Some marine species, such as the dogwhelk and the Pacific oyster, have been shown to be sensitive to less than 20 ng l^{-1} TBT, while effects on algae, bivalves, marine larvae and ecosystem recruitment have been demonstrated at 100 ng l^{-1} TBT. Analysis of water samples has revealed that levels of TBT in UK estuaries have frequently been higher than 100 ng l^{-1} and in some cases have reached concentrations in excess of 1000 ng l^{-1} (refs 22,34). There has therefore been a very significant overlap between actual TBT concentrations and levels of TBT which cause environmental damage. It was the overwhelming evidence of this overlap which prompted the UK Government to ban the use of TBT in anti-fouling paints for pleasure craft. However, the relatively slow action of the Government has not prevented the virtual destruction of many oyster-growing concerns in the south of the UK and in Scottish lochs. It is hoped that a rapid reduction in TBT levels will now ensue, allowing a recovery of these oyster-growing concerns.

Although yacht anti-fouling paints containing TBT are now banned in the UK, they are still likely to be advertised and marketed abroad in countries where they are not yet subjected to restriction. It is important that the ecotoxicological information concerning TBT reaches such countries before damage is done to the marine environment or to the local shellfish industry. It is hoped therefore that this chapter will, in some measure, contribute towards the dissemination of information on the ecotoxicology and the potential environmental impact of this highly toxic chemical.

REFERENCES

1. Waldock, M. J., Waite, M. E. and Thain, J. E. (1987). Changes in concentrations of organotins in UK rivers and estuaries following legislation in 1986. Proc. Oceans 87, Vol. 4, Int. Organotin Symp. 1352–1356. Publ. Mar. Tech. Soc. Washington DC USA.
2. Hall, L. W. Jr and Pinkney, A. E. (1984). Acute and sublethal effects of organotin compounds on aquatic biota: an interpretative literature evaluation. *CRC Critical Reviews in Toxicology* 14: 159–209.
3. Rexrode, M. (1987). Ecotoxicity of tributyltin. Proc. Oceans 87, Vol. 4, Int. Organotin Symp. 1443–1455. Publ. Mar. Tech. Soc. Washington DC USA.
4. Beaumont, A. R., Mills, D. K. and Newman, P. B. (1987). Some effects of tributyltin (TBT) on marine algae. Proc. Oceans 87, Vol. 4, Int. Organotin Symp. 1488–1493. Publ. Mar. Tech. Soc. Washington DC USA.
5. Beaumont, A. R., Newman, P. B., Mills, D. K., Waldock, M. J., Miller, D. and Waite, M. E. Sandy-substrate microcosm studies on tributyltin (TBT) toxicity to marine organisms. In 22nd Eur. Mar. Biol. Symp., Barcelona, Spain. In press.

6. Walsh, G. E., McLaughlan, L. L., Lores, E. M., Lovie, M. K. and Deans, C. H. (1985). Effects of organotins on growth and survival of two marine diatoms, *Skeletonema costatum* and *Thalassiosira pseudonana*. *Chemosphere* **14:** 383–392.
7. Beaumont, A. R. and Newman, P. B. (1986). Low levels of tributyltin reduce growth of marine micro-algae. *Mar. Pollut. Bull.* **17:** 457–461.
8. Gray, B. H., Porvaznik, M., Flemming, C. and Lanfong, H. L. (1986). Tri-*n*-butyltin aggregates and membrane cytotoxicity in human erythrocytes. Proc. Oceans 86, Vol. 4. Organotin Symp. 1234–1239. Publ. Mar. Tech. Soc. Washington DC USA.
9. Laughlin, R. B. and French, W. J. (1980). Comparative study of the acute toxicity of tri alkyltins to *Hemigrapsus nudus* and *Homarus americanus*. *Bull. Environ. Contam. Toxicol.* **25:** 802–807.
10. Davidson, B. M., Valkirs, A. O. and Seligman, P. F. (1986). Acute and chronic effects of tributyltin on the mysid *Acanthomysis sculpta*. Proc. Oceans 86, Vol. 4, Int. Organotin Symp. 1219–1225. Publ. Mar. Tech. Soc. Washington DC USA.
11. Meador, J. P. (1986). An analysis of photobehaviour of *Daphnia magna* exposed to tributyltin. Proc. Oceans 86, Vol. 4, Int. Organotin Symp. 1213–1218. Pub. Mar. Tech. Soc. Washington DC USA.
12. U'ren, S. C. (1983). Acute toxicity of bis (tributyltin) oxide to a marine copepod. *Mar. Pollut. Bull.* **14:** 303–306.
13. Bryan, G. W., Gibbs, P. E., Hummerstone, L. G. and Burt, G. R. (1986). The decline of the gastropod *Nucella lapillus* around south-west England: evidence for the effect of tributyltin from anti-fouling paints. *J. Mar. Biol. Ass. UK* **66:** 611–640.
14. Bryan, G. W., Gibbs, P. E., Burt, G. R. and Hummerstone, L. G. (1987). The effects of tributyltin (TBT) accumulation on adult dogwhelks, *Nucella lapillus*: long term field and laboratory experiments. *J. Mar. Biol. Ass. UK* **67:** 525–544.
15. Gibbs, P. E., Bryan, G. W., Pascoe, P. L. and Burt, G. R. (1987). The use of the dog-whelk, *Nucella lapillus*, as an indicator of tributyltin (TBT) contamination. *J. Mar. Biol. Ass. UK* **67:** 507–524.
16. Gibbs, P. E. and Bryan, G. W. (1986). Reproductive failure in populations of the dog-whelk, *Nucella lapillus*, caused by imposex induced by tributyltin from anti-fouling paints. *J. Mar. Biol. Ass. UK* **66:** 767–778.
17. Key, D., Nunny, R. S., Davidson, P. E. and Leonard, M. A. (1976). Abnormal shell growth in the Pacific oyster (*Crassostrea gigas*): some preliminary results from experiments undertaken in 1975. ICES Paper CM 1976/K:1 (mimeo). International Council for the Exploration of the Sea, Copenhagen.
18. Alzieu, C., Thibaud, Y., Heral, M. and Boutier, B. (1980). Evaluation des risques dus à l'emploi des peintures anti-salissures dans les zones conchylicoles. *Rev. Trav. Inst. Pech. Marit.* **44:** 301–345.
19. Alzieu, C., Heral, M., Thibaud, Y., Dardignac, M. J. and Feuillet, M. (1982). Influence des peintures antisalissures à base d'organostannique, sur la calcification de la coquille de l'huître *Crassostrea gigas*. *Rev. Trav. Inst. Pech. Marit.* **45:** 100–116.
20. Waldock, M. J. and Thain, J. E. (1983). Shell thickening in *Crassostrea gigas*: organotin anti-fouling or sediment induced? *Mar. Poll. Bull.* **14:** 411–415.
21. Alzieu, C. and Heral, M. (1984). Ecotoxicological effects of organotin compounds on oyster culture. *In: Ecotoxicological Testing for the Marine Environment.* G. Persoone, E. Jaspers and C. Claus (ed.). State Univ. Ghent and Inst. Mar. Scient. Res., Bredene, Belgium Vol. 2, 187–197.
22. Waldock, M. J. (1986). TBT in UK estuaries 1982–86. Evaluation of the environmental problem. In Proc. Oceans 86, Washington DC, Vol. 4 Organotin Symp. 1324–1330. Publ. Mar. Tech. Soc. Washington DC USA.
23. Thain, J. E. (1986). Toxicity of TBT to bivalves: effects on reproduction, growth and survival. In Proc. Oceans 86, Washington DC, Vol. 4, Organotin Symp. 1306–1313. Publ. Mar. Tech. Soc. Washington DC USA.
24. Beaumont, A. R. and Budd, M. D. (1984). High mortality of the larvae of the common mussel at low concentrations of tributyltin. *Mar. Pollut. Bull.* **15:** 402–405.
25. Laughlin, R. B., Pendoley, P. and Gustafson, R. G. (1987). Sublethal effects of tributyltin on hard clam larvae (*Mercenaria mercenaria*). Proc. Oceans 87, Vol. 4, Int. Organotin Symp. 1494–1498. Publ. Mar. Tech. Soc. Washington DC USA.
26. Smith, J. (1987). The effects of tributyltin on bivalve larvae, 72 pp. MSc. Thesis, University of Wales.
27. Beaumont, A. R., Newman, P. B. and Smith, J. Some effects of tributyltin from anti-fouling paints on early development and veliger larvae of the mussel *Mytilus edulis*. (in press).
28. Lawler, I. N. and Aldrich, J. C. (1987). Sublethal effects of bis (tri-n-butyltin) oxide on *Crassostrea gigas* spat. *Mar. Pollut. Bull.* **18:** 274–278.
29. Short, J. W. and Thrower, F. P. (1986). Accumulation of butyltins in mussel tissue of Chinook salmon reared in sea pens treated with tri-n-butyltin. *Mar. Pollut. Bull.* **17:** 542–545.
30. Alzieu, C. and Portmann, J. E. (1984). The effect of tributyltin on the culture of *Crassostrea gigas* and other species. Proceedings of the 15th Annual Shellfish Conference 15–16th May, 87–100. Shellfish Association, London.
31. Stephenson, M., Smith, D. R., Goetel, J., Icnikawa, G. and Martin, M. (1986). Growth abnormalities in mussels and oysters from areas with high levels of tributyltin in San Diego Bay. In Proc. Oceans 87, Vol. 4, Int. Organotin Symp. 1246–1251. Publ. Mar. Tech. Soc. Washington DC USA.

32. Thain, J. E., Waldock, M. J. and Waite, M. E. (1987). Toxicity and degradation studies of tributyltin (TBT) and dibutyltin (DBT) in the aquatic environment. Proc. Oceans 87, Vol. 4, Int. Organotin Symp. 1398–1404. Publ. Mar. Tech. Soc. Washington DC USA.
33. Lee, R. F., Valkirs, A. O. and Seligman, P. F. (1987). Fate of tributyltin in estuarine waters. Proc. Oceans 87, Vol. 4, Int. Organotin Symp. 1411–1415. Publ. Mar. Tech. Soc. Washington DC USA.
34. Waldock, M. J. and Miller, D. (1983). The determination of total and tributyltin in seawater and oysters in areas of high pleasure craft activity. ICES Paper CM 1983/E: 12 (mimeo), International Council for the Exploration of the Sea, Copenhagen.
35. Anon. (1986). Political fixers. *Intercity* Nov./Dec. 1986, 15–17.
36. Abel, R., Hathaway, R. A., King, N. J., Vosser, J. L. and Wilkinson, T. G. (1987). Assessment and regulatory actions for TBT in the UK. Proc. Oceans 87, Vol. 4, Int. Organotin Symp. 1314–1319. Publ. Mar. Tech. Soc. Washington DC USA.

34

The Use of a Landfill Site for Bunker C Collected from a Barge Aground at Matane, Québec: a Case Study

M. F. Khalil International Marine Environmentalists (IME) Inc., 205 est, St-Jean-Baptiste, Rimouski (Québec), Canada G5L 1Y7 and *A. Kouicem* Génilab BSLG Inc. 603, 3è rue, Rimouski-est (Québec), Canada G5L 7M9

On December 3, 1985, the oil spill in Matane, Québec, was one of the largest to happen in the St Lawrence estuary. The accident happened on a stormy night at subfreezing temperatures with wind velocity exceeding 100 km h^{-1}. *Pointe-Levy* barge was being towed by a tug-boat travelling from Montreal (Québec) to Bathurst (New-Brunswick) (Fig. 1). The barge, 92 m long, was charged with 34 000 barrels of bunker #6 oil. To keep this heavy oil from aggluti-nation, it was continuously steam-heated at 130°C during the transport. With the increasing wind vel-ocity, the cable connecting the barge to the tug-boat cracked and the barge drifted towards the south shore of the St Lawrence maritime estuary. The dominant winds are from the north-west. On the rocky bottom of the shore, three of the nine compartments of the barge were damaged, releasing around 200 metric tonnes of bunker #6 oil. The agglutinated spill spread 40 km to the east of Matane city. Approximately 150 cubic metres collected near the Matane river mouth on the west side of the marina dock.

Soon after the accident, a Canadian coastguard ice-breaker was on the scene of the accident and its helicopter located and followed the spill. The Cana-dian coastguard immediately assembled a spill response organization to assist and coordinate the assessment of the problem, the situation of the barge, and the recovery and clean-up of the spilled material.

The bunker #6 oil was trapped in large quantities of snow and ice, prohibiting conventional methods of recovery and clean-up. This mixture of oil and frozen water ended up on the shore and most was recovered using heavy equipment such as pay loaders, cranes and trucks. The location of the spill is shown in Figure 2.

A wide range of parameters can be used to describe the degree of oil contamination on the shoreline following a spill (ref. 1). According to this author, the most relevant parameter in the shoreline clean-up decision involved the measurement of the area of surface oil cover and calculation of the volume of contaminated sediments.

Advances in Underwater Technology, Ocean Science and Offshore Engineering, Volume 16: Oceanology '88

© Society for Underwater Technology (Graham & Trotman, 1988)

The Matane river mouth is situated at some hundred metres from the accident site. This river is one of the five best rivers in the Québec province for Atlantic salmon spawning and adult salmon *Salos salar* migrates in the summer from the Atlantic to the freshwater rivers for reproduction, so it was of paramount importance to locate and remove as much of the spilled oil as soon as possible. As recently reported (ref. 2), the danger of contamination of the salmon by oil is greater if the oil remains in the environment until the next summer.

The task force at Matane was formed from rep-

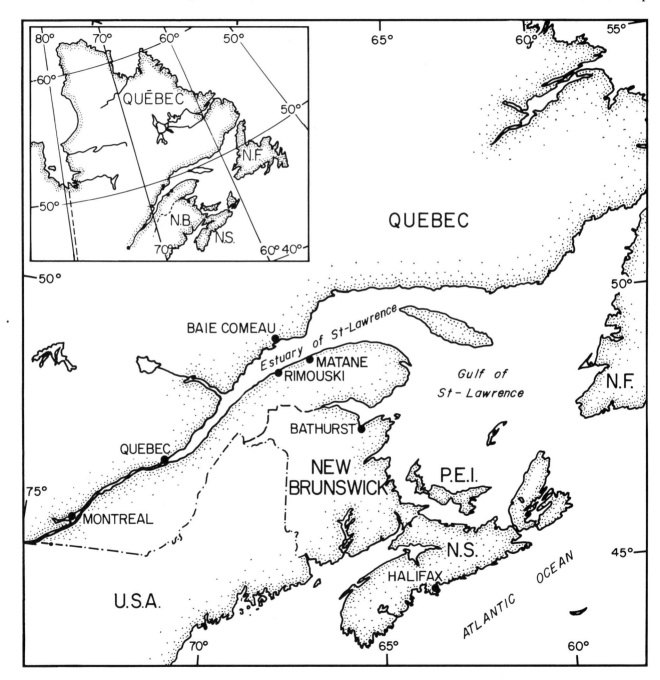

Fig. 1 Map of the St Lawrence estuary and location of Matane City.

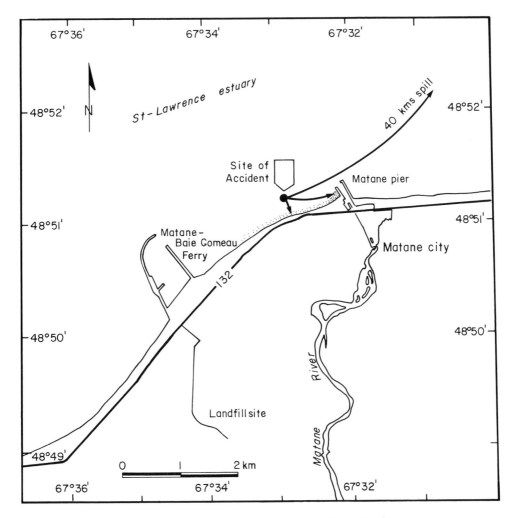

Fig. 2 Site of the accident and propagation of bunker
#6 oil.

resentatives of the Canadian coatguards, Environ-ment Canada, Environnement Québec, the oil com-pany owner of the barge, and those representing the insurance companies. Plans were made to bring clean-up and recovery equipment to the scene but several important facts concerning the spill and its implica-tions were not known. For example, the amount of damage to the barge and the possibility of pumping out bunker oil to another barge, the containment of the spill in such severe weather conditions of wind, sea state and temperature. An experience with simi-lar weather conditions occurred at Baie Verte, Newfoundland, where there was a spill of approxi-mately 500 barrels of diesel oil in March 1982. The oil was trapped under the ice until the spring thaw

in May. A study by Kiceniuk and Williams (ref. 3) showed the persistence of some of the components of polyaromatic hydrocarbons (PAH) as fluorene even after 39 months of weathering.

In an environment contaminated by oil, evaporation of the light fraction is the dominant weathering pro-cess in supratidal sand (ref. 4). After four months, concentrations of compounds less volatile than $n\text{-}C_{18}$ did not change.

In spite of the gravity of the incident at Matane, the cold weather did help in keeping the bunker oil agglutinated to ice and snow. Collecting this slush mechanically was relatively easy. Large container trucks collected the oil from the shore, and most of the oil accumulated on the pier was collected manually. One problem remained unresolved. How and where to dump such mixture of sand, ice, snow and oil? This is the subject of the present study.

Landfill site choice

The landfill site of Matane is situated approximately 3 km from the shore where the spill accumulated. The Environnement Québec ministry allowed this site to be used as a temporary measure to contain the ice, sand and oil mixture, provided the oil was removed the following summer after thawing. Part of the skimmed oil from the Matane's pier was collected in 45-gallon drums which were left unburied beside the landfill site.

Matane's disposal site: geological and hydrogeological conditions

The disposal site is formed of two distinct geological units. At the surface, sand mixed with gravel and silt covers the area. This deposit is of fluvioglacial

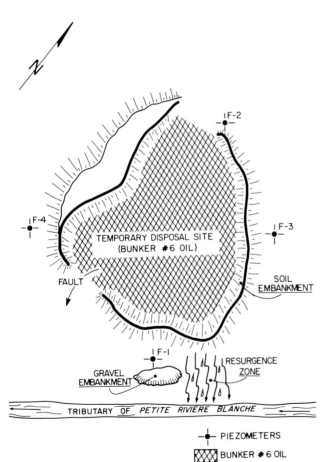

Fig. 3 Landfill site at Matane.

origin and its thickness varies between 4 and 11 metres. Under this granular soil, a cohesive deposit is present composed of clayey silt and silty clay. Just south-west of the disposal site, the clay deposit is present near the surface and partially covered by the sand-gravel top soil.

Four piezometers were installed around the site (Fig. 3) to follow the fluctuations in the groundwater levels and to permit the collection of water samples for chemical analysis. During their installation, a borehole was conducted to the bedrock. The depth reached by the boring equipment was approximately 20 m.

From the hydrogeological point of view, the surface sand-gravel forms an unconfined aquifer compared to the clayey deposit which forms a natural obstacle to ground waters. The sand-gravel unit has a coefficient of permeability $k = 10^{-3}$ cms^{-1}, the horizontal hydraulic gradient $i = 0.0075$ mm heading south-east (i.e. towards the tributary of Petite Rivière Blanche). The water migration speed estimated at 33 m year^{-1}. It was calculated that the water percolating from the site towards the tributary would need at least 333 days.

To improve this situation, we recommended that the surrounding dams be covered with non-permeable materials to help in preventing any leakage and percolation of contaminated waters. We proposed to cover the site with a non-permeable material to reduce the infiltration of water originating from precipitation and snow melt. This action would result in reducing the volume of infiltrated water from the surface and *ipso facto* weaken the hydraulic gradient and the migration speed of the water in the soil.

EXPERIMENTAL

Sampling strategy

The monitoring of the landfill site had to take in account the climatic conditions prevailing in the area. Thawing at the surface generally begins from late March to mid-May, and we expected the bulk of bunker #6 and snow which were covered with top soil to thaw completely by July 1986.

In addition to the chosen stations, core samples in the temporary disposal site were taken and used as reference of the bunker #6 concentration in the site. In addition, permeability tests were to be conducted in the soil. According to Québec provincial government regulations, permeability coefficient of 10^{-8} cm s^{-1} is acceptable. The underlying layers of soil in the disposal site are formed mainly of silt, sand and clay. The sampling strategy proposed and accepted covered a two-year period, from March to July 1986 on a fortnightly basis then from August 1986 to September 1987 on the basis of a sampling every second month.

Sampling methodology

Samples of water were taken from the four piezometers. In the tributary of Petite Rivière Blanche, three stations were identified. The first one, upstream at nearly 300 m, was taken as a reference and was not affected by the dumping site. The inclination of the water table and the running of the surface waters confirm this hypothesis. The second station was in the proximity of the dumping site; only 100 m separated the tributary from the southerly end of the dumping site. The third station was downstream the tributary and was separated from the site at least 200 m to the west. At each of the three stations, samples of water and surface sediments were collected. All water samples from these stations and the piezometers together with the sediment samples were kept at 4°C and transported to our laboratories at Rimouski (100 km). The analyses were carried out on the samples as soon as practicable.

ANALYTICAL METHODOLOGY

Infrared (IR) spectroscopy together with gas chromatography (GC) were the basis of hydrocarbon determination in the collected samples. Water from the piezometers and the tributary of the Petite Rivière Blanche were extracted with hexane by liquid:liquid extraction (ref. 5). Sediments from the river were extracted in a Soxhlet apparatus using hexane.

The analytical techniques described elsewhere (refs 6,7), were adopted using a Perkin Elmer model 521 IR spectrophotometer a Perkin Elmer model Sigma 2000 GC, equipped with capillary columns DB-5, 30 m long.

RESULTS AND DISCUSSION

During the thawing period of March–July 1986, no real percolation occurred. This was confirmed by the chemical analyses of the water in the piezometers together with the water and sediments in the tributary of Petite Rivière Blanche. The concentrations in most of the samples varied between 20 mg kg^{-1} (ppm) and the detection limits of our method. An exception to this general behaviour was the overflow incident in April 1986.

On the east side of piezometer F1, the presence of leaking water was noted. This outgoing water was running towards the tributary of Petite Rivière Blanche. On thawing, the embankment surrounding the landfill site showed some discontinuities, especially between piezometers F1 and F4 (Fig. 3). This fault resulted in the rising of underground water from the site. This running water was contaminated by an apparent hydrocarbon film on its surface.

The chemical analysis of the water and sediments in the tributary of Petite Rivière Blanche showed a net contamination. The concentration of hydrocarbons spread to nearly 100 mg kg^{-1} in both the water and the sediments. A hydraulic shovel was introduced to contain and repair the fault. A bank was filled up with available materials, creating a ditch at the outskirts of the site which was filled later by filtering sand which was available near the site.

A bed of filtering sand has been spread out on the zone of resurgence. This bed is nearly 25 m long and 5 to 7 m wide. The bed thickness could vary from 50 to 80 cm. A Caterpillar tractor type D-7 was used for cleaning the contaminated soils. The earthworks of the embankment and the bed of sand were also treated. The results were not satisfactory, owing mainly to the presence of snow and still frozen soil in the bank.

Following this intervention, the chemical analyses confirmed that no further contamination occurred in the vicinity of the disposal site. The values of hydrocarbons in the piezometers never exceeded 20 mg kg^{-1}. The tributary of Petite Rivière Blanche was sampled and analysed for hydrocarbons following the same schedule till September 1987. No hydrocarbon has been detected in the stream or its sediments.

CONCLUSION

The status of the landfill site was a temporary one. The oil company owner of the barge had the responsibility of removing the oil from the site after the thaw. It was assumed that this company had to transport the oil and burn it in a kiln or elsewhere.

We believe that modification works held at the disposal site together with the geological conditions were the cause of the non-dispersion of the bunker outside of the site. The clay deposit just south-west of the site located near the surface prevented percolation of water in that direction.

Finally the only way the contaminated water had to percolate was on the south-east side of the site. Modifications in the silty cover performed on the dams confined the disposal site at its actual location. Chemical results so far prove that the contaminated waters have not been percolating outside the site.

The permeability of the landfill site added to the possibilities of contamination of the environment by transporting the rest of the oil, the distance it has to travel, led to the conclusion that this landfill site should be kept as such. From a temporary disposal site, Matane landfill is now considered a permanent one. Adjacent to this site, a new one is now in use, as an urban waste disposal site.

REFERENCES

1. Owens, E. H. (1987). Estimating and quantifying oil contamination on the shoreline. *Marine Pollut. Bull.* **18:** 110–118.
2. Bax, N. J. (1987). Effects of a tanker accident and an oil blowout in Bristol Bay, Alaska, on returning adult sockeye salmon *Oncorhynchus nerka*. A simulation study. *Marine Environ. Res.* **22:** 177–203.
3. Kiceniuk, J. W. and Williams, U. P. (1987). Sediment hydrocarbon contamination at the site of the Baie Verte, Newfoundland, oil spill: results of a four year study. *Marine Pollut. Bull.* **18:** 270–274.
4. Strain, P. M. (1986). The persistence and mobility of a light crude oil in a sandy beach. *Marine Environ. Res.* **19:** 49–76.
5. Cauchois, D. and Khalil, M. (1974). Matière organique dissoute dans l'estuaire maritime du Saint-Laurent: Comparison et choix des méthodes. *J. Fish. Res. Board Can.* **31:** 133–139.
6. Wade, T. L., Quinn, J. G., Lee, W. T. and Brown, C. W. (1976). Source and distribution of hydrocarbons in surface waters of the Sargasso sea. In sources, effects and sinks of hydrocarbons in the aquatic environment. American Institute of Biological Sciences (ed.). In: *Proceedings of the symposium, American University,* Washington, D.C. 9–11 August 1976, 270–286.
7. Morel, G. and Courtot, P. (1981). Dosages d'hydrocarbures dans l'eau et le sédiment marin par infra-rouge et chromatographie gazeuse sur colonne capillaire. *Rev. Inst. Franç. du Pétrole* **36:** 629–666.